Oxford Avian Biology Series

Oxford Avian Biology Series

Series Editor: Professor T.R. Birkhead FRS (University of Sheffield, UK)

A new series of exciting, innovative contributions from the top names in avian biology. Topics have been carefully selected for their wider relevance to both students and researchers in the fields of ecology and evolution.

Volume 1: Avian Invasions
Tim M. Blackburn, Julie L. Lockwood, Phillip Cassey

Avian Invasions

The Ecology and Evolution of Exotic Birds

Tim M. Blackburn
Institute of Zoology, ZSL, Regent's Park, London
Julie L. Lockwood
Ecology, Evolution, and Natural Resources, Rutgers University, New Brunswick, NJ
Phillip Cassey
School of Biosciences, University of Birmingham

OXFORD
UNIVERSITY PRESS

Great Clarendon Street, Oxford OX2 6DP

Oxford University Press is a department of the University of Oxford.
It furthers the University's objective of excellence in research, scholarship,
and education by publishing worldwide in

Oxford New York

Auckland Cape Town Dar es Salaam Hong Kong Karachi
Kuala Lumpur Madrid Melbourne Mexico City Nairobi
New Delhi Shanghai Taipei Toronto

With offices in

Argentina Austria Brazil Chile Czech Republic France Greece
Guatemala Hungary Italy Japan Poland Portugal Singapore
South Korea Switzerland Thailand Turkey Ukraine Vietnam

Published in the United States
by Oxford University Press Inc., New York

First published 2009

British Library Cataloguing in Publication Data

Data available

Library of Congress Cataloging in Publication Data

Data available

Typeset by Newgen Imaging Systems (P) Ltd., Chennai, India
Printed in Great Britain
on acid-free paper by
CPI Antony Rowe, Chippenham

ISBN 978–0–19–923254–3 (Hbk.)
ISBN 978–0–19–923255–0 (Pbk.)

10 9 8 7 6 5 4 3 2 1

Preface

It is now fifty years since the publication of Charles Elton's influential monograph on *The Ecology of Invasions by Animals and Plants.* Writing in the interregnum between the conflicts in Korea and Vietnam, as the Cold War escalated, and during the period of the early nuclear weapon tests, Elton started his first chapter with the analogy of invasive species as 'ecological explosions'. More and more species were achieving enormous increases in numbers, thanks to humans interfering with the natural forces that previously held their populations in check. One such interference is to remove a species from its natural environment and release it in an area novel to it, and where it is new to the area's natural inhabitants. That type of interference was the main focus of Elton's book, and it is the main focus of this book too.

Explosion is an apt analogy for the current situation regarding exotic species in at least two additional ways. The environmental context in the late 1950s was such that Elton could write that 'we are seeing one of the great historical convulsions in the world's fauna and flora', but in the half-century since, the rate at which humans have transported species to areas beyond the limits of their native geographic distributions has increased. So too apparently has the rate at which those exotic species are accumulating in the environment. Elton's work seems all the more prescient for it. Fortunately, an ecological explosion is only one possible outcome from the translocation of a species from its native to an alien environment, and in most cases the fuse simply fizzles out. Nevertheless, we are continuing to light so many fuses that we risk setting off a real firework display.

Explosion is also apt for the literature pertaining to exotic species. Keyword searches show that, twenty years after Elton's monograph, barely a handful of papers were published on the subject in any given year. The past twenty years, however, has seen near exponential growth in the output of this field, such that hundreds of papers now appear annually. The quantity of information that has accumulated is such that we think it makes sense to restrict ourselves to synthesizing the literature relating to exotic species in one taxon, as we have done here for birds. Even this is no small undertaking.

There are distinct advantages to a taxon-specific focus, as we hope will become clear through the chapters that follow. Limiting ourselves to birds means that we also proscribe the vectors by which species are translocated beyond their range boundaries, the traits of the species translocated, the biotic and abiotic processes that are likely to influence the growth and spread, or otherwise, of exotic populations, and the mechanisms by which evolution is likely to occur at the novel location. We also gain the benefit of a taxon for which there is as good information as we have on most aspects of the invasion process, as well as on their ecology and evolution more widely. This gives us the best possible chance to construct a coherent and informative picture of how a species proceeds from native to exotic, and beyond.

The ideas that follow would not have been possible without the encouragement and inspiration of a wide range of influential and brilliant people whom we have been lucky enough to have met and worked with over the course of our careers. Particularly important in this regard have been Richard Duncan, Kevin Gaston, Paul Harvey, John Lawton, Dan Simberloff, and Mark Williamson, who have selflessly shared their thoughts and knowledge over many years. Their help has made our science better, and made us into better scientists. Others who deserve special mention include Jim Brown, Marcel Cardillo, Steven Chown, Martha Hoopes, Kate Jones, Michael Marchetti, Michael McKinney, Michael Moulton, Julian Olden, Dave Richardson, Dov Sax and Dani Sol.

We thank Alex Badyaev, Allan Baker, Barry Brook, Steven Chown, Sonya Clegg, Bob Colautti, John Ewen, Dave Forsyth, Rob Freckleton, Martha Hoopes, Jonathan Jeschke, Salit Kark, Petr Pysek, Dave Richardson, Dan Simberloff, Dani Sol, and Frank van den Bosch, who generously gave of their valuable time to read and comment on drafts of chapters, and Richard Duncan, who read the entire manuscript. Their input was in all cases constructive and insightful, and any faults with the published versions are of course our own. We would also like to thank all of those people who responded so helpfully to our requests for books, papers, figures, and data, especially Richard Duncan, Salit Kark, Mark Parnell, Nathalie Pettorelli, Colin Ryall, and Navjot Sodhi. David and Inga La Puma provided generous hospitality to P.C. and T.M.B. during repeated visits to New Jersey. We thank Ian Sherman and Helen Eaton at Oxford University Press for being such a pleasure to work with during the long period of the book's gestation.

Finally, we are grateful to Noëlle, Tabby and Henry, and Becs for their love and support.

T.M.B
J.L.L
P.C
New Brunswick
2008

Contents

1

Introduction to the Study of Exotic Birds

> For the past few years it has been my hobby to import to Tahiti birds from different parts of the world and liberate them. Up to the present time I have received and liberated about 7,000 birds of fifty-nine different species.
>
> E. Guild (1940)

1.1 Introduction

One of the pleasures of being a biologist is the excitement that comes from visiting a part of the world where you have never been before. Much of that excitement comes from the fact that, especially if it is a long distance from your home, the destination will house species and indeed entire communities that you have never previously encountered. The opportunity to observe and expand your knowledge of the natural world is what biologists like us live for, making all the drudgery of paperwork associated with academic life suddenly worthwhile.

New localities tend to be home to different species in different combinations because each of these places has a unique history. Temporal and spatial variation in environmental conditions, in the identities of the taxa that have immigrated and emigrated naturally over time, and in the interactions between them, all drive the evolution of local and regional biotas in unique directions. The result is the rich texture of biological diversity that carpets the planet, such that no two locations are identical in terms of their biological communities, and that the further two locations are from each other, the more different they tend to be. This rich texture is why we have the disciplines of biogeography, ecology, and evolution, populated with scientists working to understand pattern and process in biodiversity.

Yet, increasingly, when we arrive at a new location we are met by sights and sounds of nature with which we are familiar, either from back home or from previous visits to entirely different regions. As avian biologists, it is of course the birds that we find most striking in this regard. It is hard for any of us to recall the last time that we landed at an airport and were not greeted by the chipping of house sparrows *Passer domesticus*, or a flock of feral rock doves *Columba livia* wheeling around the terminal buildings, or the sight of European starlings *Sturnus vulgaris* or common mynas *Acridotheres tristis* strutting along the grassy verges of the car parks and slip roads. Occasionally, those species are native, and simply using the airport as a recent innovation in the structure of their local habitat. Many times, however, these species are not native, and their presence at the location barely predates that of the airport itself.

Species that are not native to a location are labelled with a variety of epithets (Falk-Petersen, et al. 2006; Lockwood, et al. 2007): exotic, non-indigenous, naturalized, alien, introduced, and invasive are some of the more common terms. Some of these terms are general and synonymous, but others have specific meanings relating to subsets of non-native species, while these definitions may differ between studies and authors. For the remainder of this book, as indeed in the title, we use the terms 'exotic' or 'exotic species' to refer to species that are not naturally present in the wild bird assemblage inhabiting a location, but have been moved beyond the limits of their normal geographic ranges by human actions (deliberate or accidental). We use this term because it is compact and because of its dictionary definition as 'originating in or characteristic of a foreign country' (and do not in this case imply anything relating to the alternative dictionary definition of exotic as 'attractive or striking because colorful or out of the ordinary'). In Section 1.3 we further discuss the issue of terminology as it relates to subsets of exotic species.

Encounters with exotic species do not stop at the exit from the airport, and each of us has had experiences with them that have shaped our careers in biology. Tim will never forget waking up after his first night spent in New Zealand, in October 1998, and listening to a dawn chorus rich with the song of European skylarks *Alauda arvensis* and yellowhammers *Emberiza citrinella*. These songs were common sounds of his youth growing up in agricultural southern England in the 1970s, but by 1998 had become much harder to hear as these species had declined under agricultural intensification (Chamberlain, et al. 2000). Yet, these species were flourishing in New Zealand, an archipelago in the southern Pacific more than 18,000 km from the land where their ancestors originated, and more than 9,000 km from the nearest point that either species naturally occurred (Cramp 1988; Cramp and Perrins 1994). Why were they surviving so well there? Did they interact with the environment and with other bird species in the same way as back home? Could their evident success in New Zealand inform conservation practices in the UK? These and other questions sparked his interest in exotic birds.

The interest in exotic birds for Julie began with a trip to Hawaii. On her first morning in Honolulu she saw and heard an abundance of birds. These birds were beautiful and so very different from the set she grew up with in Georgia in the United States, but not one was native to the islands. A walk through any of the lowland parks on the Hawaiian Islands brought sightings of yet more exotic bird species that were native to some continent very distant from Hawaii. How did all these birds get here, and why these species and not some others? How were they faring after coming into contact with each other, especially given that they shared absolutely no evolutionary history together? What effect, if any, were they having on the few remaining highly threatened native Hawaiian birds?

For Phill the interest in exotic birds was driven by an intense desire to finish his doctoral dissertation. His forays into constructing invasion pathways with freshwater microcosms had been expensive, but unproductive, and his modelling of niche-apportionment was using up all of the computers in the undergraduate and postgraduate laboratories combined. Finally, a frustrated colleague dumped a copy of *Introduced Birds of the World* by John L. Long on his desk and said 'Here, why don't you analyse this?!' These days the more time Phill spends with birdwatchers the more amusement he has spotting common mynas.

Why is it that exotic species, and birds in particular, are sufficiently ubiquitous a feature of the environment that all three of us have been so strongly influenced by them as to devote substantial proportions of our working life to studying their ecology and evolution? The reasons are threefold. First, there is a wide range of motivations for humans to transport bird species beyond the boundaries of their normal geographic ranges, and so a concomitantly wide range of bird species have been transported and subsequently introduced. Second, the reasons for avian translocation generally are not specific to certain human societies or locations (although some are, and the relative importance of different reasons varies), and so areas all around the world have experienced avian introductions. Indeed, regions with at least one exotic bird species clearly cover more of the land area of the planet than those with none (Figure 1.1). Finally, exotic birds have a long history, so that in many cases there has been plenty of time for the species to establish and spread, and in some cases come to dominate certain assemblages. It is to this history that we now turn.

1.2 A Brief History of Exotic Birds

It is generally accepted that humans evolved in sub-Saharan Africa, from where they subsequently emigrated to inhabit essentially every land area on the planet free of permanent ice or snow. The dates of the emigration are still debated, but it may have begun approximately 65,000 years ago with colonization of Asia (Quintana-Murci, et al. 1999). The current best estimates from archaeology and genetics place humans in east Asia at around this time, on mainland Australia *c.*60,000–55,000 years ago, and in North America perhaps *c.*15,000 years ago. Humans have only colonized oceanic islands much more recently, reaching Melanesia and Micronesia in the last 5,000—4,000 years, Fiji and Samoa perhaps 3,500 years ago, New Zealand in the last 1,000 years, and some exceptionally isolated islands like Mauritius and Gough only in the past 400 years or so (e.g., Diamond 1987; Anderson 1991; Milberg and Tyrberg 1993; Higham, et al. 1999; Roberts, et al.

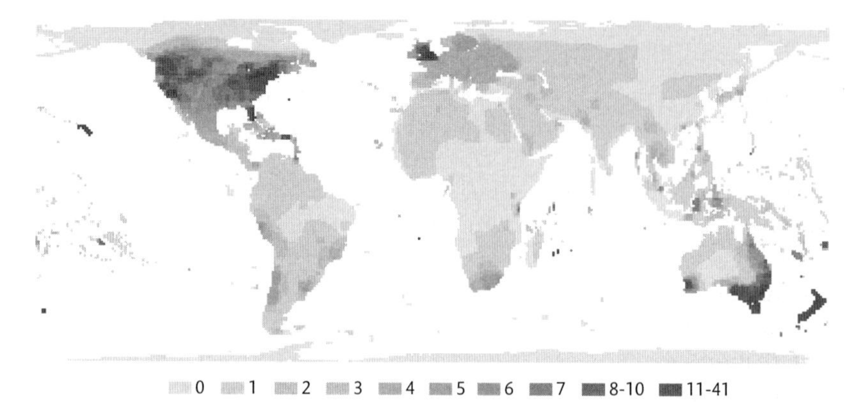

Fig. 1.1 An estimate of the number of exotic bird species found in equal area grid cells (96.486 x 96.486 km, equivalent to 1° longitude and approximately 1° latitude at the equator) across the land surface of the world. The data were compiled by M. Parnell, from the maps in Long (1981), and plotted by N. Pettorelli. The richness and extent of exotic bird species distributions are both likely to have increased since Long's work was published.

2001; Bortolini, et al. 2003; Gaston, et al. 2003; Hurles, et al. 2003; Cheke and Hume 2008).

The first evidence that humans transported other species with them on their travels (aside from the inevitable parasites) currently comes from fossils in the Bismarck Archipelago (Allen, et al. 1989; Grayson 2001). Here, bones of the grey cuscus *Phalanger orientalis* first appear in deposits on New Ireland around 19,000 years ago, at the same time as obsidian transported from New Britain by human colonizers. The cuscus was evidently transported as a food resource, to supplement the relatively impoverished terrestrial vertebrate fauna of this offshore island (Leavesley 2005).

The motive for the first avian translocations was probably the same. The earliest records of birds probably being kept in captivity come from Neolithic times, and relate to taxa such as geese, jungle fowl, and pigeons (Long 1981; West and Zhou 1989) that are still favoured as food today. The earliest evidence that such species were transported beyond their natural geographical range limits also dates from this period, with archaeological and palaeoclimatic data suggesting that red junglefowl *Gallus gallus* (the wild ancestor of the domestic chicken) had been transported north to China by *c.*8,000 years ago from their area of first domestication in south-east Asia (West and Zhou 1989). As far as we aware, however, there is no evidence that this translocation was with the purpose of introducing the junglefowl to the wild, and certainly Long (1981) does not record it as established there.

Grayson (2001) suggests that the first instance of a non-domesticated bird species being translocated for the purpose of exotic introduction may be the flightless rail *Nesotrichis debooyi* (see also Olson 1974). Grayson suggested that this species was introduced from Puerto Rico, where it is known palaeontologically, to several other islands in the Caribbean, including the Virgin Islands. Wetmore (1918) found it to be common in kitchen midden deposits on St Croix indicating that it was a preferred food item of the prehistoric human inhabitants there. If it was an exotic introduction, therefore, it was likely to have been transported and released for consumption. Nevertheless, it is hard to distinguish this hypothesis from alternatives such as that the prehistoric inhabitants reared them in captivity. Given that the introduction happened so long ago and our evidence is restricted to archaeological surveys, it is also possible that *N. debooyi* was actually native to all islands in the greater Puerto Rican area, which would have been joined by land bridges at the height of the last glaciation, but that evidence of a pre-human presence on many of these islands is lacking. Whatever the rail's biogeographic or introduction history, however, *N. debooyi* is now extinct on all the islands on which it once occurred.

A more well-documented account of widespread avian translocation comes from the colonization of islands in the Pacific by the Lapita people. The earliest sites attributable to this civilization, characterized by its distinctive style of pottery, date to the period of *c.*3,300–3,500 years ago, and are found in Near Oceania. However, by 2,900–3,500 years ago this civilization was present on islands in Vanuatu, New Caledonia, Fiji, Tonga, and Samoa (Hurles, et al. 2003). There followed a pause of 500 to 1,000 years before islands in eastern Polynesia were colonized. Hurles, et al. (2003) note that there is little doubt that the Lapita people introduced several species to Remote Oceania, including red junglefowl. It follows that this species is likely to have accompanied the Lapita people on their entire colonization of Oceania from 3,300–3,500 years ago. Red junglefowl were free ranging on many of these islands at the time of first European contact, indicating that they had escaped or been released from captivity on repeated occasions (Long 1981), although we do not currently know how soon following first colonization this occurred on any island in the region.

While much of the evidence for early translocation concerns edible species, food was not the only reason why prehistoric peoples moved bird species beyond the limits of their native geographic ranges. Most notably, pre-Columbian cultures in Central America used feathers extensively in decorations, such as headdresses, garments, and ceremonial objects (Haemig 1978). The brightly coloured feathers of tropical species such as quetzals and macaws were particularly prized for this purpose, but their geographic location meant that some civilizations could only acquire such feathers through trade. Clearly this trade sometimes involved live birds. For example, skeletons of the scarlet macaw *Ara macao* have been found associated with archaeological sites scattered from southern Utah (in present-day USA), to northern Mexico, hundreds of kilometres north of their natural geographic range limits in Veracruz, south-eastern Mexico (Minnis, et al. 1993).

In at least one case, there is good evidence for the deliberate release by a pre-Columbian society of a tropical bird species beyond its natural range for the purposes of establishing an exotic population. Haemig (1978) recounts the history of the great-tailed grackle *Quiscalus mexicanus* in the central highlands of Mexico, as described in the writings of the sixteenth-century Benedictine monk Bernard de Sahagún. The Aztecs called this bird Teotzanatl, which translates as "divine or marvelous grackle". Haemig (1978) then quotes Sahagún to explain why:

> It is named *Teotzanatl* because it did not live here in Mexico in times of old. Later, in the time of the ruler Auitzotl it appeared here in Mexico. For he commanded that they be brought here from the provinces of Cuextlan and Totonacapan. It was made known especially that those which came here were to be fed... And when they were still esteemed, no one might throw stones at them. If anyone stoned them, they chided one another... 'What are you doing over there? Do not shout at, do not stone the lord's birds!' Sahagun ([1577] 1963) cited in Haemig (1978)

Subsequently, Haemig (1979) suggested that a deliberate prehistoric release may also explain the highly restricted distribution of the painted jay *Cyanocorax dickeyi* in the highlands of western Mexico. The closest relative of the painted jay is the very similar white-tailed jay *Cyanocorax mystacilis*, found over 4,000 km to the south in lowland Ecuador. Haemig (1979) proposes that these two are actually the same species, and that the Mexican population derives from releases of birds traded up the South and Central American coasts. Man-made artefacts of South American style found in western Mexico support the existence of trade between these regions. Whether or not this hypothesis is correct can of course now be determined using genetic analyses unavailable when Haemig wrote his article, although as yet the relevant test has not been performed.

Despite increasing evidence of a long history of avian introductions, most exotic bird species derive from the period of the great European diaspora in the eighteenth to the twentieth centuries (Long 1981; di Castri 1989; Crosby 1993; Mack, et al. 2000; Chapter 2), and especially from the activities of British settlers. The three regions with the greatest number of introductions are, in order, Australasia, Pacific islands, and the Nearctic. These are, of course, all regions where settlement in the most recent centuries has been dominated by Europeans (Blackburn and Duncan 2001b). The predominance of the British in producing exotic species becomes apparent when one considers that roughly 40% of all known introduction events concern releases in just four geopolitical areas: Hawaii, New Zealand, USA, and Australia (see also Chapter 2). The last three of these were British colonies for greater or lesser periods of time, while the fourth, Hawaii, includes the British union flag on its State flag.

In many cases, initial enthusiastic but uncoordinated attempts by Europeans to establish exotic species gave way to organized attempts to enrich entire faunas. The aim of the earliest such organizations was to bring animals and plants from around the world to enrich the homelands from which the colonists had departed. The first systematic introductions were probably carried out by the Zoological Society of London, which had amongst its goals "the intention of bringing species from . . . every part of the globe, either for some useful purpose or as objects of scientific research", as early as 1847 (McDowall 1994). However, as McDowall (1994) further noted, "the Society also had other objectives, and having few funds for acclimatisation, made little progress in that area". Nevertheless, it was not long before 'acclimatization societies' were formed specifically for the purpose of naturalizing exotic species. The first of these was started in France in 1857, followed shortly after by the first British society. It was the neo-European colonies that took up the task of acclimatization most enthusiastically, however.

The apogee of this activity was probably to be found in New Zealand, as relayed in accounts by Thomson (1922), Lever (1992), and McDowall (1994). Here were located more than half of the acclimatization societies formed around the world after 1863, and with near comprehensive coverage of the entire country (Lever 1992; McDowall 1994). These societies both answered the call to send interesting species back to the 'Old Country', and strived to import new and useful species into New Zealand to the benefit of colonies that had themselves only recently become established. The societies counted amongst their active members Premiers, cabinet ministers, and men such as James Hector, Frederick Hutton, Thomas Kirk, and Walter Buller, whose names are prominent in the history of New Zealand science. The records they kept of their activities provide a valuable resource for understanding the invasion process.

Europeans deliberately attempted to establish exotic bird species for a broad range of motives. Food was again an important one, but the wide range of domesticated species farmed by Europeans meant that many such releases were less important as a source of meat than for the 'sporting' opportunities that hunting them provided. In New Zealand the introduction of birds for hunting took on the added element of escaping the rigid European class system. The European colonists of New Zealand, for example

> recalled the sport which was forbidden to all but the favoured few, but which they had often longed to share in . . . the grouse on the heather-clad hills, the pheasants in the copses and plantations, the . . . partridges in the stubbles . . . and there rose up before their vision a land where all of these desirable things might be found and enjoyed. (Thomson 1922)

Thus, "in early colonial New Zealand, there was nothing to prevent them from going hunting or fishing, except that the right species were not present" (McDowall 1994). Yet, "New Zealand should swarm with game. There is no destructive animal or reptile: the climate and soil are perfection . . . and there is the finest cover and perpetual profusion of the finest foods for everything from jack-snipe to elephants" (Hursthouse 1857). New Zealand now plays host to a wide range of game bird species, from California quail *Callipepla californica* to wild turkey *Meleagris gallopavo* (Heather and Robertson 1997), and, by 1922, Thomson was able to write that "so-called acclimatisation societies are today only angling and sporting clubs". Indeed, game birds are amongst the most frequent subjects of introduction attempts and are over-represented in lists of exotic species worldwide (Chapter 2), despite frequent difficulties in getting such species established (Long 1981; McDowall 1994; Duncan, et al. 2001).

However, sport was far from the only motive for naturalization, and more practical aims were also entertained, such as reducing the ill effects of insect pests of agriculture on the livelihoods of growing European populations. Julius von Haast of the Canterbury Acclimatisation Society noted that "at the beginning the "sporting" side played a subsidiary part. The Society was concerned not so much to bring out birds for massacre, as to import those that would be useful to the community in the destruction of insects and other pests" (von Haast 1948). The reality of relying on birds as biocontrol agents has been questioned, given that most early introductions to New Zealand were evidently of sporting species (McDowall 1994), but it is none the less the case that biological control has motivated some avian naturalization events. For example, black swans *Cygnus atratus* were introduced to Christchurch, New Zealand, to control water cress, an exotic plant that itself was feared would hinder the establishment of exotic trout (McDowall 1994). New Zealand also imported several passerine species, including dunnock *Prunella modularis*, European starling, and Australian magpie *Gymnorhina tibicen*, to control the "blasting plague of insects, which crawled over the country in vast hordes" (Drummond 1907).

New Zealand is certainly not the exception in terms of colonists employing birds as potential biocontrol agents. Some of the exotic bird species introduced for biocontrol have come to dominate the avifaunas into which they were released, such as the great kiskadee *Pitangus sulphuratus* on Bermuda. Other species were considered successful enough in this role that they were repeatedly introduced across several locations, as is the case for the common myna. There are also scattered examples of exotic bird species being introduced to control other elements of the natural world that colonists found noxious. For example, the Chimango caracara *Milvago chimango* and turkey vulture *Cathartes aura* were evidently introduced to Easter Island and Puerto Rico, respectively, to act as scavengers (Lever 2005).

Exotic bird introductions were not just for functional reasons. As the epigraph to this chapter shows, some people adopted naturalization as a hobby, apparently for little reason other than to see what could persist at their location (e.g., Guild 1938; 1940). Some introductions were also motivated by aesthetic amenity and nostalgia. For example, many exotic species have brightly coloured plumage, or musical song, and seem likely to have been released for those reasons alone. The liberation of golden pheasants *Chrysolophus pictus* and silver pheasants *Lophura nycthemera* in Europe are probable examples of this (Long 1981). McDowall (1994) found it hard to see why the common nightingale *Luscinia luscinia* would have been introduced to New Zealand for reasons other than its beautiful song. Lever (1992) quotes the 1881 annual report the South Australian Acclimatisation Society on the introduction of songbirds and relates that these birds were released "in the hope that they may…impart to our somewhat unmelodious hills the music and harmony of English country life". Later in his book, Lever quotes the same Society as saying that exotic "goldfinches now beautify the hedgerows" of South Australia. Across the Tasman Sea, such species were desired because they "would greatly contribute to the pleasures of the settlers of New Zealand and help to keep up those associations with the Old Country which it was desired should be maintained" (Wellwood 1968). Yet, as McDowall (1994) notes in the title to his chapter 24, such species were "nostalgic, but often a nuisance".

Most current major religions stress the importance of good deeds to ensure good karma and eternal life. The precise nature of these good deeds varies across religions and sects, but at least in the eastern philosophies of Buddhism and Taoism, releasing animals back into the wild is one way in which good karma can be accrued. There is thus good business to be done in countries where these religions predominate, selling wild-caught animals for release. Agoramoorthy and Hsu (2007) estimate that on the island of Taiwan alone, religious adherents spend around US $6 million annually to set free around 200 million wild animals. Severinhgaus and Chi (1999) reported that 29.5% of Taiwanese people (of all religions) participate in prayer animal release, and that 6% of the 68,538 prayer birds available for sale in Taiwan were exotic. Agoramoorthy and Hsu identified seventy-five species of exotic birds from their own personal observations in the wild in Taiwan. Of these species, eighteen were reported to be breeding and eight were known to be successfully established and expanding (Agoramoorthy and Hsu 2007). Given that the ritual release of birds and other species is likely to be most common in east and south-east Asian countries for which good inventories of exotic species are relatively poor, the number of exotic bird species established in these areas may be greatly underestimated by compilations such as Long (1981) and Lever (2005).

Exotic populations derive not only from the planned release of individuals, but also from unplanned releases. These may be unplanned but deliberate, or just plain accidental. Examples of the former include the liberation of cage birds by owners incapable of, or unwilling to, look after their pets, or by the bird breeders and pet dealers themselves, under the influence of economic fluctuations in the cage bird market (see section 2.2.3). The Hispaniolan amazon *Amazona ventralis* was apparently introduced to Puerto Rico when birds were liberated after authorities there refused to permit the importation of a consignment brought from Hispaniola by boat (Forshaw 1973). A recent example of a species establishing an exotic population after hitching a lift on human transport is the house crow *Corvus splendens* in the Netherlands (Ottens and Ryall 2003). From a natural distribution in the Indian subcontinent, the house crow has established exotic populations across the Indian Ocean, from Australia to Africa (Nyári, et al. 2006). In many of these areas the crow is a pest of crops and injurious to native bird species (Long 1981), and in South Africa an eradication programme is being mooted. The Dutch population of house crows has recently started to breed (Ryall 2003).

Finally, exotic bird species have also been introduced for the purpose of their conservation. At the start of the twentieth century, the greater bird-of-paradise *Paradisaea apoda* was thought to be at risk of extinction within its native range of New Guinea because their feathers were highly valued by the millinery trade. Sir William Ingham purchased the island of Little Tobago in the Caribbean as a sanctuary for the species, and equipped an expedition that translocated there almost fifty individuals from the Aru Islands off New Guinea. The species persisted on Little Tobago for a considerable period of time (Long 1981), but is now apparently extinct. Exotic populations of some other species, while not originally established for such purposes, do now have a conservation role. Examples include the mandarin duck *Aix galericulata* and golden pheasant, both of which have exotic populations in the United Kingdom but are now threatened with extinction in their native Asia (Madge and Burn 1989; Gibbons, et al. 1993).

Overall, the history of avian translocations is one of continual change. The original motives of direct and easy access to food and feathers shifted to visions of the enhancements to the natural environment that abundant game and pretty birds with an appetite for pests would bring. Eventually, deliberate releases would become legislated against in many parts of the world (see e.g., McDowall 1994), and most of the later establishments were the result of accidents or unplanned liberations. The identities and characteristics of the species translocated changed accordingly. The locations associated with translocations also changed, as the sites of the dominant roving civilizations shifted. In Chapter 2, we discuss some of the consequences of these changes for global patterns in the transport and introduction of exotic bird species. Before addressing pattern and process in avian invasions, however, it is necessary to provide a framework within which the causes and consequences of such invasions can be analysed, and, it is hoped, understood. It is this framework—the invasion pathway—that we now elucidate.

1.3 The Invasion Pathway

Successful establishment at an exotic location is the end point of a process that begins in the native range. For a species to change from one with purely native populations—populations in areas where the species evolved or to which it spread without the direct intervention of humans—to one with free-living exotic populations, some of the individuals in the native population must negotiate a process that presents a number of hurdles, at each of which a species may fall for a range of different reasons. We term this the 'invasion pathway', following Lockwood, et al. (2007). It can be usefully illustrated as a sequence of stages (Figure 1.2).

The concept of the invasion process as a series of stages dates back at least to Williamson and Brown's (1986) analysis of British invasions, as part of the Scientific Committee on Problems of the Environment (SCOPE) programme on the ecology of biological invasions. Williamson and Brown noted that the

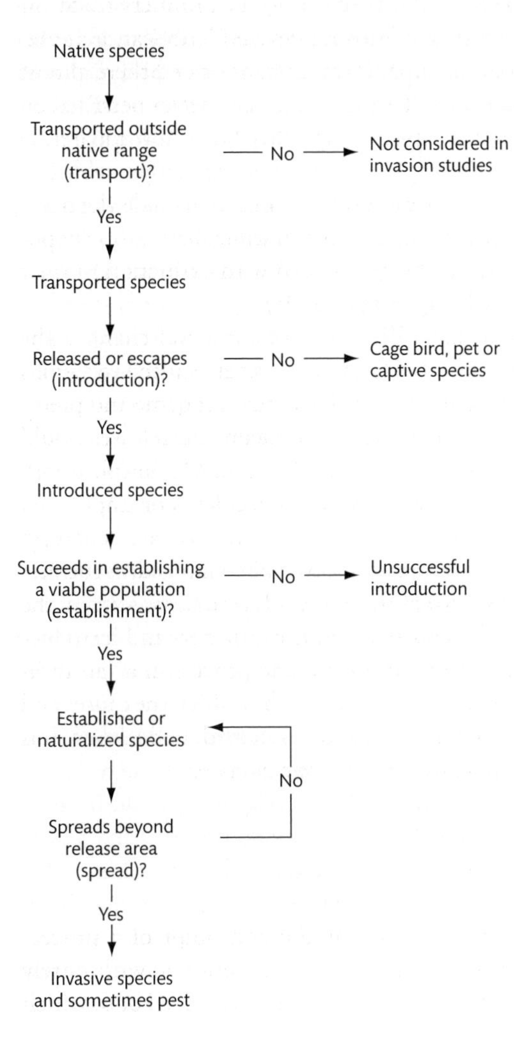

Fig. 1.2 Schematic representation of the invasion pathway, the transitions in the process of a human-caused invasion that we recognize in this book (following Duncan, et al. 2003; Sol, et al. 2005a).

proportion of introduced species becoming established, and the proportion of established species becoming pests, both tended to be around 0.1, and that this figure was relatively consistent across different taxonomic groups (see also Lodge 1993). Later, Williamson (1993) added a third proportion—the proportion of those imported that subsequently appear in the wild—and suggested that this also tended to lie close to 0.1 for different taxonomic groups. These proportions formed the basis of the widely cited 'tens rule' (see Williamson 1996), which for many years was the only accepted generalization in invasion ecology. A key point of this rule is that since only 10% of imported species make it into the wild, only 10% of those introduced species establish in the exotic location, and only 10% of these established species go on to become pests, only 1 in 1,000 imported species go on to become pests. Thus, the vast majority of species that find their way out of their native range thanks to human activities are deemed harmless. Whether or not we believe the 'tens rule', and the support for it provided by birds is patchy at best (Figure 1.3; see also Williamson 1996; Cassey, et al. 2004b; Jeschke and Strayer 2005), its importance lies in the explicit treatment of an invasion as a series of stages.

The number of stages that the invasion pathway is broken into varies between authors, but four stages are common to most recent treatments (e.g., Lockwood 1999; Kolar and Lodge 2001; Duncan, et al. 2003; Cassey, et al. 2004b). These can be ascribed to Williamson (1996), who identified the stages in terms of the types of species that had passed through them. These species are imported (brought into the country, contained), introduced (found in the wild but not yet with a self-sustaining population), established (with a self-sustaining population), and pest (with a pronounced negative economic and/or ecological impact). Clearly, an exotic species that is a pest is also necessarily an imported, introduced, and established species. It is more standard now to think of the invasion pathway as the series of stages through which an invasive species has had to pass. We adopt a scheme where a successful invader must pass through four stages termed transport, introduction (release or escape), establishment, and spread. The species that successfully pass through each of these stages are termed transported, introduced (or released if deliberate), established or naturalized, and invasive, respectively (Figure 1.2).

Another way to think about this pathway is as a series of filters (cf. Richardson, et al. 2000). All species have the potential to become exotics but the ones that actually make it are those that can pass through the filters imposed by the invasion pathway. Thus, some species are excluded because they do not have the desired characteristics (e.g., dull plumage), or because their geographic distributions do not locate them in the right place to be collected for translocation by humans (e.g., eastern Siberia). Some species may be excluded because their rarity means that sufficient individuals cannot be captured to found a new population, or because they cannot be kept alive in captivity. Still more may be excluded because they

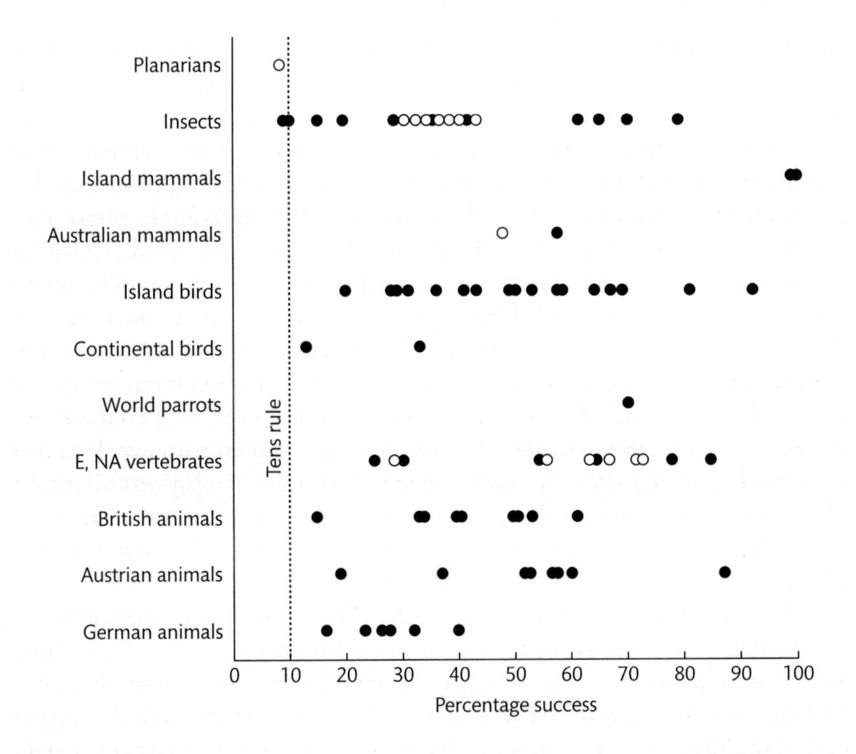

Fig. 1.3 Evidence for the 'tens rule', as applied to data for the establishment (filled circles) and spread (open circles) stages of invasion. Each point refers to a study of a given taxon at a given location. If the rule applied, most points would lie close to the vertical dotted line, indicating that 10% of introduction events resulted in establishment, or that 10% of established species became invasive. Reprinted with permission from Jeschke and Strayer (2005). © 2005 National Academy of Sciences, USA, where data sources are listed.

cannot survive in the exotic location, or because they survive but cannot achieve a positive population growth rate. From this list it should be clear that a filter can act at different points in the invasion pathway, and that different types of filters act at different points.

There are several key advantages that accrue through thinking about an invasion as a series of stages. First, with explicit acknowledgement of the invasion pathway comes the recognition that different stages present diverse and distinct challenges to the species entrained. At each stage, different filters act representing different social and biological factors (Kolar and Lodge 2001), and different characteristics may be advantageous in allowing species to navigate them. Traits that predispose a species to enter and survive the transport stage, for example, may not dispose the species to succeed following introduction. Second, if there are indeed identifiable traits that predispose a species to enter and survive one stage, it is obvious that the set of species being fed into the next stage is unlikely to be a random subset of all possible species.

It follows from these first two points that the characteristics that drive invasion success should ideally be studied at each stage in the invasion pathway separately. This approach avoids the possibility that cross-stage interactions produce misleading conclusions about the correlates of success by conflating the characteristics of species that have failed in the invasion pathway at different stages for dramatically different reasons (Prinzing, et al. 2002; Cassey, et al. 2004a; Blackburn and Jeschke 2009). Thus, while some studies have considered whether the characteristics of species that succeed in establishing are a non-random subset of all other species (e.g., Ehrlich 1986), such comparisons are likely to be uninformative about the true determinants of establishment success. The reason is that they compare the characteristics of species that have succeeded in establishing with the characteristics of species that have failed at earlier stages in the invasion pathway (Figure 1.2), and so have never had the chance to establish. This latter set includes species that have never even been considered for introduction. For example, if only large-bodied species are transported and released, it is inevitable that successful introductions will be large-bodied, even if body size has no bearing on establishment success at all. This is analogous to the problem of source pool specification in studies of the composition of island faunas (e.g., Graves and Gotelli 1983; Schoener 1988; Gotelli and Graves 1996). Only by analysing sets of species that have been exposed to any given stage on the invasion pathway can the characteristics driving success and failure be correctly assessed.

A third advantage of the invasion pathway model is that it helps to clarify the point(s) in the process where different factors may act to promote or retard an invasion. For example, it is entirely clear that features of the recipient community, such as species richness or evenness, cannot prevent an invasion at the transport or introduction stages. Conversely, features such as aspects of a bird's life history may act at several stages, but the invasion pathway model helps clarify the potential for synergistic or antagonistic effects of such factors at different stages.

Fourth, the invasion pathway model also helps to clarify the types of characteristics that are likely to influence success at each stage. To address this point, however, we need first to consider what the different types of such characteristics actually are.

The factors that may determine whether or not a species becomes an exotic invader can be divided into three categories: species-, location-, and event-level characteristics (Blackburn and Duncan 2001b; Duncan, et al. 2003; Sol, et al. 2005a; see also Williamson 1999; 2001). Species-level characteristics are intrinsic traits of the species concerned, such as body mass, reproductive rate, or geographic range size. Location-level characteristics are features of the location in which the introduction and, if the introduction proceeds to establish, post-establishment spread occurs. Examples include climate, structure and composition of the native biota, and insularity. Event-level characteristics are those that are associated with, and often unique to, specific cases of the invasion pathway. Examples of these include such things as the number of individuals released, the condition or composition (i.e., age, sex) of the individuals, time of year, date of introduction, and weather. Species-level and location-level characteristics are factors that will be identical (or nearly identical) for every invasion that involves a given species, or to a given location. Event-level characteristics, in contrast, are factors that may differ between different invasion events involving the same species to the same location. They can also be viewed as factors that attempt to quantify the idiosyncrasy of invasion events.

Modelling an invasion as a series of stages helps to delineate where in the invasion pathway different types of characteristics are most likely to influence success. Notably, the factors that influence transport and introduction will generally be species-level, and to a lesser degree, location-level characteristics. Certainly location may influence the likelihood that a species is introduced, because of variation in local laws or adherence to international treaties. Nevertheless, it is the characteristics of the species themselves that determine the value of the species to importers and exporters, or the likelihood that a species will hitchhike successfully. Conversely, the interaction of species traits with the recipient environment and idiosyncrasies of the introduction event means that the establishment and spread stages of invasion will be much more dependent on features of the specific population released, and hence on event-level characteristics. Species- and location-level effects are of course likely to play a role in moderating or overcoming the effects of such idiosyncrasies, but it is at these later invasion stages that event-level characteristics are most important.

Not all introductions of a given species have the same establishment outcome, and likewise nor do all introductions to a given location. Moreover, the introduction of a given species to a given location does not necessarily result in the same invasion outcome every time. Indeed if these conditions were generally true, it would be very easy to make predictions in invasion ecology (Williamson 2006)! This degree of uncertainly holds even for species that we tend to think of as common 'winners' in the bird invasion game. For example, the European starling did not succeed in establishing a viable population on its first introduction to New York, but did following a subsequent release (Wood 1924). The successful establishment could have been due to subtly different effects of the precise location (e.g., the presence or absence of specific predators) or the species introduced (e.g., circannual propensity for dispersal), or not so subtle effects of the given event (e.g., numbers released, their age, or sex ratio). Whatever the reason, however, it follows that whether or not establishment occurs is a feature of the specific population introduced. That is also true of spread following establishment. In contrast, transport and introduction are features of the species, because a species either is transported (and introduced) or it is not. Therefore establishment and spread should be analysed at the population level, and transport and introduction at the species level.

Up to this point, we have considered how thinking about the invasion pathway as a series of stages helps in understanding the causes of avian invasion success. It clarifies that different stages make different demands on species, that we need to assess the challenges addressed by each stage separately, and that this separation helps to identify how and where in the process different types of factors will have the greatest influence. However, these benefits are tempered by the methodological context within which the analysis of avian invasions must generally be performed. We now need to consider how the types of data that we have available affect our analyses of the invasion pathway.

1.4 Analysing the Invasion Process

> One thing clearly emerges from any study of the history of acclimatisation and it is this, that no scientific method was followed in selecting species for introduction.
> *Annual Report of the Otago Acclimatisation Society* (1964).

There is a variety of methodological approaches that biologists can take when developing and testing hypotheses (Diamond 1986; McArdle 1996; Gaston and Blackburn 1999). The most powerful is the controlled manipulative experiment, where the effect of a process on a biological system is compared to a control system that is (ideally at least) identical in every way except for the process of interest. Unfortunately, this approach has several shortcomings with respect to the study of avian invasions. The most important is that ethical considerations preclude most experimental treatments that involve the deliberate introduction of exotic species, or the deliberate lethal removal of vertebrates for scientific purposes only. It is also difficult to see how manipulative experiments could be performed that would inform on some of the key aspects of transport, introduction, or invasion, such as selectivity (Chapter 2) or causes of spread (Chapter 6). These shortcomings of manipulative experiments, however, are circumvented by experiments in nature (Diamond 1986). These experiments are changes in the natural environment brought about by human activity that provide indirect opportunities to study the consequences of such changes to systems. Human-mediated avian invasions represent a classic example of this kind of experiment. The long history of such invasions and the large amount of historical data that are available as a result, represents an invaluable resource for analysing the invasion process (see also

Richardson, et al. 2004; Fridley, et al. 2007). Moreover, these avian experiments in nature concern spatial and temporal scales, and patterns of change, that are more relevant to natural processes than would be those achievable by any manipulative experiment.

However, historical data, and indeed experiments in nature more generally, are not without their shortcomings. It is important for readers of this book to be aware of these issues going forward as they should temper interpretation of the information we present, for both better and worse. One problem with experiments in nature is that they lack control treatments. This situation makes it harder to assess what processes are actually driving changes in the study systems, or how the system would have changed in the absence of the human intervention. For example, if we observe higher establishment success for bird populations introduced to competitor-poor relative to competitor-rich locations we may infer that establishment is easier in the former, but without experimental controls it is difficult to establish whether it is actually the number of competitors (and not some other feature of the environment with which competitor richness is correlated) that is driving this result. Similarly, we may observe a decline in native species' abundances following the invasion of an exotic competitor, but without controls it is difficult to establish conclusively that the two are causally linked, and that the declines would not have happened even in the absence of the invader.

A second problem with experiments in nature, such as historical invasions, is that the manipulations to systems are not designed with the aim of testing the effects of specific processes. This situation means that the data tend to be horribly confounded with respect to different processes, and hence that untangling the causes of their effects is difficult. For example, Sol (2000a) found that establishment success for bird species introduced to Hawaii was significantly higher than for bird species introduced to the continental United States (43.7% vs 13.3%). However, the sets of bird species introduced to each location also differed in composition. The difference in establishment success may thus be due to the invasion potential of the different bird species introduced to the different locations, rather than to differences in the invasibility of the different locations. Indeed, Sol (2000a) was able to show that there was no difference in establishment success between these two regions when only bird species introduced to both were compared: most species either failed in both places or succeeded in both places. We return to the issue of confounding processes, and suggest some solutions to it, at the end of Chapter 2.

The exotic bird data also suffer from a set of problems in addition to those common to other experiments in nature, that influence the interpretation of analyses (see also Sol, et al. 2008). These problems pervade many of the subsequent chapters, especially those that specifically address the various steps on the invasion pathway.

First, studies of avian invasions vary in their definitions of invader. For example, Pranty (2004) provides a list of 207 non-native bird species that have been reported in the wild in Florida. However, he also identifies subsets of this list for which the species' occurrence has been verified by archived photographic or specimen evidence ($N = 95$), and for which breeding has been reported in Florida outside captivity ($N = 68$). Pranty goes on to note that it is uncertain whether or not the individuals of some reported species (e.g., red-legged honeycreeper *Cyanerpes cyaneus*) originated from captivity or as natural vagrants. This example illustrates that what constitutes a genuine introduction is often unclear, and noise may be added to analyses by the inclusion of species on the basis of the escape or release of the odd individual, or on vagrancy from other (perhaps established exotic) populations. This issue is rarely considered explicitly in invasion studies and the work we cite in this book is no exception.

Similarly, definitions of location and event also differ across studies. For example, Moulton, et al. (2001b) studied galliform introductions to the Hawaiian Islands and New Zealand. Their unit of analysis was the individual island for Hawaii, but the entire country for New Zealand. Duncan and Blackburn (2002) subsequently pointed out that introductions to New Zealand were carried out by acclimatization societies based in different administrative regions. They suggested that it was more appropriate to analyse galliform introductions by region within New Zealand, with the result that the number of introduction events differed markedly between the two studies.

Variability in definitions of invader, event and location means that no two analyses of any stage in the invasion pathway concern identical samples, in terms either of size or composition. Indeed, we ourselves cannot say exactly how many species have been subject to introduction attempts (although the total number is now certainly >500), how many introduction events this involves (certainly >2,000), or how many species have established exotic populations as a result (perhaps >250). This problem is compounded by incomplete data for various factors of interest for specific analyses, which modifies samples further. Readers will undoubtedly notice the variation in sample size across the analyses we review here, and for the new analyses we present. This variation is inescapable in the circumstances. Nevertheless, it is possible to think of all these analyses as concerning samples of the same universe of invasion events, and hence that they provide us with information about the characteristics of this universe. In that sense, one benefit of having them all presented in a single book is that it gives the reader an unprecedented opportunity to glimpse this universe and to draw their own conclusions about the factors that influence the invasion process.

Second, many, if not most, studies of the invasion pathway in birds analyse data that are out of date to a greater or lesser degree. For example, in his summary of the exotic avifauna of Florida, Pranty (2004) noted that only eighteen species are officially considered as established breeding species (American Birding Association 2002), but that a further nineteen species are known or presumed to breed in the State. Pranty's list of Floridian exotics also includes more species of Neotropical origin than a previous analysis of global introductions (Blackburn and Duncan 2001a). Outside of Florida, and in other regions that have dense human populations (including groups of dedicated birdwatchers), we continue to record new exotic bird establishment events. For example, in 1995, the ashy-throated parrotbill *Paradoxornis alphonsianus* (Timaliidae) was recorded for the first time in Italy, at Brabbia Marsh Regional Reserve <www.surfbirds.com/blog/birdingitalynet/515>. Subsequently a second species, the vinous-throated parrotbill *Paradoxornis webbianus*, was also recorded from that same marsh. These populations apparently derive from escaped cage birds, number more than 100 individuals, and are considered to be self-sustaining. Yet, the most recent catalogue of exotic bird species (Lever 2005) includes no parrotbill species. Similarly, in Tokyo and its vicinity, seventy-one exotic bird species were observed in the period 1961–81. During this period the increase in the number of new exotic bird species reported to be breeding in the wild averaged one per year (Narasue and Obara 1982). Eguchi and Amano (2004) list forty-three exotic bird species as having been recorded breeding in Japan at least once, whereas a similar list published in 1990 included only twenty-four species. Eguchi and Amano acknowledge the poor quality of the historical data, but note that some of the additional species they list are likely to be recent colonists. At least three of these species are not included in Lever's (2005) catalogue. Finally, Leven and Corlett (2004) list nineteen species that they consider to have colonized Hong Kong following human-mediated introduction. All species were first recorded in the twentieth century, ten are missing from Lever (2005), and the accumulation of species shows no signs of slowing (Figure 1.4).

Clearly there are far more exotic bird species than current catalogues would lead one to believe, and it is recently established species that are most likely to be missing from them. Moreover, if more recent introductions are more likely to be unplanned or accidental (see Chapter 2), it will be harder to assess the frequency and scale of ongoing introductions (Eguchi and Amano 2004).

Arguably, relying on catalogues of historical invasions, such as Long (1981) and Lever (1987; 2005), has the advantage that the populations listed have been in the non-native location long enough that the outcomes of the invasion events are known with a degree of certainty. For many recent introductions, such as the parrotbills in Italy mentioned above, the longer-term fate of the population remains unclear. Using older data at least removes an element of uncertainty from the analysis of introductions. Nevertheless, it is equally true that the omission of more recent invasion events has the potential to bias analyses. For example, treatments of non-randomness in introduction (Chapter 2) will suffer from a failure to include all introductions, especially in considering how the composition of introductions

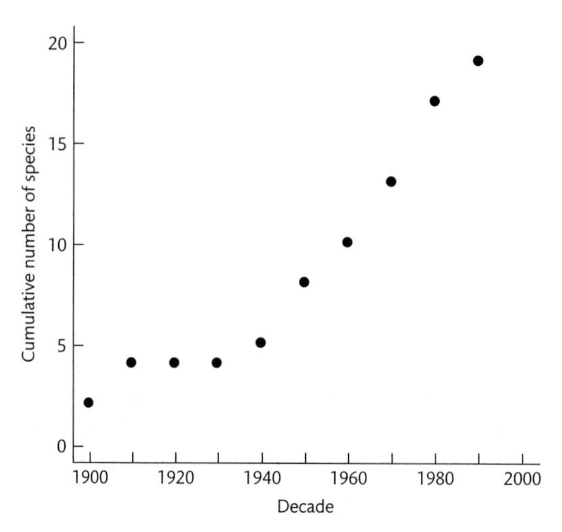

Fig. 1.4 Cumulative number of exotic bird species recorded as established in Hong Kong. From data in Leven and Corlett (2004).

changes over time (section 2.2.3). These analyses will lose generality as a result. It would be of extreme practical utility to assemble and continually update systematic, global data on exotic bird populations to rectify this problem (cf. <www.issg.org/database>).

The third problem with bird invasions is that many of the data we would like to have for a comprehensive understanding of the invasion process are simply lacking. This problem is especially evident for the early invasion stages (see Chapter 2). We have very little systematic information on which species are transported beyond their geographic range limits, in what numbers, and which species are held in captivity and not subsequently released. Those data that we do have tend to be heavily proscribed in terms of taxon (e.g., Cassey, et al. 2004b), location (e.g., Nash 1993), or time (e.g., Thomsen, et al. 1995). Early introductions, at least in the historical period, tended more to be the outcome of planned attempts to establish certain desirable species in certain locations. Therefore some care was taken to record important aspects of the introduction process, such as the number of individuals introduced (Chapter 3). Even so, Phillips (1928) was moved to write that the 'real importance should be emphasised of properly recording all these bird introductions. In the past there has been more often than not an absolute neglect to record such facts in available places... The result is that probably 90 per cent of these biological experiments are lost to science'. More recent introductions increasingly are unplanned side effects of the cage bird trade (Chapter 2), and so are even less well documented than this (Eguchi and Amano 2004). Historical data on avian invasions are thus riddled with imperfections. Nevertheless, these imperfections are outweighed by the fact that the data exist at all.

1.5 Précis

However one might feel about arriving at a new place and being met at the airport by a familiar exotic flora and fauna—and parallels have been drawn with the negative feelings evoked by finding the same fast food companies selling the same products in every city in every country around the world (Cassey, et al. 2005a)—it is none the less true that the presence of such species provides a biologist with golden opportunities. These are opportunities that an increasing number are taking, such that the field of invasion ecology is rapidly burgeoning (Figure 1.5; Lockwood, et al. 2007). It is fifty years since the publication of Charles Elton's (1958) now classic monograph on the ecology of invasions by animals and plants, and more than a decade since the first modern synthetic review of the field, by Mark Williamson (1996). Even in the dozen years since the publication of Williamson's book the study of invasions has taken huge strides forward.

So where does the study of exotic birds fit into the invasion ecology picture? A review of geographical and taxonomic bias in invasion ecology by Pyšek, et al. (2008) identifies three trends that speak to this question. First, given the absolute number of exotic birds that have established in the wild, species within this taxonomic group have received disproportionately limited research attention within invasion ecology. This result was surprising to us since there is the perception that in other fields, such as conservation biology, birds are too often studied given their limited contribution to the entire list of the Earth's species. In fact, the paucity of studies of exotic birds is probably not as great as it appears in Pyšek, et al. (2008),

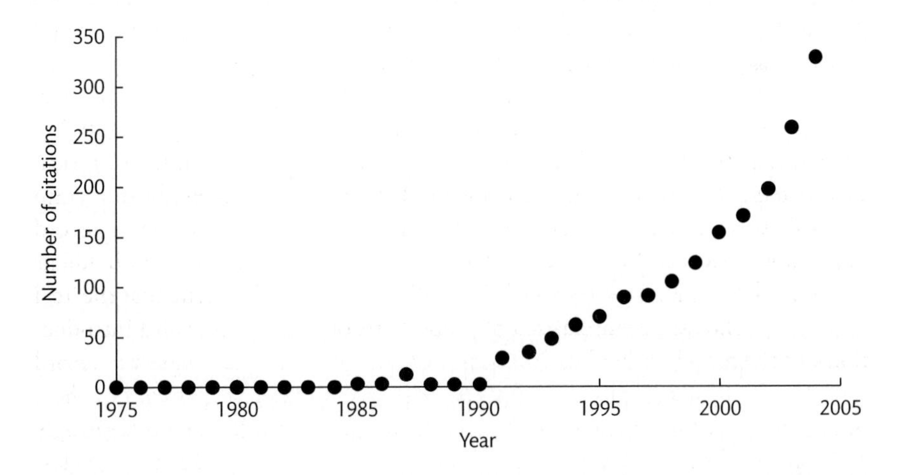

Fig. 1.5 Number of citations that included the terms 'invas*' and 'ecolog*' returned from a search of the Science Citation Index between the years 1975 and 2005. Reprinted with permission of John Wiley & Sons Ltd from Lockwood, et al. (2007).

because the authors only address detailed species-level studies, whereas many exotic bird studies are multi-species comparative analyses. Nevertheless, there is still a bias against autecological studies of exotic birds.

The remaining two trends go a long way towards explaining the first. Pyšek, et al. (2008) show that there is a profound geographical bias in where invasion ecologists conduct their research. Predictably, Europe and America are over-represented in terms of research locations, although this may be justified to some extent because these areas also seem to hold a large relative number of exotic species. Oceanic islands, on the other hand, have very high numbers of exotic species but are uncommon locations for invasion ecology research. Given that exotic birds have preferentially been introduced to islands (Chapter 2), the shortfall of research on birds may be a consequence of the propensity for oceanic islands to be ignored within invasion ecology. The third trend brought to light by Pyšek, et al. (2008) is that research energy seems to be focused on exotic species that are perceived to produce the most economic or ecological harm.

Exotic birds have never been widely considered as 'pests' and thus the failure of invasion ecologists to focus on them may also be the result of the attitude that birds are curious little additions to urban areas but not terribly detrimental to native biodiversity. In fact, bird introductions have been dominated by purposeful attempts. Exotic birds serve as the foundation for our multi-billion dollar domestic poultry industry. The pet bird trade is well established and ornamental species grace our many wildlife and zoological gardens worldwide. Pimentel, et al. (2000; 2001) estimated the economic losses to exotic species in the United States, United Kingdom, and four other countries. While the manner in which they arrived at their estimates for losses can be debated, it is the comparative impacts of birds relative to other species that is most important here. Pimentel and colleagues suggest a total of US $2.4 billion per year in damages due to exotic birds in the six countries examined, through effects including damage to buildings, damage to crops, and the spread of poultry diseases. This is less than 1% of the total cost of US $314 billion estimated for all exotic species in the six countries. Arguably, then, in comparison with the costs of the direct effects (and their control and amelioration) of other recently established non-native taxa (plants, mammals, insects, diseases), the impacts of exotic birds are a minor inconvenience.

Nevertheless, while US $2.4 billion may be a small proportion of the total cost of exotic species in the six countries studied by Pimentel, et al. (2000; 2001), it is still a significant sum in its own right. Moreover, birds apparently contribute disproportionately to the total, since they produce 0.76% of the cost but comprise only around 0.2% of all exotic introductions to these countries (Pimentel, et al. 2001). In Australia, at least fourteen bird species have established widespread mainland populations of which nine (64%) are considered to be either moderate or serious pests. The annual cost through agricultural loss as well as managing these species (control and research) is estimated at US $12 million (Bomford and

Hart 2002). Thus, it seems to us naive to ignore the economic impacts of exotic birds compared with other vertebrate taxa.

In the broader context of invasion ecology, this book provides a needed push towards a greater focus on avian invasions and their influence on native biodiversity. In return, we believe that the study of exotic birds can bring deeper insight into our broad understanding of invasion ecology. As the previous section demonstrates, historical data on avian invasions are riddled with imperfections. Nevertheless, relative to most other plant and animal taxa, the historical data for birds are exceptionally good! This is especially true for the introduction stage of the invasion pathway: as the composition of this book demonstrates, studies of establishment make up a relatively large fraction of the bird invasion literature. Moreover, birds have long been the subject of disproportionate attention by naturalists, due to their beauty, diversity, and visibility, and this attention has been carried over into the scientific sphere. There is thus a huge bank of supplementary information on their distribution, life history, ecology, and evolution, which can be used to augment the invasion data with a view to understanding the processes that drive species along the invasion pathway. We hope this book will prompt a greater focus on birds within invasion ecology so as to achieve both theoretical and practical goals.

In Chapter 2, we assess the first two steps on the invasion pathway: transport and introduction. We start by summarizing why what we know about the history of avian invasions leads us to expect non-randomness in the transport and introduction processes. We proceed to review evidence for non-randomness in the types of birds that get transported and introduced, where they are transported and introduced to, and, finally, changes in the identity of these bird species through time. We consider non-randomness in transport and introduction together because we have little information on them as separate processes. We finish the chapter by outlining the significant consequences transport and introduction have for the study of subsequent invasion stages.

Chapters 3 to 5 focus on the next step on the invasion pathway: establishment. Establishment is the stage that has been the focus of most invasion studies in birds, and so offers a large amount of information to summarize. The factors that influence establishment success have been hard to identify in other invasive taxa because, for most species, we do not have a comprehensive list of which species attempted to establish but failed (Lockwood, et al. 2007). Inasmuch as the historical record of bird introductions provides this information, these chapters fill a substantial gap in what drives establishment success for any exotic species. However, the research on this topic has not always been congruent in its findings, and considerable disagreement exists over the causes of successful establishment.

One point on which most writers agree is that the numbers of individuals contributing to an introduction event (i.e., propagule pressure) is an important contributory factor. In Chapter 3, we present reasons why we would expect this, which relate to the problems of survival that face small populations of any organism, before laying out the evidence for a relationship between propagule pressure and establishment success. We finish the chapter by considering some of the other associations of propagule pressure, especially those which may affect the conclusions of studies of establishment that do not account for it.

In Chapter 4 we consider the role of species-level traits. Previous treatments of such traits have failed to find consistent associations between species traits and establishment success. However, we find that framing the issue in terms of the small population problem again provides useful insights into the role of species-level traits, and allows some general conclusions to be drawn about the roles of population growth rates, the predisposition to Allee effects, and the ability to cope with novelty, in aiding establishment. Our approach to sorting through the influence of species' traits on establishment success should provide a suitable framework for similar explorations within other exotic taxa. We suspect that the more widespread confusion over the importance of species' traits to establishment success is a by-product of failing to view the influence of these traits within the context of the small population paradigm.

Chapter 5 reviews the influence of the recipient location on the probability of establishment success. This subject is another that has a long history within invasion ecology research but that has, in general, produced no clear consensus on what makes a site more or less invasible. This confusion has reigned in the study of avian invasions too, and indeed many readers of this book (if they remember anything about bird invasions at all) will recall heated arguments over the role of competition in determining establishment success. We review and update (dare we say resolve) this argument in Chapter 5. We also consider the array of other biotic interactions that can influence establishment success such as predation, parasitism, and mutualistic interactions. Beyond the influence of species interactions, there is a clear role for the biophysical environment in determining the success of exotic bird introductions, which we review in Chapter 5 as well.

Once an exotic species has successfully established a self-sustaining population, it may spread beyond its initial point of introduction and establish meaningful interactions with species native to this location. In Chapter 6 we review the relatively large literature on models of geographic range expansion (spread) in invasive species. In particular, we note in this chapter the tremendous bias towards a few, very successful and widespread exotic bird species, on our understanding of exotic species range expansion. For the same reasons that exotic birds have already contributed to our understanding of the invasions (i.e., there is detailed information on their numbers and locations), we suggest that with the right research effort they can also help us gain a clear picture of why some exotic species do not expand their ranges once established, or only do so after a long lag period.

In the second part of this book, our focus shifts away from the invasion pathway and onto some of the ongoing research opportunities that exotic species present. As the chapters in recent edited volumes (Sax, et al. 2005b; Cadotte, et al. 2006) prove, exotic species can provide important insights into (and reciprocally be enlightened by) ecology, evolution, and biogeography. Notably, exotic species allow us to study change processes in real time, rather than infer past events; records of the time, place, and characteristics of species introductions allow rate processes to be measured; exotic species derive from so many different species and taxa, and occur in so many different environments, that the full panoply of biological questions is open to investigation using them (Sax, et al. 2005c).

In Chapter 7 we consider how exotic birds interact with native species and how they serve to reshape global biodiversity patterns. Both exotic and native species are distributed unevenly across the environment, such that some areas house more species, and other areas house fewer. The origins of these distributions for exotic and native bird species are undoubtedly very different, yet they share several common features, such as species–area relationships on islands, and latitudinal gradients. We examine whether the same processes produce the same patterns in each set of species, and what this tells us about the causes of distribution patterns in native species, and also in exotics. We then consider the associations that exotic species forge in their recipient communities through their biotic interactions with native species, including native birds.

In Chapters 8 and 9 we consider the long-term outcomes of exotic bird establishment—namely, the propensity for, and pattern of, evolution in exotic bird populations. We show that exotic birds have occasionally lost substantial amounts of genetic variation via the introduction process, but that just as many (if not more) populations have not lost genetic variation. It is difficult to pin down the influence that the loss or gain of genetic variation has on establishment success; this is a research area that is ripe for exploration. The same can be said of the opportunities to generate basic insights into evolution that is provided by exotic birds. We present evidence that exotic birds have evolved in their phenotypes over the relatively short time spans in which they have been established in their new locations. Some of the shifts observed are entirely consistent with macroevolutionary patterns observed amongst native bird species; others are more curious and require us to reconsider some long-held views of evolutionary dynamics. Yet, again, we see great potential for research on the evolution of exotic birds to provide insight into basic evolutionary theory and the role of evolution in the impacts of all invasive species.

Finally, in Chapter 10, we present some overall conclusions from the work we have reviewed, together with our thoughts on prospects for the future. In particular, we summarize the major lessons that have emerged so far from the study of exotic birds, and discuss what those lessons might tell us about invasions more generally. We also highlight some of the important questions that remain to be tackled.

There is one last point that we think needs to be made before we proceed to the main body of the book. The processes that structure ecological assemblages are incompletely understood and controversial (e.g., Lawton 1999; Hubbell 2001; Brown, et al. 2003; Chave 2004; Simberloff 2004; Gaston and Chown 2005; West and Brown 2005). Exotic species are pervasive, and studying their evolutionary ecology and the consequences of their successful establishment is one way in which an understanding of these processes may be advanced. The invasion process itself is also of intrinsic theoretical and practical interest, in terms of understanding the ways in which species spread, and how the likelihood of such spread might be reduced in certain cases. In these respects, invasive species are an opportunity to be exploited (Brown and Sax 2004). Nevertheless, it is an opportunity that we do not think will be impaired by a growing environmental awareness to stem the flow of invading exotic species. Biological invasions have been (and continue to be) a genuine threat to the livelihoods, way of life, and life itself, of populations and species in every biome on earth (see Simberloff 2003). There is nothing to suggest that the rate of invasion is slowing down, and hence that these are problems that will even level off any time soon, let alone simply go away. It is possible for scientists to study these processes with objectivity, but we should not confuse objectivity with neutrality. We would all three of us rather be greeted at airports around the world by the sight and sound of native species.

2

Transport and Introduction

> A hundred years of faster and bigger transport has kept up and intensified this bombardment of every country by foreign species, brought accidentally or on purpose, by vessel and by air, and also overland from places that used to be isolated.
>
> C. S. Elton (1958)

2.1 Introduction

The brief history of avian introductions presented in the previous chapter reveals that, owing to human intervention, a large and varied set of bird species now exist as exotic populations in areas removed from their native geographic ranges. The set of species that humans have attempted to translocate (and have not successfully established) to new regions is even larger. Yet, the key reasons for translocation (primarily game hunting, commercial interests, ornamentation, or unplanned cage bird releases) are relatively few, while the characteristics of the species introduced for these different reasons are likely themselves to differ. For example, species introduced as game birds will tend to be 'larger and tastier' than the average species, and will tend to have behavioural characteristics that make them interesting 'sport'. Ornamental and escaped cage bird species will tend to have showy plumage or musical songs, as these characteristics are likely to enrich the environment of their human owners. Ornamental species may also tend to be large, as the point of that bright plumage is lost if the bird is not obvious to the observer. Thus, it seems likely that the characteristics of species introduced to exotic locations will be a limited set of the traits expressed by all bird species, and also a non-random subset of these traits.

If exotic species are not random with respect to their biology, we would also expect them to be non-random with respect to their phylogeny. Evolution means that closely related species ought to resemble each other, because they will share many features of life history and ecology through common descent rather than through convergent or parallel evolution (Felsenstein 1985; Harvey and Pagel 1991; Freckleton, et al. 2002). It follows that the desirable characteristics that lead to the introduction of a species are also likely to be shared by its relatives, that the introduction of those relatives will be similarly desirable, and hence that exotic species should be clustered into a limited selection of bird taxa.

The characteristics of exotic species will be further constrained by the coincidence of human geography and avian biogeography. Naturalization occurs when a species is moved sufficiently far to escape or be released beyond the natural boundaries of its geographic range. Early on in the history of naturalization, such movements were associated with human colonization events. The greatest period of colonization in human history in terms of numbers of colonists was undoubtedly the period 1850–1930, when millions of emigrants left Europe, primarily for the 'neo-Europes' of the Nearctic and countries in the southern temperate zone (Crosby 1993). There are many recorded exotic bird introductions from this period (Figures 2.1a–b; Jeschke and Strayer 2005), and we would expect their distribution to correlate with the distribution of peoples from the great European diaspora. However, naturalization is a two-way process, and useful species can be moved to Europe as well as from it.

Given that we expect introduced birds to be a non-random subset of all bird species, this non-randomness will be expressed in a variety of ways: in location (of both origin and introduction), phylogeny (species identity), and characteristics (life history and behaviour). Moreover, broad characteristics of species are not distributed at random with respect to phylogeny (Harvey and Pagel 1991) or location (Gaston and Blackburn 2000), while phylogeny and location are also not independent (Kark and Sol 2005), and so these three principal aspects of non-randomness should all themselves be associated with respect to exotic species. It should also be clear that all of this non-randomness is a consequence of the factors

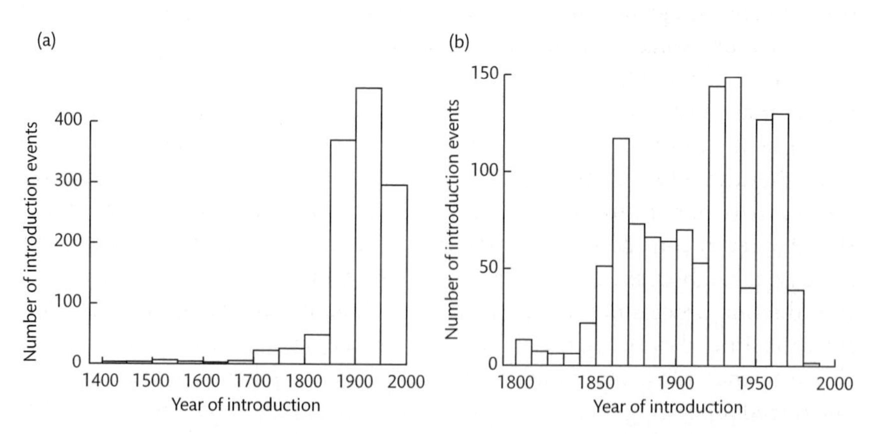

Fig. 2.1 (a) Frequency distribution of year of introduction for the 1,243 introduction events since AD 1400 analysed by Blackburn and Duncan (2001b) for which an estimate of introduction date was available. (b) Frequency distribution of dates of introduction for the 1,127 introduction events in (a) that occurred after AD 1800.

that influence which species become entrained in the transport and introduction stages of the invasion pathway.

Unfortunately, transport and introduction are the hardest stages in the invasion pathway for which to obtain the rigorous data required quantitatively to assess the causes of entrainment. One reason for the difficulty is that the annual global trade in birds is enormous: over 1 million individuals of more than 1,000 species are legally traded annually worldwide (Gilardi 2006). Accidental introductions may be facilitated by increasing volumes of trade, and hence be linked to areas importing large numbers of bird individuals and species for pets or aviculture. Ascertaining the identities of all individuals across every shipment to assess what does or does not get transported, and what does or does not subsequently escape or get released, is impractical at best. And this is for birds—one of the most conspicuous, watched, and well-studied taxa in the world.

Another reason for the difficulty in assessing which species are involved in transport and introduction is that most data refer to historical events. Records of introductions are particularly good from the nineteenth and twentieth centuries when acclimatization societies were importing and introducing bird species (Chapter 1). Yet, what information there is on trade refers only to the most recent decades, and the identities of species in trade are likely to have changed markedly over time. Notably, the current rarity of some species in their native ranges may be a consequence of excessive past trade. An example is Spix's macaw *Cyanopsitta spixii* for which trapping was the proximate cause of the rarity that lead to its eventual extinction in the wild (Juniper and Parr 1998; Birdlife International 2000; Pain, et al. 2006). Attempts to protect threatened species include the imposition of trade bans. Altogether, this evidence suggests that species traded in recent times may differ somewhat from those traded in the past.

The lack of suitable information on transport and introduction separately means that most studies of these early invasion stages concatenate the two (e.g., Jeschke and Strayer 2006). This approach obviously makes it impossible to distinguish where in the early stages of the invasion pathway those native species that are not introduced are filtered out of the process. Thus, we begin this chapter by considering non-randomness in the geography, phylogeny, and biology of the bird species that get transported and introduced, and how the introduction process influences the non-randomness observed. We then consider the limited evidence that currently exists on the characteristics of species that influence whether they succeed in passing through each of the transport and introduction stages individually. Finally, because non-randomness in what gets introduced is linked to the first two stages on the road to becoming a successful invader, we discuss how this non-randomness affects the types of questions that can be addressed, and approaches that need to be taken, in analysing introduction data.

2.2 Patterns in the Transport and Introduction of Birds

2.2.1 Non-randomness in Phylogeny

If introduced species are a non-random subset of all birds, and tend differentially to come from groups with certain desirable characteristics, exotic birds should be clustered with respect to phylogeny. Thus, species chosen for introduction should derive from fewer genera, families, and orders than a same-sized random sample of bird species. As far as we are aware, no test of this prediction exists in the literature, and so we provide such a test here. The basis of this test was the list of 426 bird species that Blackburn and Duncan (2001a; 2001b) identified as having been the subject of introduction events around the world. We randomly sampled 426 species from the 9,702 species listed in Sibley and Monroe (1990; 1993), and counted the numbers of genera, families, and orders in this sample. Repeating the process 1,000 times gives an estimate of the expected distribution (mean and range within which 99% of simulations fall) of number of higher taxa that 426 species would represent were they a random sample of all birds. The random sampling technique produces expectations of 325 (99% range 306–343) genera, 77 (99% range 68–86) families, and 20 (99% range 17–22) orders, on average, being represented in 426 randomly chosen bird species. The 426 introduced bird species actually comprise 228 genera, 49 families, and 14 orders, which are fewer taxa at all levels of the hierarchy than expected by chance alone. We can conclude that the exotic bird species that have been chosen for introduction are clustered within a relatively limited set of bird taxa. The obvious next question is which bird taxa is it that have disproportionately contributed to the list of avian introductions?

Several studies have assessed this question, with broadly congruent conclusions (Table 2.1). The null hypothesis against which the representation of any higher taxon in the list of introductions is judged is that it is represented by no more species than would be expected were the list a random sample of all bird species. The studies in Table 2.1 have exclusively focused on the taxonomic level of the family, although there is no fundamental reason why analysis should not be at higher (e.g., order) or lower (e.g., genus) levels. Analysis of global introduction data reveals that families with significantly more introduced bird species than expected are the Phasianidae, Passeridae, Psittacidae, Anatidae, and Columbidae (Blackburn and Duncan 2001b). These five families contain more than half of all introduced bird species, despite comprising fewer than 15% of all extant bird species. Other families with relatively high representation on the list of introductions included Rheidae, Odontophoridae, and Fringillidae (Table 2.1). This taxonomic make-up strongly reflects the introduction of birds to new locations primarily for the purposes of providing hunting (game birds, wildfowl, pigeons, New World quails) or for their aesthetic qualities (parrots, game birds, and wildfowl again),

Table 2.1 Taxa identified as significantly over-represented amongst invaders in studies of taxonomic non-randomness. Lockwood (1999) and Lockwood, et al. (2000) assessed non-randomness amongst established species at the global scale, using different analytical methodologies. Blackburn and Duncan (2001b), Duncan, et al. (2006), and Blackburn and Cassey (2007) assessed non-randomness amongst introduced species, at the global scale, for New Zealand, and for Florida, respectively. Kark and Sol (2005) analysed non-randomness in birds introduced to regions with Mediterranean climate systems around the globe; the results presented here were in the original manuscript but cut from the published version, and were kindly supplied by S. Kark.

Family	Lockwood 1999	Lockwood, et al. 2000	Blackburn and Duncan 2001b	Duncan, et al. 2006	Kark and Sol 2005	Blackburn and Cassey 2007
Struthionidae		✓				
Rheidae	✓	✓				
Casuariidae		✓				
Phasianidae	✓	✓	✓	✓	✓	
Odontophoridae	✓	✓				
Anatidae	✓	✓	✓	✓	✓	✓
Psittacidae	✓	✓	✓		✓	✓
Columbidae	✓		✓		✓	
Ciconiidae						✓
Passeridae	✓	✓	✓		✓	✓
Fringillidae					✓	

although some instances of the introduction of species in this latter group reflect escapes (e.g., ruddy duck *Oxyura jamaicensis* in the United Kingdom; Hughes 1996), or accidental or unplanned releases (Eguchi and Amano 2004).

2.2.2 Non-randomness in Space

The observation that life's diversity peaks in tropical regions has been described as ecology's oldest pattern, dating back at least to the early years of the nineteenth century (Hawkins 2001). The species richness of birds conforms to this general rule (Orme, et al. 2005), albeit with some interesting lacunae. Notably, the tropical regions of the New World are by far the richest in bird species, and hold almost as many bird species as the Afrotropical and Indo-Malaysian regions combined (Table 2.2). Were bird introductions random, we would expect that most introduced bird species should be tropical in origin, with a particularly high representation of Neotropical species.

Table 2.2 Numbers of species with breeding populations in each of eight biogeographic regions from Thomas, et al. (2008), together with the number and percentage of those species that have exotic populations from the data used by Blackburn and Duncan (2001a). Note that species can have breeding populations in more than one region, so that the first two numerical columns do not sum to the total number of species or the total number of exotic species, respectively.

Region	Number of species	Number that have an exotic population	%
Australasia	1691	64	3.78
Antarctica	52	1	1.92
AfroTropics	2091	75	3.59
IndoMalaysia	1825	80	4.38
Nearctic	876	52	5.94
Neotropics	3628	69	1.90
Oceania	262	8	3.05
Palearctic	1745	92	5.27

In fact, it is the Nearctic and Palaearctic that have the highest percentages of their native bird species introduced elsewhere (Table 2.2). Numerically dominant as the region of origin for introductions is the Palaearctic, producing ninety-two exotic bird species (Table 2.2). In these terms the Neotropical avifauna has been the least successful, with a lower percentage of bird species introduced elsewhere even than the Antarctic, with its single introduced species (the king penguin *Aptenodytes patagonicus,* introduced to Norway in the 1930s: Long 1981). A test for equality of proportions on the data in Table 2.2 reveals that the proportion of bird species with introduced populations differs across regions ($\chi^2 = 61.6$, d.f. = 7, $P < 0.001$; this result is not altered by excluding the two regions depauperate in both species and introduced populations). Thus, we can conclude that the introduction of bird species to novel locations has not been random with respect to their region of origin.

This non-randomness extends also to the origin of introduced bird species with respect to latitude. Blackburn and Duncan (2001b) plotted the frequency distribution of latitudinal midpoints for the native ranges of introduced species. This histogram shows that most introduced species have low latitudinal midpoints to their native ranges, and that the mode in latitudinal midpoints is at low southern latitudes (Figure 2.2a). This broadly matches the frequency distribution of latitudinal range midpoints for all bird species (Figure 2.2b). In other words, the richness of both groups peaks near the equator. Nevertheless, the patterns of latitudinal midpoints for all bird species and introduced bird species differ

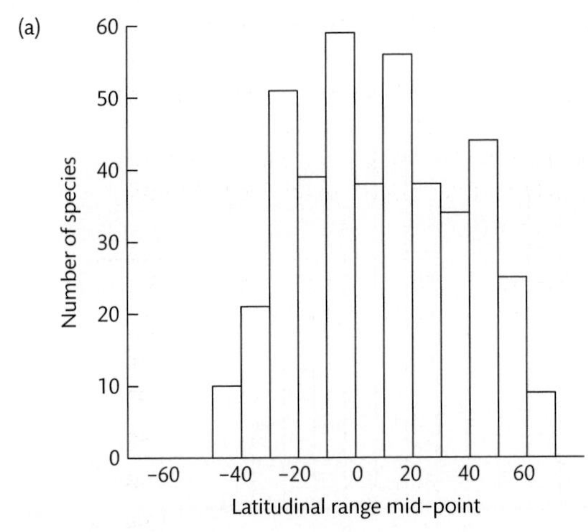

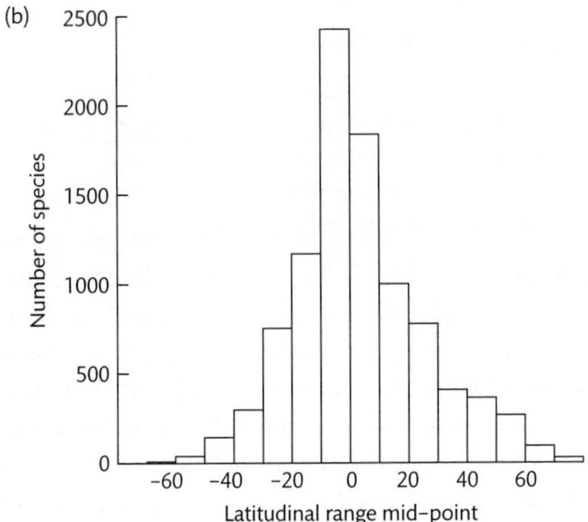

Fig. 2.2 The frequency distribution of native latitudinal range mid-points for (a) bird species that have been introduced somewhere in the world. Reprinted with permission of John Wiley & Sons Ltd from Blackburn and Duncan (2001b); and (b) all bird species (from data used by Orme, et al. 2006). Negative latitudes are in the southern hemisphere.

substantially in detail. Notably, the frequency distribution for introduced bird species is much less markedly peaked, and shows a relatively even spread of mid-points from around 30°S to 50°N. In contrast, bird species in general show a much higher equatorial peak in numbers. Thus, more introduced bird species seem to hail from northern mid-latitudes than expected given the overall pattern in the geographic distribution of bird species.

Land area is distributed across the globe with a distinctly northern bias—roughly two-thirds of land lies in the northern hemisphere. Yet, the frequency distribution of latitudes recipient of bird species introductions reveals that relatively few introduction events have occurred above 30°N (Figure 2.3). The mode in location of introduction is clearly at low latitudes in the northern

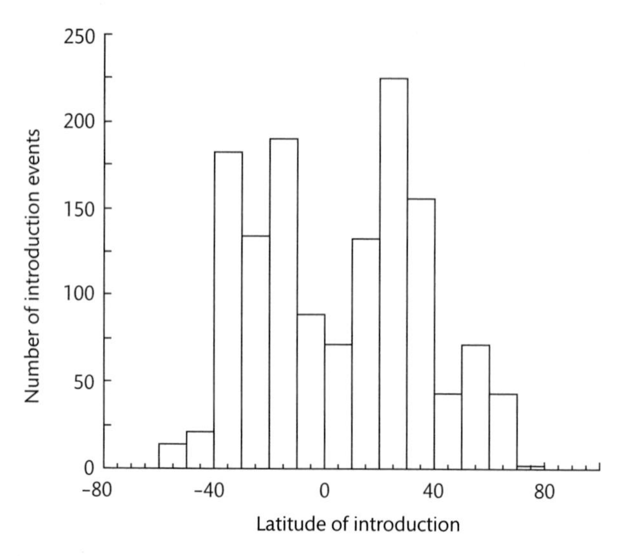

Fig. 2.3 The frequency distribution of latitudes of introduction for avian introduction events. Note that because the exact latitude was unknown for most introductions, each was assigned a latitude equal to that of the mid-point of the country in which the introduction occurred. Negative latitudes are in the southern hemisphere. Reprinted with permission of John Wiley & Sons Ltd from Blackburn and Duncan (2001b).

hemisphere, with peaks in both hemispheres between 20° and 40°. These peaks are largely a consequence of introductions to different island groups (Blackburn and Duncan 2001b). Notably, the large numbers of bird introductions to low northern latitudes is a consequence of the geographic location of the Hawaiian and Caribbean islands, on which many bird species have been liberated. Introductions to New Zealand and other Pacific islands, Australia, and the islands of the Indian Ocean largely determine the position of the southern hemisphere peak in introduction locations. Globally, more than two-thirds of all avian introductions have been to islands, even though islands make up only a small fraction of all land area around the globe. The bar between 30° and 40°N largely represents introductions into North America and the Mediterranean basin.

By grouping introductions into biogeographic regions, this predominance of introductions to islands is further emphasized. The greatest numbers of introductions are to Australasia and Oceania, with 286 and 271 introduction events respectively (Table 2.3). One hundred and twenty of the Australasian introductions concern New Zealand (Duncan, et al. 2006). A further 125 introduction events have occurred on islands in the Indian ocean, and

Table 2.3 Region of introduction x region of origin for 1,378 introduction events analysed by Blackburn and Duncan (2001b). The numbers by region of origin are used to reference region of introduction. For clarity, intra-regional introduction events are in bold.

	Region of introduction											
Region of origin	**1**	**2**	**3**	**4**	**5**	**6**	**8**	**10**	**12**	**13**	**14**	**Total**
Afrotropics [1]	**20**	0	22	17	15	1	24	7	33	13	0	152
Antarctica [2]	0	**0**	0	0	0	0	0	0	0	1	0	1
Atlantic [3]	0	0	**1**	0	0	0	0	0	0	0	0	1
Australasia [4]	1	0	0	**103**	1	1	1	4	44	6	9	170
Caribbean [5]	0	0	0	0	**8**	0	0	0	1	0	0	9
Central/S. America [6]	1	1	0	1	35	**14**	0	22	32	6	0	112
Holarctic [7]	0	2	0	5	0	1	0	4	3	13	0	28
Malagasy [8]	2	0	1	0	0	0	**32**	0	0	0	0	35
Multi-regional [9]	13	2	6	26	5	10	26	14	26	22	15	165
Nearctic [10]	2	0	1	17	17	5	0	**48**	30	16	0	136
New World [11]	0	0	0	0	6	1	0	3	2	0	0	12
Pacific [12]	0	0	0	0	0	0	0	0	**16**	0	0	16
Palaearctic [13]	8	0	3	70	3	6	4	49	23	**57**	3	226
South-east Asia [14]	10	0	7	47	7	9	38	29	61	27	**80**	315
Total	57	5	41	286	97	48	125	180	271	161	107	

ninety-seven on islands in the Caribbean. For continental areas, the tally is led by the Nearctic and Palaearctic, with 180 and 161 introduction events, respectively. Despite the predominance of introductions to islands, however, birds have been introduced to all major regions of the world, and to most latitudes free of ice (Duncan, et al. 2003).

While locations of origin and locations of introduction are not random, these two aspects of exotic bird geography are also not random with respect to each other (Blackburn and Duncan 2001b). A relatively high proportion of introduction events actually occur within the biogeographic region of origin of the species concerned. For example, Blackburn and Duncan (2001b) showed that at least 379 out of 1,378 avian introduction events (27.5%) were to the biogeographic region of origin (Table 2.3). This number is likely to be a slight underestimate, as it is often not clear from which region the introduced individuals of widespread species were collected. Just considering bird introductions to New Zealand, one-quarter (30/120) of the species derived from other locations in the Australasian region. However, more than one-third (41/120) of the species introduced came from the Palaearctic, the original home for a high proportion of the human settlers (Duncan, et al. 2006). These figures confirm the impression that settlers would source species with which they were familiar from the 'Old Country', but also suggest the importance of trade routes. The recent colonial history of New Zealand (King 2003) attests to the likely importance of trade and transport to and from both Europe and Australia, and so it is little surprise that the majority of bird species introduced to New Zealand also originated in these two biogeographic regions. This coincidence has potentially important consequences for our understanding of the causes of introduction success (Chapter 3).

Finally, Kark and Sol (2005) showed that non-randomness due to taxonomy and to location of introduction are also not random with respect to each other, for birds introduced to regions with Mediterranean climate systems around the globe. Thus, Psittacidae were mainly introduced to California, Columbidae to Australia, and Passeridae to the Mediterranean basin itself.

2.2.3 Non-randomness in Time

While the history of recorded avian introductions dates back at least 3,000 years (Chapter 1), it is only in the last two centuries that the introduction of bird species to exotic locations has become a common event (Figure 2.1a). Notably, the frequency of introduction attempts increased dramatically in the middle of the nineteenth century with the establishment of acclimatization societies (Figure 2.1b; Jeschke and Strayer 2005). The rate at which human-mediated introductions have been occurring over the last few centuries has in many cases been many times greater than the rate at which natural colonizations occur (Gaston, et al. 2003). As an example, we can consider the history of the avifauna of St Helena, in the Atlantic Ocean. This island was first discovered in 1502, and by 1588 had already experienced its first avian introduction. An account of the island published in the early nineteenth century noted that

> Partridges, pheasants, pigeons, and other birds have been introduced ... The partridges are pretty numerous, and several coveys of them are seen among the bare rocky hills, where it does not appear that there is any thing for them to eat. Here too we met with some beautiful ring pheasants and rabbits, which, together with Guinea hens, were introduced by the Governor. (Anon. 1805)

At least thirty-five exotic bird species have so far been introduced to St Helena (Long 1981; Lockwood, et al. 1996), of which nine have apparently established viable populations. This gives a rate of introduction of approximately one species every fourteen years, and a rate of successful introduction of around one species every fifty-five years since the island was discovered. In contrast, the minimum geological age for the island is probably of the order of 7 million years (the date of the last volcanic eruption: Baker, et al. 1967), and St Helena is known to have had at least twenty-two native bird species (of which twelve are now extinct: Blackburn, et al. 2004). That gives a rate of successful natural colonization of one species every 320,000 years—assuming that all native species derived from separate colonization events. Even assuming that 95% of all successful naturally colonizing bird species have subsequently gone extinct (following Gaston, et al. 2003), the rate of exotic introduction is still roughly 300 times greater than the natural colonization rate. Many other locations have experienced similarly high rates of exotic bird introduction (e.g., 120 species introduced to New Zealand in <250 years; Duncan, et al. 2006). Although the frequency of avian introduction events around the world appears to have declined in recent years, as discussed in the previous chapter (Section 1.4) this is likely to be as much a consequence of the timing of publication of catalogues of attempts as any real slow down.

Time has influenced the taxonomic composition of introduced birds, and consequently led to temporal changes in the biological characteristics of introduced species. Figure 2.4 shows how the proportion of introductions in the six bird orders with most introduction events varied across the period 1800–1950. A high proportion of early introductions concerned game birds (Galliformes), but this order's representation in introduction events has clearly declined over time.

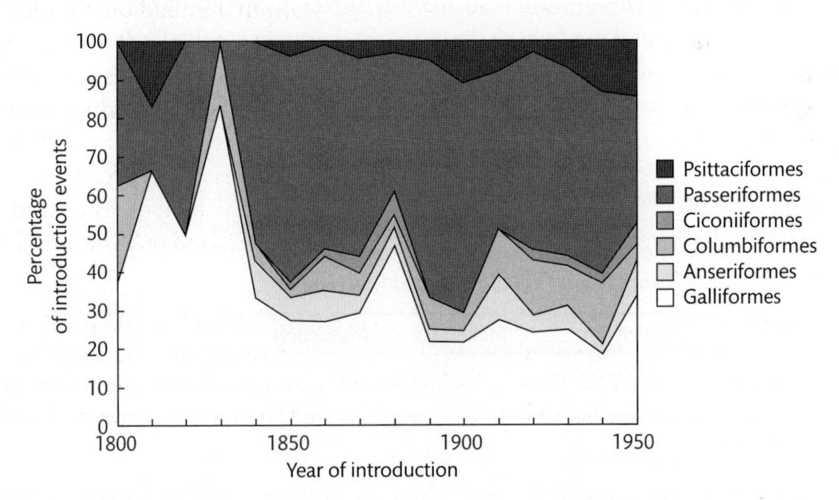

Fig. 2.4 The proportional taxonomic composition of introduction events in the period 1800–1950. Only the six bird orders with more than thirty introductions for which an estimate of date of introduction was available in the period are included. From data analysed by Blackburn and Duncan (2001b).

In contrast, the proportion of events that relate to parrots (Psittaciformes) has increased steadily since the 1850s. Passerine introductions appeared to peak in the second half of the nineteenth century, while the proportion of introductions concerning wildfowl (Anseriformes) has varied little since the 1840s. These patterns are also reflected in the mean date of introduction for these orders (Table 2.4), which varies from 1885 for Galliformes to 1940 for Psittaciformes, with Passeriformes and Anseriformes between these extremes. The probability that the spread of introduction dates across these six bird orders does not differ is very low (Kruskal-Wallis test, $\chi^2 = 72.1$, d.f. = 5, $P < 0.001$).

The temporal variation in the taxonomic composition of bird introductions between 1800 and 1950 seems likely to reflect changes in the reasons for the introductions. Introductions early in this period would have consisted more of attempts by acclimatization societies to naturalize species considered beneficial to the colonists in their new environment. The early predominance of galliform introductions probably reflects this predilection. In contrast, following the disbanding of these societies in the first half of the twentieth century, an increasing proportion of introductions were likely to be a consequence of the deliberate or accidental release of species from the cage bird trade (e.g., Richardson, et al. 2003; Eguchi and Amano 2004; Leven and Corlett 2004). The increasing relative frequency of parrot introductions in the later part of the period 1800–1950 may reflect this change. Indeed, Carrete and Tella (2008) record that nearly 99% of the 21,315 birds recorded for sale on the Spanish pet market concerned passerines (61%) and parrots (38%).

Relevant in this latter regard is an analysis by Robinson (2001) of the effects of changes in avicultural markets on the supply and price of captive birds. Robinson took all her examples from the order Psittaciformes. She noted that high prices or a rapid increase in demand for a species leads to increases in production of

Table 2.4. Mean date of introduction ± standard error for events concerning the six bird orders with most introduction events. *N* = sample size From the data used in Blackburn and Duncan (2001b).

Order	Mean date of introduction	N
Galliformes	1885 ± 4.8	339
Columbiformes	1898 ± 7.4	99
Anseriformes	1902 ± 11.8	71
Passeriformes	1907 ± 2.2	557
Ciconiiformes	1925 ± 7.1	31
Psittaciformes	1940 ± 3.7	96

the species via breeding programmes (Figure 2.5). However, this quickly leads to oversupply, and a decline in the price that the species can command. The result of price collapse is to leave breeders with birds that are costly to keep but that sell for little. One likely outcome is that breeders release the unwanted birds, and Robinson (2001) provides anecdotal evidence that the numbers of abandoned parrots grew rapidly in the last decade of the twentieth century as a consequence of oversupply. Carrete and Tella (2008) found that increasing invasion success for exotic bird species in Spain was positively associated with the number of individuals available on the market.

2.2.4 Non-randomness in Specific Characteristics

Since the traits of species are in part shaped by the effects of the environment in which they live, it is no surprise to find that specific characteristics of bird species exhibit spatial non-randomness. For example, species living at high latitudes tend to have larger geographic range sizes (Orme, et al. 2006), population sizes (Gaston and Blackburn 1996), and population densities (Currie and Fritz 1993), while small geographic range size also appears to be associated with tropical mountain ranges (Hawkins and Diniz-Filho 2006; Orme, et al. 2006). Clutch size is well known to vary latitudinally, both across (Moreau 1944; Lack 1948; Skutch 1949; Kulesza 1990; Cardillo 2002) and within (Lack 1947; Bell 1996; Dunn, et al. 2000) species. Mean body mass also tends to be higher in species living at high latitudes (Cousins 1989; Blackburn and Gaston 1996a; Gaston

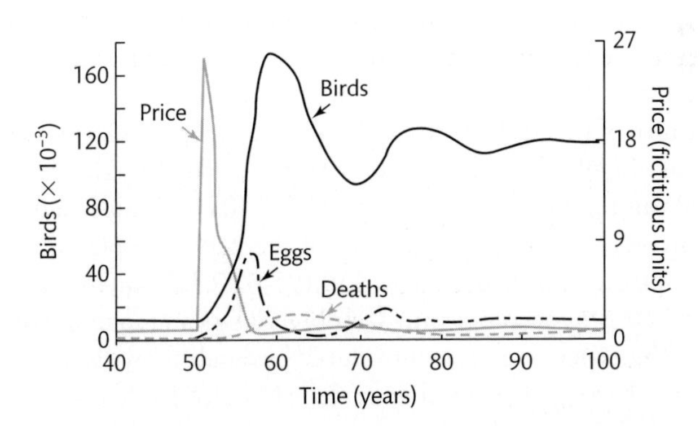

Fig. 2.5 Output of a model to assess an expansion in the market for cage birds, from 10,000 to 100,000 bird lovers. The market was expanded at the fifty-year point, leading to an increase in the price of birds. This provides the incentive for breeders to increase their production, which rapidly brings the price down again but leaves breeders with an excess of unprofitable birds. Reprinted with permission of Cambridge University Press from Robinson (2001).

and Blackburn 1996; Olson, et al. 2009) and in desert environments (Olson et al. 2009). Since many physiological and life-history characteristics are strongly correlated with avian body mass (e.g., Bennett and Harvey 1987; Trevelyan, et al. 1990), it follows that these traits are likely to show spatial variation as well (e.g., White, et al. 2007). Thus, since the origin of introduced bird species is not random with respect to either phylogeny or space, it may also be non-random with respect to specific characteristics.

The over-representation amongst introduced birds of species from northern latitudes, and from families such as the Phasianidae, Anatidae, Psittacidae, and Columbidae, suggests that introduced species may tend to be relatively large-bodied. Cassey (2001a) showed that this is indeed the case. The geometric mean body mass for all land birds has been estimated at 50.5g (Gaston and Blackburn 1995), but the equivalent estimate for introduced land birds is significantly larger at 116.6g (Cassey 2001a). Similar findings were reported by Jeschke and Strayer (2006) for introductions between Europe and North America, and by Blackburn and Cassey (2007) for bird introductions to Florida. The latter study also showed that selectivity by character is not only a consequence of the previously demonstrated selectivity by family, as exotic species tend to be large-bodied even accounting for the family membership of exotic species. In a related vein, wildfowl that have been introduced somewhere in the world have larger body masses than those that have never been introduced (Blackburn and Duncan 2001b).

The over-representation amongst introduced birds of species from northern latitudes also suggests that such species may tend to have relatively large geographic range sizes. We tested this hypothesis for the first time for all introduced bird species, using a recently compiled global database on the geographic ranges of all extant bird species (Orme, et al. 2005). The geometric mean geographic range size of the 425 introduced bird species with extant native ranges is 2.17 million km^2, and thus substantially greater than the mean for all bird species of 583,000 km^2. The significance of this difference can be assessed by permutation test, selecting 425 species at random from the global avifauna, calculating the geometric mean range size of this sample, and repeating 1,000 times to produce a distribution of expected outcomes. Ninety-nine per cent of the 1,000 sample means calculated using this approach fall in the range 430,195–759,327 square kilometres, and thus are nowhere near encompassing the observed mean range size of introduced bird species. Thus, species that are introduced have substantially larger native geographic ranges than expected by chance alone (see also Jeschke and Strayer 2006; Blackburn and Cassey 2007).

The significantly greater than expected native range sizes of introduced bird species could result from the bias in the taxonomic composition of such species, if introduced species tended to come from families with larger than average range sizes. This effect does not account for the pattern in wildfowl, as wildfowl species that have been introduced somewhere in the world have larger geographic range sizes than those that have never been introduced (Blackburn and Duncan 2001b). The effect of taxonomic composition can be assessed more generally by repeating the permutation test described above, this time replacing the native range size of each species with the native range size of a species chosen at random from the same family. This approach constrains the family-level composition of the list of randomly chosen species in each iteration of the model to be the same as that of the species actually introduced. Ninety-nine per cent of the 1,000 sample means calculated this way fall in the range 469,638–781,661 square kilometres. Thus, the random expectation is again nowhere near encompassing the observed mean range size of introduced bird species of over 2 million km^2. The same is true for the more restricted set of species introduced to Florida (Blackburn and Cassey 2007). Introduced birds thus still have substantially larger native geographic ranges than expected by chance alone even when controlling for the potential bias resulting from taxonomic non-randomness.

The larger than expected native range sizes of introduced bird species suggests that widespread species are more likely to be entrained in the invasion process. However, the influence of range size on the probability of introduction may be confounded with an effect of population size. Population size is the strongest predictor of introduction probability in wildfowl (Blackburn and Duncan 2001b), while population size and geographic range size are highly correlated across wildfowl species (Gaston and Blackburn 1996), as indeed they are in many other bird assemblages (e.g., Gaston and Blackburn 2000; Blackburn, et al. 2006). The influence of range size on introduction probability in wildfowl is greatly reduced when controlling for population size. Multivariate analysis suggests that large population size and large body size are the best predictors, of those tested to date, of introduction probability in wildfowl (Blackburn and Duncan 2001b).

Further evidence for trait selectivity in the set of introduced bird species comes from comparison of those species of British bird that have, versus those that have not, been introduced somewhere in the world (Blackburn and Duncan 2001b). Once again, population size is the strongest predictor, out of those tested, of introduction probability: species that are abundant in Britain are more likely to have been introduced. Species with broad geographic ranges in Britain are also more likely to have been introduced, while resident species are more likely to have been introduced than species that only visit Britain for the summer or winter. Body size has no influence on the probability that a British bird species has been introduced somewhere in the world.

One criticism of Blackburn and Duncan's (2001b) analysis is that Britain was almost certainly not the source of all individuals involved in the introductions analysed. Thus, conclusions about the importance of certain characteristics in driving introduction probability may be biased if these characteristics differ in Britain relative to the actual region from which the introduction originated. Blackburn and Duncan (2001b) argued that this should not affect their conclusions, as geographic range sizes and abundances tend to be positively correlated across regions (e.g., species widespread and abundant in Britain will also be widespread and abundant across Europe; Gaston 1994; Gregory and Blackburn 1998), while the species are likely to have similar body sizes and migratory strategies wherever it was in their geographic range that individuals were collected for introduction. Nevertheless, to address this problem, they repeated their analyses comparing bird species that have or have not been introduced from Britain to New Zealand, for which Britain is known to be the source of the introduced individuals. The results were essentially identical: abundant, widespread, resident species are more likely to have been the subject of introductions (Blackburn and Duncan 2001b). These analyses suggest that character selectivity also occurs independently of non-randomness due to geography (see above), as it occurs even in a geographically restricted subset of introduced species (Duncan, et al. 2003). More evidence of character selectivity comes from a study by Jeschke and Strayer (2006) of bird species introduced between Europe and North America (in either direction), in comparison to species in both avifaunas that were not introduced. Multivariate models suggested that carnivorous species were less likely to be introduced than herbivores, but that introduction probability was more likely for species with human affiliation (see also Chapter 5), that were introduced for hunting, fecund and long-lived. Human affiliation and long lifespan were also significant in models that controlled for phylogenetic autocorrelation. Finally, Newsome and Noble (1986) analysed the characteristics of species introduced to Australia. This set of species appears to include high proportions of ground-nesting species, of species that have largely vegetarian diets, and of species that use grassland, cultivated, or suburban habitats (Duncan, et al. 2003). Unfortunately, however, in the absence of information on the prevalence of such species in the global avifauna as a whole, whether or not these proportions are actually higher than expected awaits a more formal analysis.

2.3 Correlates of Transport and Introduction Separately

The studies reviewed above that consider non-randomness in the characteristics of introduced bird species provide important information on the early steps of the invasion process. Nevertheless, these studies concatenate two stages on the invasion pathway as defined in Chapter 1—transport and introduction. Thus, they do not distinguish at which of these steps those native species that are not introduced actually falter. It follows that they also cannot identify the reasons for failure at the different stages, and thus cannot determine where on the pathway different filters might act. As far as we are aware, only one study to date has performed the requisite analyses to address these points.

Cassey, et al. (2004b) analysed information on transport, introduction, and establishment success for species in the order Psittaciformes. Thomsen and Brautigam (1991) estimated the global trade in parrots to be in the region of US $1.4 billion annually, and associated with this big business are relatively high-quality data concerning which species have been transported. Moreover, parrots are large bodied by the standard of birds, frequently noisy, colourful, and gregarious, and so are readily observed by both amateur birdwatchers and members of the public alike. Regional (e.g., Johnston and Garrett 1994; Garrett 1997; Sol, et al. 1997; Pithon and Dytham 2002; Pranty 2004) and global (e.g., Long 1981; Lever 1987, 2005) data on the identities of parrot species introduced and established are also of relatively high quality. These data facilitate the analysis of characteristics associated with the first three stages of the invasion pathway for Psittaciformes. Cassey, et al. (2004b) assessed the influence on the likelihood of passing through each of these stages of variables relating to the abundance and distribution, ecological flexibility, life history, and global movement of parrots. This last category concerned the types of trade (if any) to which the various species were subjected.

Unsurprisingly, species known to be traded on international markets were significantly more likely to be transported beyond the boundaries of their geographic range. However, the same was also true of species recorded in local trade (e.g., species traded within the boundaries of their geographic range). Species known from captivity and known to be kept as pets were also more likely to be transported. Interestingly, Cassey, et al. (2004b) found no relationship between probability of transport and whether or not a species was listed on the Convention on International Trade in Endangered Species (CITES) Appendix I. This suggests that if such species were originally the subject of heavy trade, then that trade has decreased in response to listing. However, it also suggests that, if so, then listing has not reduced trade enough to produce the expected negative relationship between Appendix I listing and probability of transport! These same associations all held when probability of transport was substituted by probability of introduction (Cassey, et al. 2004b).

Transport was more likely for parrot species with wide geographic distributions, and hence for species that are more likely, on average, to be encountered by bird trappers and collectors. Species that are recorded as pests in their native ranges, such as the galah *Eolophus roseicapillus* in Australia, were also more likely to be transported. Conversely, range-restricted species were less likely to be transported, as were species with high-threat status. This last relationship may less reflect the effectiveness of threat listing than it does the fact that threatened species tend to have small geographic ranges and/or low population sizes, and so are less often encountered by trappers. Multivariate analysis revealed only two significant independent predictors of transport probability—geographic range size and fledging time. Parrots with wide geographic distributions, and that spend more time in the nest, are more likely to pass through the transport stage on the invasion pathway (Cassey, et al. 2004b).

Very similar results are obtained for probability of introduction. Introduction was more likely for parrot species with wide geographic distributions, and for species with large population sizes. Introduction was less likely for range-restricted species and for species with higher-threat status. Note that these analyses only concern those species that have been transported, and so they are not a simple consequence of the effect of threat status on transport probability. That rarer species less often make it out of captivity and into an exotic environment may reflect the greater value to owners, breeders, or traders of such species (Cassey, et al. 2004b). The corollary, as identified in Robinson's (2001) analysis of the economics of the cage bird trade, is that commoner (more widespread and abundant) species may tend to be those that cost money to keep but command little to sell. Parrots can vary in price from a few US dollars for common species such as budgerigars *Melopsittacus undulatus*, to several thousand dollars for rare or long-lived species such as macaws (Robinson 2001). Common species may be exactly those more likely to be abandoned as unwanted.

Amongst the life-history traits, introduction was more likely for parrot species with broad diets and sexually dichromatic plumage, and less likely for species undertaking population movements. Body size did not predict probability of introduction alone, but did enter the multivariate model for this stage, along with native population size and sexual dichromatism. Widespread, large-bodied species lacking sexual differences in plumage colouration are more likely to pass through the second stage on the invasion pathway, and so make it from captivity into the exotic environment. The body-size effect seems to run counter to Robinson's (2001) economic analysis, where she suggests that oversupply from parrot breeders more rapidly occurs for fast-breeding *r*-selected species, which will also tend to be small bodied (Bennett 1986). If this oversupply is a major influence on which species get released, we might expect to detect a negative effect of body size. Nevertheless, the cost and commitment required to keep individuals of large, long-lived species may be more relevant here.

Cassey, et al. (2004b) also assessed the degree of phylogenetic association in the probability of transport and introduction using Moran's *I*. Moran's *I* is a measure of autocorrelation that is most familiar to ecologists through its use in quantifying the strength of similarity between observations due to their spatial proximity, but it can also be used to quantify phylogenetic autocorrelation (Gittleman and Kot 1990). Proximity here is measured in terms of taxonomic classes, such that all species within a genus, for example, are assumed to be separated by the same taxonomic distance. Cassey and colleagues found that what autocorrelation there was in the probabilities of transport and introduction was negative. This result suggests that high probabilities of transport and introduction are not aggregated within certain taxa, but rather spread more evenly than expected across at least some parts of the parrot phylogenetic tree. These findings are in interesting counterpoint to evidence showing clumping in the taxonomic associations of species when transport and introduction are concatenated (section 2.2.1 above). There, transport and introduction were found to be taxonomically aggregated, in fewer families than expected, and specifically in a limited number of families with properties desired by those driving the import and liberation of exotic birds. The probability of transport plus introduction is high for parrots in general (e.g., Lockwood 1999; Lockwood, et al. 2000; Blackburn and Duncan 2001b; Duncan, et al. 2006), but is not obviously clustered within parrot taxonomic groups.

The lack of positive autocorrelation in the probabilities of transport and introduction for parrots suggests at least a couple of possible interpretations. One is that probabilities of transport and introduction are primarily driven by distribution and abundance, both of which are traits known to be relatively weakly phylogenetically conserved in birds (Gaston and Blackburn 1997; Blackburn, et al. 1998; Gaston 1996; Webb, et al. 2001; Webb and Gaston 2003; 2005; but see Böhning-Gaese, et al. 2006): knowledge of the rarity of one species allows little to be inferred about the likely rarity of a relative. The lack of positive phylogenetic autocorrelation in distribution and abundance may explain the similar lack of autocorrelation exhibited by probabilities of transport and introduction (Cassey, et al. 2004b). Alternatively, this latter result could be a consequence of the locations from which parrots are collected for transport. If collectors tend to focus on specific locations to acquire parrots, then those caught may comprise species coexisting at those locations. These species may tend to represent a relatively broad taxonomic spread of parrots, since close relatives may be more likely to share niche characteristics and so suffer competitive exclusion (cf. 'Darwin's naturalization hypothesis': Darwin 1859; Daehler 2001; Diez, et al. 2008; Proçhes, et al. 2008). If so, this could explain the relatively broad taxonomic spread of transported parrots. A knock-on effect to introduction may account for the effect at this stage. An analysis of non-randomness in the geography of transport and introduction for parrots is missing from Cassey, et al. (2004b) but would allow the merit of this second suggestion to be assessed.

2.4 What Do We Learn about the Early Stages of Invasion from Studying Introduced Birds?

Invasion is a multi-stage process, and we have argued previously that the best way to understand the process in its entirety will be to study each stage separately (Section 1.3). Data to analyse the characteristics of species that pass through the transport and introduction (release or escape) stages separately are rare, but those data that are available suggest that concatenating these stages may not be a significant handicap to a broad understanding of which species in a taxon make it as far as introduction.

A consistent thread that runs through all the analyses of character selectivity in transport, introduction, and the two combined, is that the availability of a species contributes substantially to its success. Large geographic range size and/or large population size are significant predictors of transport and introduction probability in bird species with populations in Britain and which have been introduced elsewhere in the world, British bird species introduced to New Zealand, bird species introduced to Florida, introduced wildfowl species, and in parrots. For the British bird species, residents are also more likely to have been transported and introduced than species present in the country for only part of the year.

Becoming entrained in the invasion process requires, first, being collected within the native range. Availability will clearly increase the likelihood that this happens. Species with wide ranges are more likely to overlap the locations in which collectors operate, while more abundant species (and species with wide ranges tend also to be disproportionately abundant: Gaston and Blackburn 2000) will be more commonly encountered within those ranges. Cassey, et al.'s (2004b) analysis of parrots suggests an interesting lacuna, whereby the length of the fledging period further contributes to this availability. Many parrot species are captured on, or removed from, the nest. For example, Martuscelli (1994) found that 41/49 nests of the red-tailed parrot *Amazona brasiliensis* surveyed in south-east Brazil had been robbed. Clearly, a long period spent in the nest is likely to increase the likelihood of capture in these circumstances. That said, there is no correlation between fledging time (data from Cassey, et al. 2004b) and the proportion of nests lost to harvesting in the studies collated by Pain, et al. (2006; $r = -0.04$, $N = 14$, $P > 0.05$).

The probability of being introduced is likely to increase with the numbers present in captivity (Cassey, et al. 2004b), because the risk of accidental escape is higher for more commonly kept species, because the economics of the cage bird trade dictate that such species are more commonly deliberately abandoned (Robinson 2001), or because species desired for acclimatization may be imported in large numbers to improve the chances of success (see Chapter 3). Species easily available for collection are likely to make it into captivity in larger numbers. In addition to these probabilistic arguments, abundant, widespread species may be more likely to be captured, transported, and introduced if it is species that are most familiar that people most desire to introduce (Duncan, et al. 2003).

Superimposed on availability are effects of location and taxonomy. There are many widespread, abundant species that have never been the subject of introduction attempts because they either belong to taxa lacking the qualities deemed suitable by human society, or reside in the wrong areas. Historically, it was species living in areas from which European settlers departed, and to which their emigration subsequently took them, that have featured predominantly in lists of exotic introductions. Game birds were initially favoured too. Nevertheless, these aspects of selectivity are not constants. The reasons for introductions have changed over history, and escaped or liberated cage birds, notably parrots, have increased in prevalence in recent decades (Figure 2.4). Temporal analyses of selectivity in location have not yet been performed, but may be expected also to show changes as general increases in global commerce have opened up new opportunities for trade in birds. Such changes may also reflect the taxonomic trends identified in Figure 2.4.

The influences of taxonomy and location on transport and introduction nevertheless do not produce the effect of availability. Selectivity due to correlates of availability is present in taxonomic (wildfowl, parrots) and geographic (Britain) subsets of the introduction data. It is the widespread and abundant species in favoured taxa in favoured regions that end up being introduced. We may speculate whether it is just a coincidence that these species also happen to be widespread compared to the average bird (section 2.2.4), or whether this is in part a consequence of the same factors that led the human inhabitants of European regions to be successful invaders across much of the globe (Diamond 1998).

2.5 Conclusions

In general, it seems that the set of bird species transported and introduced around the world reflects the intersection of biological characteristics with societal demands. Those demands vary across space and time, and, thus, so do the characteristics with which they intersect. Not all instances of transport or introduction are deliberate acts in response to these demands. Nevertheless, be they deliberate releases, accidental escapees or opportunistic stowaways, the helping hand of human society is evident in all of them.

One consequence of this helping hand is that species that transit the first two stages of the invasion pathway are not necessarily a set that we would perhaps imagine a priori would make good colonizers. Although it is difficult to quantify the relative colonization ability of different avian taxa, one way in which this can be assessed, for land birds at least, is by the extent to which a group has managed to colonize oceanic islands. Presumably, such groups have been more readily able to meet the challenges of locating new habitable areas, across large expanses of uninhabitable areas. We assessed this using a database of the bird species inhabiting 220 oceanic islands around the world at the time of first human colonization (Blackburn, et al. 2004). This includes a total of 1,537 bird species, and 7,166 island x species combinations. Excluding seabirds, the bird families with the highest proportions of all island species in this database are the Psittacidae, Columbidae, Rallidae, Corvidae, and Fringillidae. The bird families with the most separate island populations (island x species combinations) are Columbidae,

Rallidae, Ardeidae, Passeridae, and Fringillidae. While there is a good degree of overlap between families that oceanic island occupancy identifies as good colonizers and that have been the subject of many human-mediated introductions (e.g., Psittacidae, Columbidae, Passeridae; Table 2.1), some families of good natural colonizers (e.g., Rallidae, Corvidae) are notably lacking in human-introduced species. Indeed, the analyses presented by Duncan, et al. (2006) suggest that Corvidae have been introduced to New Zealand less often than expected by chance.

Since invasion is a sequential process, each stage in the pathway sets the species pool for the next step. Not all introduced species establish viable exotic populations, and so the characteristics of established species must be a subset of those introduced. Whether or not these characteristics are a random subset is a question we will explore in detail over the course of the next three chapters. Either way, however, this has some important consequences for how we should analyse and/or interpret the data presented in subsequent chapters (see also Sol, et al. 2008).

First, non-randomness in transport and introduction imposes patterns on the subsequent invasion stages, which if not taken into consideration may bias conclusions about them. For example, studies on the types of locations that favour exotic bird establishment (e.g., Sax 2001) need to take account of the fact that introductions are non-random with respect to location. Geographic patterns in establishment may simply reflect geographic patterns in introduction (Chapter 7). Similarly, analyses of the taxonomic composition of established species (Lockwood 1999; Lockwood, et al. 2000) need to account for the fact that introduced species themselves show taxonomic structure (Blackburn and Duncan 2001b), and likewise for analyses of specific characteristics such as body size. More generally, non-randomness in transport and introduction emphasizes the need to specify the correct null hypothesis in analyses of subsequent invasion stages (Prinzing, et al. 2002; Duncan, et al. 2003; Cassey, et al. 2004a; see Gotelli and Graves 1996 for a general discussion of null models). A significant issue here is that the source pool of samples (be they species, locations, or events) for any comparison is correctly specified. For example, the null hypothesis for the effect of body size on establishment success is not that the body sizes of successfully established species do not differ from those of a random sample of all bird species, but rather that they do not differ from those of a random sample of introduced bird species.

An analysis on the consequences of specifying different source pools was carried out by Cassey, et al. (2004a) using the same parrot data set they explored with respect to transport and introduction. They found that different definitions of the source pool of species for analysis of characteristics related to establishment led to different patterns of significance in a wide range of predictor variables. For example, successfully established parrot species have larger than expected native population sizes than other parrot species. However, this result pertains because widespread parrot species are more likely to be transported and introduced. Comparing established species just to species that are introduced but which fail to establish reveals that it is in fact parrots with smaller native population sizes that tend to establish successfully. An interesting aside here is that very similar results were obtained for analysis of the introduction stage, regardless of whether the pool was specified correctly or incorrectly. This result supports the conclusions drawn above that concatenation of the transport and introduction stages may not be a significant handicap to a broad understanding of which species in a taxon progress as far as introduction into exotic environments (though we still advocate that the stages should be analysed separately wherever possible).

A further issue with respect to bias involves temporal changes in the reasons for introductions. As noted above, many introductions in the nineteenth and early twentieth centuries were a consequence of planned releases by acclimatization societies. These societies often kept detailed records of the species and circumstances of the release, including useful information on numbers of individuals liberated, and on outcomes. In the later part of the twentieth century, there is inevitably a decline in the quality of information surrounding introductions with the apparent increase in the prevalence of unplanned or accidental releases. If the causes or consequences of successful invasion are changing with the type of introduction, our ability to detect any such changes is greatly diminished.

The second consequence of the non-randomness in transport and introduction is that it can result in problems of confounding between traits relevant to understanding transit through subsequent invasion stages (Duncan, et al. 2003; Sol, et al. 2008). Notably, it can confound event-level, species-level, and location-level factors, making it harder to establish the relative influence of each of these levels on the causes of success or failure. For example, species that are common in Great Britain were introduced to New Zealand in larger numbers (Blackburn and Duncan 2001b). However, species that are common in Britain also possess certain life-history characteristics that distinguish them from rare species (Nee, et al. 1991; Blackburn, et al. 1996; Gregory and Gaston 2000). If commoner species were also more likely to succeed in establishing in New Zealand, it would be difficult to judge whether this was down to species-level life-history traits, or to the event-level characteristic of number of individuals introduced. Care thus needs to be taken when analysing subsequent invasion stages to identify possible confounding interactions and control for them statistically (Duncan, et al. 2003). Unfortunately, this has not always been the case in published analyses. Furthermore, these analyses may be affected by the third consequence of non-randomness in transport and introduction, that of non-independence.

An important assumption of standard statistical tests is that the errors around any relationship are independently distributed (Sokal and Rohlf 1995)—that is, the deviation of any one point from the relationship does not provide information about the deviation of any other. However, this assumption may frequently be violated in studies of the invasion process because of the issue of non-randomness. Consider an analysis that explores the effect of life-history traits on the outcomes of a series of introduction events. The success or failure of a set of these events may be correlated because they all concern species introduced to one location (which may, for example, be a particularly amenable environment), or because they all concern the same species (which may, for example, be an especially good invader). Regardless of the causes of the outcomes observed, the outcomes would be correlated because of the clustering of events, either by location or by taxon. The error terms associated with the outcomes would therefore also be correlated, and so not independent as standard statistical tests require. This situation is equivalent to the issue of pseudo-replication in experimental design (Hurlbert 1984), which can lead to biased parameter estimates and elevated type I error rates. In effect, the true sample size is less than the number of introduction events, because some of those events do not provide independent sources of information on the relationship between introduction success and the life-history traits.

This issue of non-independence needs to be taken into account when analysing later stages in the invasion pathway (Blackburn and Duncan 2001a; 2001b; Duncan, et al. 2003; Sol, et al. 2008). One way to do this is explicitly to model the non-independence in any statistical analysis. A variety of techniques now allow non-independence to be incorporated, including generalized linear mixed modelling (Goldstein 1995) and generalized estimating equations (Waclawiw and Liang 1993). An alternative is to restrict comparisons to certain locations or taxa, although in many cases this will result in limited sample sizes for statistical inference. Unfortunately, many analyses of the invasion pathway were published before these issues were widely understood. This issue needs to be borne in mind over the coming chapters.

3

The Role of Contingency in Establishment Success

I returned and saw under the sun, that the race is not to the swift, nor the battle to the strong, neither yet bread to the wise, nor yet riches to men of understanding, nor yet favour to men of skill; but time and chance happeneth to them all.

Ecclesiastes 9: 11

3.1 Introduction

In the previous chapter, we considered those characteristics of species that determine whether or not they are transported beyond the limits of their native distributions, and the characteristics that then influence which of the species chosen to be transported are subsequently introduced into a novel recipient environment. Despite the relatively small number of studies that have addressed these questions (in birds as well as other taxa), it is clear that those species that progress through these first two steps on the invasion pathway are a restricted set of all birds, in terms of taxonomy, geography, and availability. It is these species, and this set of characteristics, whose ability to establish a viable exotic population in the new environment is then tested. Whether or not any of these traits makes a difference to successful establishment is a key question in invasion research (e.g., O'Connor 1986; Ehrlich 1989; Williamson 1996; Williamson and Fitter 1996; Brooks 2001; Cassey 2002b; Duncan, et al. 2003; Vázquez 2006; Blackburn, et al. 2009).

However, it is not just the characteristics of the species that are likely to influence whether or not a bird species that is introduced becomes an established exotic. To these species-level traits must be added features of the location. Just as it seems likely that some species make better establishers than others, so too is it likely that some locations are more amenable to receiving exotic species, regardless of the identities of the species introduced. Attempts to identify such location-level characteristics make a substantial contribution to the invasion literature (e.g., Elton 1958; Brown 1989; Ehrlich 1989; Pimm 1989; Case 1996; Williamson 1996; Keane and Crawley 2002; Shea and Chesson 2002; Cassey 2003; Duncan, et al. 2003; Lockwood, et al. 2007). Nevertheless, the characteristics of species and location together are insufficient for a complete understanding of introduction success. For example, the successful introduction of European starlings *Sturnus vulgaris* to Central Park, New York, in 1890 had been preceded in 1872–3 by a failed introduction of the same species to the same location (Wood 1924). Such examples suggest that there is an additional element to establishment success, one that varies from introduction to introduction. There is a growing appreciation in invasion biology regarding the importance of such event-level factors in establishment success. It is these that we consider in this chapter, before going on to address the effects on establishment of species-level traits in Chapter 4, and location-level characteristics in Chapter 5.

If the introduction of a given species to a specific location sometimes succeeds and sometimes fails, we can ask what it might be that differs between these respective events that determines the different outcomes? One possibility is that the differences are the simple consequence of blind chance. One introduction may coincide with unseasonably bad weather, or all the females liberated may suffer predation. In monomorphic species, no females may be introduced at all. Certainly there is a stochastic element to most, if not all, events concerning natural systems, and so failures may be inherently unpredictable, as has often been argued (Gilpin 1990; Lodge 1993; Williamson and Fitter 1996; Williamson 1999; 2006). On the other hand, perhaps some introductions succeed where others fail because, in the introduction lottery, the odds are stacked in their favour. If so, it makes sense to consider how chance might affect introduction success, and what it might be that improves the odds in favour of a positive outcome. In this we might reasonably exploit experience from another area of biology where understanding the persistence or otherwise of populations is of paramount interest: conservation biology.

3.1.1 Small Populations and the Game of Chance

Caughley (1994) distinguished between two different paradigms in the science of conservation biology. The 'declining-population paradigm' concerns understanding and ameliorating the processes by which populations are driven to extinction, such as the 'evil quartet' (Diamond 1984; 1989) of overkill, habitat destruction, invasive species and chains of extinction. Alternatively, the 'small-population paradigm' deals with the risk of extinction inherent in low numbers, which largely concerns issues that arise from the population dynamics and population genetics of such populations.

Exotic species are typically introduced in low numbers relative even to the population sizes of threatened species (Figure 3.1). For example, if exotic bird populations are assessed under the International Union for Conservation of Nature (IUCN) Red List criteria (Mace and Lande 1991; Birdlife International 2000) immediately after introduction, almost 80% would fall within the critically

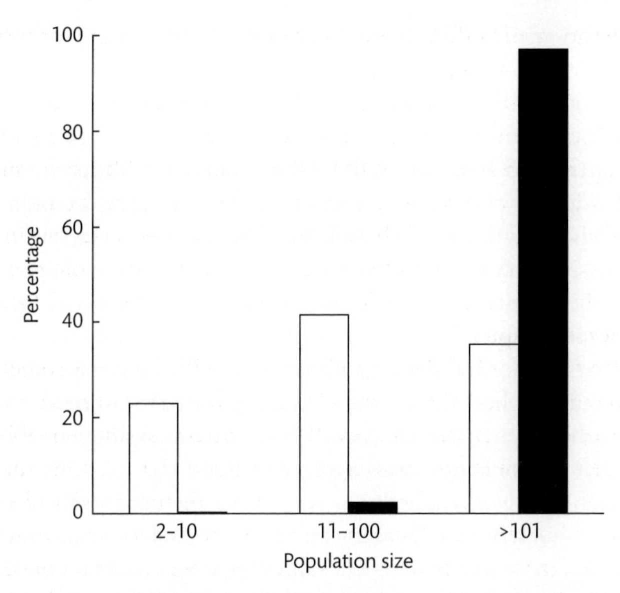

Fig. 3.1 A comparison of the percentage of population sizes (number of individuals) of 987 bird species defined as globally threatened by BirdLife International (2000; data as used in Blackburn and Gaston 2002; filled bars), and 875 exotic bird introduction events around the world (from data supplied by D. Sol; open bars) in different population size classes. It is clear that most exotic birds are liberated in low numbers relative to the population sizes of most globally threatened species, albeit that many of these latter species may be at risk because of declining populations due to external threatening processes.

endangered category (<250 individuals). Recent empirical analyses have suggested that population sizes required to prevent extinction (Brook, et al. 2006; Traill, et al. 2007) and reduce the risk of adverse genetic factors (Spielman, et al. 2004) are much larger than those involved in the introduction of most avian species. For example, Traill, et al. (2007) estimate the median minimum viable population size for bird species to be around 3,700 individuals—around fifteen times the size of most exotic bird populations immediately after introduction. The parallels between the small-population paradigm for conservation and the problem faced by invasion biologists in understanding what determines the extinction (establishment failure) or otherwise of an exotic species are abundantly clear. Theory and practice in the small-population paradigm is, and indeed has been, a fertile area in which for invasion biologists to prospect for ideas (e.g., Pimm 1989; 1991; Grevstad 1999; Sakai, et al. 2001; Lockwood, et al. 2005; Taylor and Hastings 2005; Freckleton, et al. 2006a).

Clearly, a declining population will become extinct if that decline is not checked. This is simply to say that an exotic population that cannot maintain a positive growth rate in the novel environment, whether because of characteristics of the species, the location, or both, is doomed to fail. However, even if a small population has a positive average growth rate and could under normal circumstances persist more or less indefinitely at a location, it may become extinct because of the effects of demographic or environmental stochasticity on that growth rate (Caughley 1994; Melbourne and Hastings 2008). Demography can influence growth rate because individuals within a population vary in their age, sex, fecundity, and survival. The precise growth rate of a population thus depends on its specific composition, and this can vary substantially (and stochastically) between generations or years, especially in a small population. The environment can also affect population growth rates, for example through the effects of poor conditions (e.g., a succession of drought years) on fecundity or survival. Just as a casino can be bankrupted on the spin of a roulette wheel, despite the odds being biased in its favour, so unfavourable demographic or environmental events can drive to extinction an introduced population that we would otherwise expect to persist.

A considerable body of research has considered the effects of stochasticity on population dynamics, principally using modelling approaches, but sometimes supported with data (reviewed by Lande, et al. 2003; Fagan and Holmes 2006). The primary conclusion to be drawn from such models is that understanding these effects is not straightforward: the likelihood that a population will persist depends on the variability in population vital rates, and the relative contributions of demography and the environment to that variability (Freckleton, et al. 2006a; Melbourne and Hastings 2008). That said, some broad generalizations can be proffered.

First, it seems likely that exotic bird populations will not be near carrying capacity, nor likely to be limited by density-dependent processes, at least during the establishment phase. It follows that the most informative models for them will be based on these assumptions, and concern sexually reproducing populations.

Second, the effect of demographic stochasticity on population growth rate decreases as population size increases. Most models suggest that demographic stochasticity is only a significant risk for very small populations. For example, the stochastic birth–death model simulated by Grevstad (1999) suggests that demographic stochasticity is only important in influencing the persistence of populations consisting of fewer than a few dozen individuals, and even then only when the population growth rate is itself low (Figure 3.2). Caughley and Gunn (1996) suggest that its effects are negligible by the time a population reaches 300 individuals. Lande's (1993) stochastic models of density-independent population growth show that time to extinction under demographic stochasticity increases almost exponentially with population size, as long as population growth rate is positive. However, recent work by Melbourne and Hastings (2008) suggests that the risks of demographic stochasticity may have been underappreciated at larger population sizes. This is because demographic variance due to predictable variation in vital rates amongst individuals (e.g., due to body size), and to variation in sex ratios, may have previously been erroneously attributed to environmental stochasticity.

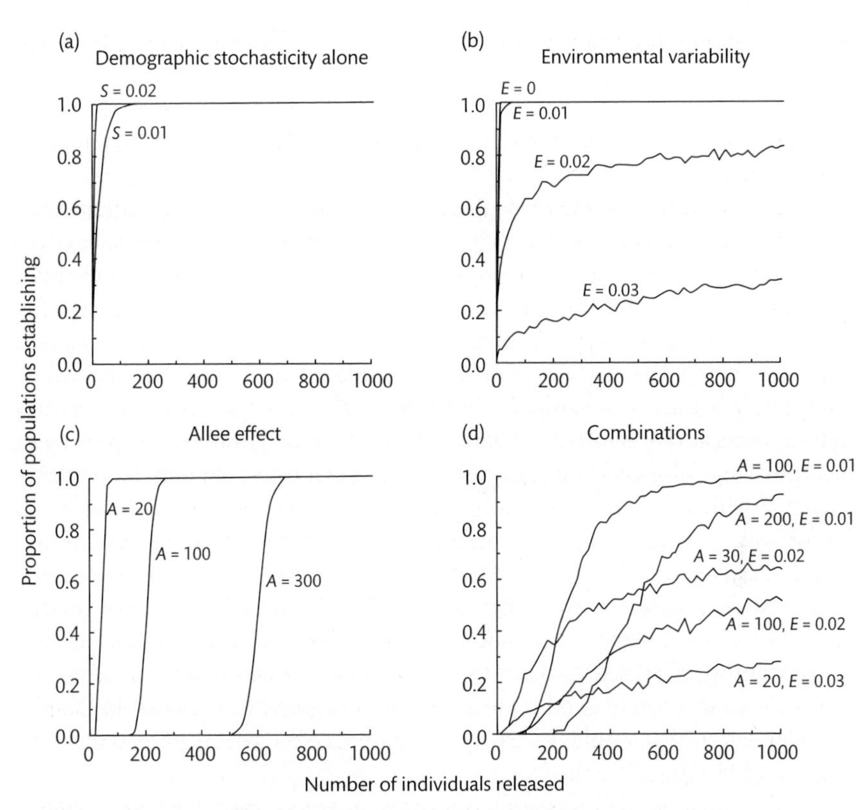

Fig. 3.2 The influences of (a) demographic stochasticity, (b) environmental variability (in addition to demographic stochasticity), (c) Allee effect (in addition to demographic stochasticity), and (d) combinations of the above, on the relationship between the number of individuals released and the proportion of simulated populations establishing. Except where indicated otherwise, fecundity = 200 ± 50 eggs per female, *s* (expected survival rate of eggs) = 0.02, *E* (standard deviation in year-to-year survivorship as a measure of environmental variability) = 0 and *A* (intensity of the Allee effect) = 0. Reprinted with permission of the Ecological Society of America from Grevstad (1999).

Third, environmental stochasticity produces variation in population growth rates that is largely independent of population size (Caughley and Gunn 1996): the proportionate effect on a population of bad weather, for example, does not depend on the number of individuals in the population (although it may be reduced by a population's extent). Thus, it follows that demographic stochasticity is a less important cause of extinctions than environmental stochasticity once the population size is sufficiently large (Lande 1993). Nevertheless, while the effect of environmental stochasticity on population growth rate may be size-independent, it is clear that a proportional decline in a smaller population is more likely to take that population down to a size where demographic stochasticity matters (and into the 'extinction vortex': Gilpin and Soulé 1986). Indeed, environmental stochasticity can greatly decrease establishment success if the variability it induces in population growth rates is high relative to the mean rate (Lande 1993; Grevstad 1999; Figure 3.2).

Environmental stochasticity in growth rate may also reduce a population to the point at which Allee effects (Allee 1931; 1938) become important. An Allee effect is a positive relationship between any component of individual fitness and either numbers or density of conspecifics, hence also termed positive density dependence (Stephens, et al. 1999; Courchamp, et al. 2008). It is often expressed as a reduction in population growth rate at low population density, such as may occur for example because it is harder for individuals of a sexually reproducing species to find mates at low densities (Dennis 1989). A variety of modelling approaches incorporating Allee effects suggest the existence of a threshold population density, below which persistence is greatly reduced (e.g., Grevstad 1999; Dennis 2002; Taylor and Hastings 2005). How well defined this threshold is depends on the presence of stochasticity in population growth rates (Figure 3.2). Interestingly, as well as decreasing the persistence of populations initially above the threshold density, environmental (or indeed demographic) stochasticity can increase the probability of persistence for populations below the threshold density via the reverse process: a run of good years for population growth can boost the density of a population that would otherwise succumb to Allee effects (Grevstad 1999; Taylor and Hastings 2005).

Small populations will also suffer from a loss of genetic variation (Brook 2008). Sampling effects mean that introduced populations are likely to contain fewer alleles than the native population from which they were taken, while genetic drift within the introduced population will cause further alleles to be lost by chance. Levels of inbreeding in small populations may be high, which can in turn lead to the loss of heterozygosity, the expression of deleterious alleles, and consequently inbreeding depression in some (though not all) species. These processes will continue to act as long as a population remains small. Evidence of their action in exotic species is dealt with at length in Chapter 8. They all have the potential to reduce the mean fitness of a population, and hence elevate the likelihood that it will go extinct (reviewed by Frankham, et al. 2002; Frankham, et al. 2004). Note, however, that genetic variation within small, introduced populations may be higher than within-population variation in the native range, if the individuals introduced are sampled from a range of distinct genetic lineages across the native range (Chapter 8).

Clearly, then, there is a variety of reasons why small populations of threatened species provide a headache for conservation managers, which may equally jeopardize the establishment success of an exotic species, given that these tend to be liberated in low numbers (Figure 3.1). Moreover, it is also clear that many of these problems decrease as population size increases. It follows that we might expect the number of individuals introduced to be an important determinant of whether or not establishment is successful (Newsome and Noble 1986; Ehrlich 1989; Veltman, et al. 1996; Williamson 1996; Dean 2000; Lockwood, et al. 2005; Hayes and Barry 2008).

3.2 Propagule Pressure and Establishment Success

3.2.1 Evidence for an Association

The total number of individuals of a species introduced at a given location is generally termed its propagule pressure (Williamson 1996), or sometimes introduction effort (Veltman, et al. 1996; Duncan 1997; Blackburn and Duncan 2001b). Since not all of these individuals are necessarily introduced in one go, propagule pressure can be viewed as having two components: the number of introduction events (propagule number) and the number of individuals per introduction (propagule size) (Pimm 1991; Carlton 1996; Veltman, et al. 1996; Lockwood, et al. 2005). Propagule pressure is the sum over all introduction events of the number of individuals liberated, or the product of propagule number and mean propagule size. Propagule pressure is very clearly related to the theoretical and empirical concepts of minimum viable population size in conservation biology (Terborgh and Winter 1980; Traill, et al. 2007).

At least sixteen studies have analysed the effect of propagule pressure on exotic bird species establishment success (Table 3.1). Almost all of them find a significant positive relationship between these two variables. The only slight dissenters from this unequivocal association are studies by Newsome and Noble (1986) and Drake (2006). Drake found that a measure of propagule pressure did not explain significant variance in establishment success for ring-necked pheasant *Phasianus colchicus* introductions to the USA when included with a measure of hybridicity (see section 8.4.3). However, Drake divided propagule pressure by the numbers of years over which releases occurred, which may mask effects of release size or number. Newsome and Noble (1986) found that propagule pressure did not significantly influence establishment success for Australian species of bird introduced within Australia, although it did for species introduced from outside Australia. Establishment success was notably high for Australian species introduced in low numbers. However, there are two reasons why these results may be biased. First, the within-Australian introductions combine data from species introduced from one part of Australia to another with natural colonizations, and possibly also reintroductions to the historic range. Second, all introduction events resulting from escapes, where propagule pressure is not actually known, were assumed to concern low numbers (<20) of individual birds. These two problems may cause the influence of propagule pressure to be underestimated in this set.

Table 3.1 Studies of the effect of propagule pressure, size, or number on establishment success in birds. + = significant positive relationship, NS = relationship not statistically significant.

Study	Details	Propagule size	Propagule number	Propagule pressure	Notes
Dawson 1984	Introductions to New Zealand			+?	Unpublished BSc thesis cited in Williamson (1996). It is unclear whether propagule size or pressure is analysed
Newsome and Noble 1986	Introductions to Australia of non-native and Australian species			+ for non-natives NS for Australian species	There were no large introductions of species native to Australia. See text for more details
Pimm 1991	Game bird introductions to North America	+			Relationship sigmoidal
Veltman, et al. 1996	Introductions to New Zealand		+	+	The minimum adequate model includes both
Duncan 1997	Passerine introductions to New Zealand		+	+	The minimum adequate model includes propagule pressure
Green 1997	Introductions to New Zealand			+	
Sol and Lefebvre 2000	Introductions to New Zealand			+	
Cassey 2001	Introductions to New Zealand	+	+		Across species and using independent contrasts to account for phylogenetic relatedness

Duncan, et al. 2001	Introductions to Australia	+	+	Across species and using independent contrasts. Also significant effect of number of release sites
Moulton, et al. 2001	Galliform introductions to the island of Hawaii		+	Medians were 179 individuals introduced for successful species, and 14 for failures
Duncan and Blackburn 2002	Game bird introductions to New Zealand		+	
Sol, et al. 2002	Global introductions	+		Significant in a minimum adequate model, with taxonomic order as a random effect
Brook 2004	Introductions to New Zealand and Australia	+	+	The best model identified using information theoretic model selection is in terms of propagule number alone
Cassey, et al. 2004c	Global introductions	+	+	Results for propagule number not reported in the paper
Møller and Cassey 2005	Introductions to New Zealand		+	
Sol, et al. 2005b	Global introductions		+	Same data as in Cassey, et al. (2004c)
Drake 2006	Pheasant introductions in the USA		NS	Propagule pressure is divided by the number of years of releases and fitted in a model with a measure of hybridicity
Duncan, et al. 2006	Introductions to New Zealand		+	

Most of the analyses of propagule pressure for exotic birds concern introductions to New Zealand. This is a consequence of the unusually thorough and clear historical compendium of the early introduction events provided by Thomson (1922; see also Chapter 1), although even for New Zealand there is a certain amount of disagreement as to how many species have actually been introduced (Duncan, et al. 2006). Nevertheless, regardless of the number of species included in the analysis, all studies of New Zealand introductions that assess the effect of propagule pressure identify it as a primary determinant of establishment success. Figure 3.3 shows how the proportion of all bird introductions to New Zealand fall into each of three propagule pressure categories (≤20, 21–100, >100 individuals introduced) in comparison to the proportion of successful introductions falling in those categories (see also Williamson 1996). It clearly shows that while most introductions involve small numbers of birds, most successful introductions relate to species introduced in larger numbers. This conclusion also holds for subsets of these species, including game birds (Duncan and Blackburn 2002) and passerines (Duncan 1997).

Because propagule pressure is a composite variable, a high value for it can be obtained through an increase in either propagule size or propagule number separately, or in both together. The relative influence of propagule size and propagule number in driving the effect of propagule pressure has not been well studied, although Cassey, et al. (2004c) showed that propagule size and propagule number are positively correlated ($r = 0.71$, $N = 300$) for avian introductions around the world. So far, only Cassey (2001b) has presented results for the effects on avian establishment of propagule size and number separately. His results indicated that propagule size and number independently helped explain observed variation in which birds introduced to New Zealand established self-sustaining populations.

Several further studies have addressed the relative predictive powers of propagule number and propagule pressure, although it must be borne in mind that these variables are not independent. All find that propagule number and pressure

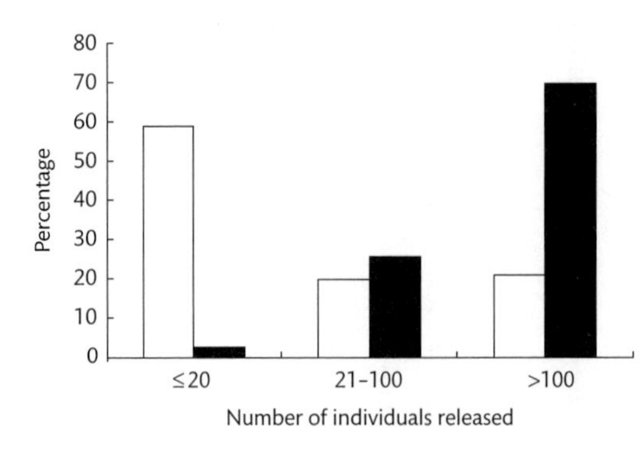

Fig. 3.3 The percentage of those bird species introduced to New Zealand in different categories of propagule pressure (open bars), and the percentage of those species in each category that successfully established (filled bars). From data used by Duncan, et al. (2006).

are consistently positively related to establishment success (Table 3.1), although the relative contributions vary. For example, Duncan (1997) found that a stepwise multiple regression model for establishment success in New Zealand passerines included propagule pressure but not propagule number. In contrast, Brook (2004) combined data for bird introductions to New Zealand and Australia, and modelled success using the total number of individuals introduced (N), number of introduction events (E), and an interaction (I). He then used an information theoretic approach to distinguish amongst the seven different models possible with these three variables. The best model was in terms of number of introduction events alone (E), although N + E is also reasonably well supported (ΔAIC = 1.4) and the full model N + E + I is ranked third (Table 3.2). The number of individuals introduced (N) alone faired relatively poorly as an explanation for establishment success in this study. The minimum adequate model identified by Veltman, et al. (1996) to explain establishment success in New Zealand birds included both propagule number and pressure. Finally, to complete the set of all possible outcomes, Duncan, et al. (2001) found that neither propagule number nor pressure entered into the minimum adequate model for establishment success in birds introduced to Australia! Rather, it was the number of introduction sites that mattered (see also Veltman, et al. 1996). It would be interesting to revisit these analyses using propagule size instead of pressure. For New Zealand, at least, the results from Cassey (2001b) show that both predict success.

Table 3.2 Model selection results relating the likelihood of successful establishment by exotic birds in Australia and New Zealand to the log-transformed total number of individuals introduced N, the log-transformed number of release events E, and an interaction term I calculated as N x E. Log*L* = maximized log-likelihood, *K* = number of parameters, AIC_c = small-sample value Akaike's Information Criterion, Δ_i = difference between model AIC_c and that of the best model, and w_i = Akaike weight of the model. From Brook (2004).

Model	Log*L*	*K*	AIC_c	Δ_i	w_i
E	-52.13	2	108.3	0	0.444
N + E	-51.79	3	109.8	1.4	0.219
N + E + I	-51.04	4	110.4	2.1	0.159
E + I	-52.12	3	110.4	2.1	0.156
I	-55.48	2	115.1	6.7	0.016
N + I	-55.44	3	117.1	8.7	0.006
N	-60.25	2	124.6	16.3	0.000

Although it is clear that establishment success broadly increases with different partial indices of propagule pressure (e.g., size and/or number), the exact shapes of these relationships remain little explored. Two exceptions are Dawson's (1984) study of bird species introduced to New Zealand (cited in Williamson 1996), and Pimm's (1991) analysis of game bird introductions. Notably, Pimm (1991) showed that a sigmoidal function well fitted the relationship between establishment success and propagule size (Figure 3.4). Success asymptotes, such that above around 300 individuals, increasing the size of the propagule does not increase the probability that it will establish. Even at the larger sizes, however, success is low for these introductions, flattening out at around 15% of events in each class. Other studies have also found that establishment success is low for game birds (Duncan, et al. 2001; Cassey, et al. 2004c), perhaps because recreational hunting excessively elevates their mortality rates (although one might reasonably heed Phillips (1928) when he wrote 'It would be more interesting than instructive, perhaps, to attempt to account for failures in planting game and song birds. Wherever an expensive enterprise fails, sportsmen's journals are found surging with ready-made explanations that have not the slightest scientific foundation.'). Dawson's (1984) data also show that success is an approximately sigmoidal function of propagule pressure, although he apparently did not formally model this.

We used the data on which the analyses of propagule size in Cassey, et al. (2004c) and Sol, et al. (2005b) were based to investigate the shape of its relationship to establishment success in more detail. These data included 550 introduction events for which a precise estimate of propagule size was available. We assigned these events to a series of bins of increasing numbers introduced, and calculated the proportion of events in each bin for which the introduction succeeded in

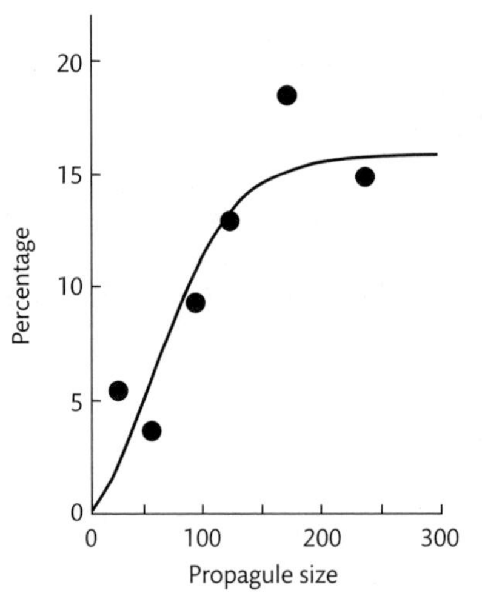

Fig. 3.4 The relationship between propagule size and the percentage of successfully established game bird introductions to North America. Introductions are grouped into bins at intervals of roughly fifty individuals. Reprinted with permission from Pimm (1991). © 1991 University of Chicago.

establishing. Figure 3.5 shows the results. When all events are included, percentage success seems to be a sigmoidal function of propagule size up to around 200 individuals, beyond which per cent success declines as more individuals are introduced. However, the substantial proportion of higher numbers introduced that concern game birds largely explains the reduced success at higher propagule sizes. Game birds are frequently bred in specialized game farms for release in large numbers, have higher propagule pressures than species in other bird families (Cassey, et al. 2004c), and, as noted above, also have relatively high failure rates. When these introductions are excluded, the sigmoidal pattern is retained, but now per cent success plateaus at a higher percentage level, at around 50–100 individuals introduced. Moreover, per cent success does not decline at higher propagule sizes (the relatively low proportion for the highest propagule size category can perhaps reasonably be discounted as it is based on just six events).

Further investigations into the form of the relationship between propagule size and introduction success would be useful to establish the generality of the sigmoidal pattern found by Dawson (1984), Pimm (1991) and in Figure 3.5. There is, nevertheless, some evidence that the exact pattern will vary by taxon and region. Cassey, et al. (2004c) showed that establishment success increased with propagule

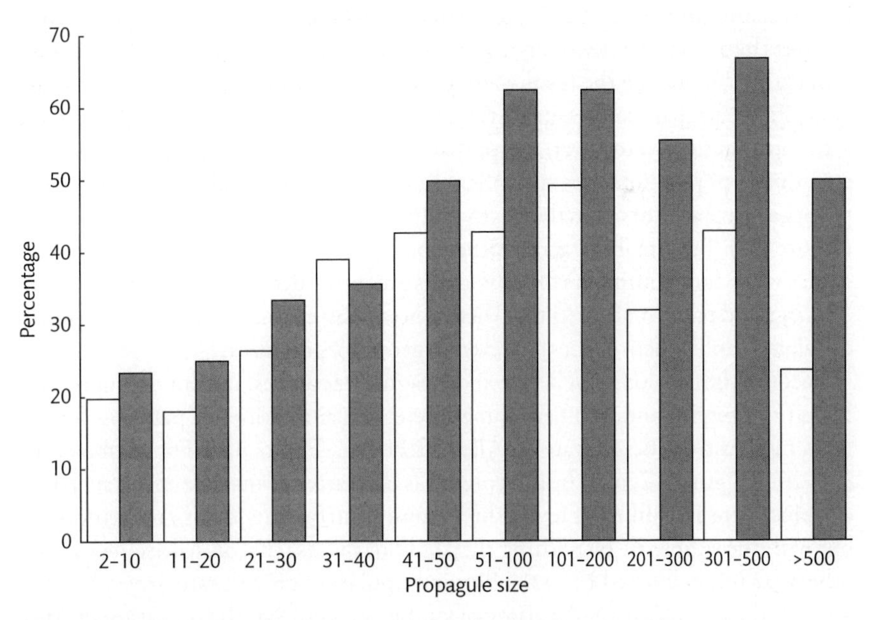

Fig. 3.5 The percentage of bird species introduced to locations around the world in different categories of propagule pressure that succeeded in establishing. Open bars are for all species (N = 550), filled bars are with species in the game bird families Cracidae, Numididae, Odontophoridae, and Phasianidae excluded (N = 266). From data supplied by D. Sol.

size for all of the ten bird families and 11/13 geographic regions for which sufficient data were available (albeit that not all the correlations were significant) but with widely varying regression coefficients. Thus, the increase in success for each increment in propagule size differed across families and across regions, resulting in different mean levels of success (Cassey, et al. 2004c).

3.2.2 Implications of the Association

The consistent positive association between metrics of propagule pressure and establishment success all point to small population size being as much of a hurdle for the establishment of exotic bird species as it is for the persistence of threatened species. Demonstrating mechanism from pattern is rightly notorious for its difficulty, and the patterns revealed in the studies of avian propagule pressure should not be over-interpreted. Nevertheless, they exhibit some features that we feel would merit attention in terms of understanding the small-population paradigm—as well as planning the reintroduction of conservation-dependent species (Cassey, et al. 2008).

First, there are clear thresholds in the influence of propagule pressure on establishment success (Figures 3.4 and 3.5). This effect is most obvious for larger propagule pressures, but at least in Figure 3.5 there also seems to be a lower threshold. If fewer than twenty individuals are introduced, then it is very likely that population extinction will be the result—roughly 80% of all introductions with precise quantitative propagule pressure estimates fail, or three-quarters of all non-game bird introductions. However, the probability of extinction declines rapidly with introduced population size, such that above around 100 individuals, increasing propagule size appears to make relatively little difference to establishment success (Figure 3.5). Presumably, factors other than population size determine the outcomes when larger numbers of individuals are introduced (see Chapters 4 and 5). This sigmoid pattern suggests that there is no consistent marginal benefit, in terms of avian establishment success, to ever-larger propagule sizes.

Second, the relationships between propagule size and establishment success for birds (Figures 3.4 and 3.5) bear some interesting similarities to patterns derived from models that incorporate an Allee effect (e.g., Figure 3.2). For example, the pattern in Figure 3.5 could be interpreted as indicative of an Allee effect, such that establishment is unlikely if fewer than around thirty individuals are introduced, but with the simple all-or-nothing threshold in success that an Allee effect would otherwise imply blurred by variability in population growth rate driven by variation in the environment. A variety of studies suggest that the operation of Allee effects can be detected on the basis of thresholds in the probability of establishment (or extinction) as a function of population size (Grevstad 1999; Dennis 2002; Leung, et al. 2004). That success plateaus at a propagule size of around 100 individuals also fits with the observation that the effects of environmental stochasticity are independent of population size (Caughley and Gunn 1996), once the population is large enough to escape Allee effects (or demographic stochasticity; see below).

Also interesting in this regard is Brook's (2004) study of bird introductions to New Zealand and Australia. Brook assembled data on the population dynamics of the species in their native ranges to parameterize a simple population model of exponential growth in a stochastic environment, where stochasticity is expressed through a random component to variation in the population growth rate. He used this model to calculate, for each species, the population size required for >50% probability of establishment. He then predicted the introduction outcomes for these species based on the premise that successful introductions would tend to be founded by larger populations than his estimated population size for >50% success. His predicted outcome was correct in 83% of cases—even though some of the numbers used in parameterization clearly relate to declining native populations and so are unlikely to represent a 'true' intrinsic growth rate for the species. Brook's predictions are relatively accurate despite no account being taken of Allee effects, perhaps because of the apparently small threshold population size for such effects in birds. Thus, whether the assumed threshold for Allee effects is thirty individuals, as Figure 3.5 might imply, or zero individuals, as Brook (2004) implicitly assumes, may make little difference to the predicted propagule size above which establishment is more likely than not.

The data currently available for bird introductions would certainly benefit from more detailed investigation to assess whether Allee effects can be distinguished from the possible influence of demographic stochasticity. Both of these processes have their strongest effects at very small population sizes (indeed, elements of demographic stochasticity and Allee effects are rolled together by Courchamp, et al. 2008), but decline in importance as population size increases (although see Melbourne and Hastings 2008). Thus, while the relatively rapid increase in establishment success with propagule size, such that success increases to a plateau at around 100 individuals introduced (Figure 3.5), matches models with Allee effects (e.g., Grevstad 1999), it is also congruent with model findings that the effects of demographic stochasticity should only matter for populations of a few dozen individuals (e.g., Lande 1993; but see Melbourne and Hastings 2008). Detailed studies of the population dynamics, behaviour, genetics, and ecology of the exotic bird populations in question may help to distinguish the two (Lande, et al. 2003). Allee effects may be especially important for social or colonial species, and it has been argued that the degree of sociality may reflect the severity of the Allee effect for a species (Courchamp, et al. 1999; Stephens and Sutherland 1999). If so, we might expect differences in the influence of propagule pressure on establishment in social versus solitary bird species. We are not aware of a published test of this suggestion, but there is some evidence that sociality per se influences success (see Chapter 4).

Third, it would be interesting to investigate the relative influences of propagule size and propagule number on establishment, given that avian studies to date show that either, both, or neither may be the primary determinant of success (see also Drake, et al. 2005). Several small introductions may increase mean population fitness and adaptive potential relative to a single large introduction because of sampling effects on the genetic composition of the population (DeSalle 2005), and so explain situations when propagule number is a better predictor of success than propagule size (e.g., Brook 2004). Alternatively, whether propagule pressure should be concentrated in one large introduction event, or distributed across many small introductions, may depend on the relative magnitudes of environmental stochasticity and Allee effects. For example, Hopper and Roush (1993) argued that number of individuals per introduction should be the best predictor of success if Allee effects apply, but number of introductions if environmental stochasticity is the most important determinant of success (see also Haccou and Iwasa 1996; Haccou and Vitunin 2003).

Grevstad (1999) modelled the probability that establishment will occur in at least one population if 1,000 individuals are introduced, divided into varying numbers of introduction events, r, of varying numbers of individuals, n (where r and n are integers, and $rn = 1{,}000$). In these models, the effect of demographic stochasticity on probability of establishment was negligible except for species with very low reproductive rates. Varying the strength of environmental stochasticity and the Allee effect produced the following generalizations. When Allee effects are small, introductions should be made as small as possible: propagule number is favoured over propagule size. When Allee effects are intermediate in size, and there is considerable environmental stochasticity, propagules should be intermediate in size and number. When Allee effects are strong, introductions should always be large. Finally, when environmental stochasticity and Allee effects are both small, the optimum introduction size is flexible above the number required by the Allee effect. Overall, there is a general positive relationship between the strength of the Allee effect and the size of the propagule required, while high levels of environmental stochasticity or strong Allee effects mean that different introduction strategies can greatly influence success or failure (Grevstad 1999; see also Hopper and Roush 1993; Drake and Lodge 2006).

The temporal and spatial distribution of the different introduction events is also likely to influence establishment success (Lockwood, et al. 2005). Haccou and Iwasa (1996) showed that success in fluctuating model environments depends on the timing of the introduction, and so is higher if introductions are sequential rather than simultaneous. The advantage of sequential introductions was reduced if propagule pressure was very high (Haccou and Iwasa 1996), and also depended on the degree of autocorrelation in the environment (Haccou and Vitunin 2003). This result pertains because the most important factor in success is the duration of runs of 'bad luck' in the conditions experienced by an incipient exotic population.

For example, if environments are negatively autocorrelated, so that a bad year at a site is likely to be followed by a good year, then it makes sense to free exotic individuals using sequential introductions. This tactic increases the likelihood that an introduction event coincides with good conditions. Conversely, if the environment is positively autocorrelated such that bad years follow bad, or if there is no or only weak autocorrelation, then it makes more sense to spread the risk of failure by distributing introductions across different, independently varying sites (Haccou and Vitunin 2003). Adverse conditions in one location are then less likely to affect populations in other locations (Hanski 1989).

Overall, therefore, whether and how introduction events should be distributed in space and time to maximize establishment success will vary, depending on the magnitude of environmental fluctuations, their temporal and spatial autocorrelation, whether or not Allee effects apply, and the reproductive potential of the species. Untangling the relative influence of these effects will be difficult. Nevertheless, variation in the form of the relationship between propagule pressure and success across bird taxa, and indeed more widely, and variation in the importance of its components, may provide significant opportunities in this regard. For example, it would be interesting to attempt to identify characteristics of the location that explain when (for a given propagule pressure) either many small or fewer large propagules is the better strategy to maximize establishment success. Insight may also be gained from exploring how this variation might relate to the influence of evolutionary or ecological differences between taxa (see Chapter 4). The implications for conservation reintroductions should be obvious.

3.3 Other Associations of Propagule Pressure

Given that propagule pressure is a consistent and usually strong predictor of establishment success, a pertinent question is whether propagule pressure is associated with other factors thought to be important for exotic bird establishment. If so, it could bias perceptions of the types of species that make good invaders, or the types of locations that are easy or difficult to invade. Thus, here we consider the evidence that propagule pressure varies systematically with characteristics of the species introduced and the location of introduction. We finish by noting two important further consequences of variation in propagule pressure for avian introductions.

3.3.1 Are Some Species Consistently Introduced in Larger Numbers than Others?

An obvious correlate of propagule pressure is likely to be the degree of planning that went into the introduction. In general, a planned and organized release ought to concern more individuals than accidental releases. We used the data of Cassey, et al. (2004c) to test for an influence of planning on propagule size. Planning does indeed make a significant difference to the number of individuals released (Table 3.3). Deliberate releases concern between nine and 250 times as many individuals on average as escapes, depending on the measure of central tendency. Introductions that comprised escaped and released individuals also tended to have low propagule size. This result probably stems from these release events occurring without concern for eventual establishment in mind, as for example with the release of superfluous cage birds (e.g., Robinson 2001; Chapter 2).

For British bird species that had been introduced to New Zealand, Blackburn and Duncan (2001b) compared components of propagule pressure to characteristics of the species in their British native range. They found that resident species, and species with larger population and geographic range sizes in Britain were introduced to New Zealand more often and in greater total numbers. Multivariate analysis identified British population size and migratory category as the strongest independent correlates of both propagule number and propagule pressure in New Zealand. Planned releases involve more individuals but, as already noted in Chapter 2, numbers introduced are greatest for species that are readily available for capture, transport, and subsequent release.

Table 3.4 lists the top ten exotic bird species, as judged by their median propagule size across introduction events for which this has been reported precisely, and for which at least five such estimates were available. The list is dominated by game birds, as might have been expected given that many of the game bird introductions derive from individuals bred in specialized game farms for release in large numbers (Long 1981; Lever 2005). Unsurprisingly, Galliformes also has the highest mean propagule size, averaged over all quantified introduction events for the orders (Table 3.5). However, propagule sizes are highly skewed, with most

Table 3.3 The median and arithmetic mean propagule pressure for different manners of introduction (both = some individuals deliberately and some accidentally released) for the 525 introduction events for which precise estimates of number of individuals introduced and information on manner of introduction were available. Propagule pressure differs significantly across introduction modes (Kruskal-Wallis test, $\chi^2 = 24.5$, d.f. = 2, $P < 0.001$). N = number of introduction events.

Manner	Median	Mean	N
Deliberate	50	3273.0	503
Accidental	6	12.8	12
Both	5.5	28.2	10

Table 3.4 The top ten bird introductions, based on ranking by median propagule size from introduction events for which precise estimates of number of individuals introduced were available. N = number of introduction events from which median is calculated (minimum for inclusion = 5).

Order	Species	Median	N
Galliformes	Kalij pheasant *Lophura leucomelanos*	1091	7
Galliformes	Red junglefowl *Gallus gallus*	1047.5	10
Galliformes	Black francolin *Francolinus francolinus*	864	13
Galliformes	Common quail *Coturnix coturnix*	750	6
Galliformes	Reeves's pheasant *Syrmaticus reevesii*	548	15
Galliformes	Grey francolin *Francolinus pondicerianus*	366	13
Galliformes	Red-legged partridge *Alectoris rufa*	331	12
Galliformes	Grey partridge *Perdix perdix*	248	27
Ciconiiformes	Chestnut-bellied sandgrouse *Pterocles exustus*	137	5
Passeriformes	European skylark *Alauda arvensis*	129	10

Table 3.5 The median and arithmetic mean propagule pressure in different orders, for the 550 introduction events for which precise estimates of number of individuals introduced were available. N = number of introduction events.

Order	Median	Mean	N
Galliformes	85.5	5763.4	282
Tinamiformes	172.5	1431.9	10
Apodiformes	375.0	375.0	1
Ciconiiformes	25.5	214.8	18
Passeriformes	31.0	65.2	162
Psittaciformes	4.0	55.75	8
Strigiformes	49.5	53	6
Anseriformes	8.0	52.2	41
Craciformes	25.0	25.0	2
Struthioniformes	24.0	24.0	2
Columbiformes	12.0	24.1	15
Gruiformes	16.0	16.0	1
Coraciiformes	4.0	4.0	2

species introduced in low numbers most of the time (Figure 3.1; Figure 3.6). Median propagule sizes for these orders are much lower, most notably for the Galliformes (Table 3.5). This metric also reveals that Anseriformes tend to have been introduced in surprisingly low numbers, which conclusion is also supported by Figure 3.6. Nevertheless, propagule sizes do differ significantly across the different orders (Figure 3.6). Thus, analyses of the association between propagule sizes and the characteristics of species will need to account for differences in average numbers introduced across taxa.

Cassey, et al. (2004c) used global data on avian introductions and propagule size to test whether availability correlates with propagule size across introduction events more widely. The best measure of availability we have for these species is their geographic range size. While we lack abundance data for these species, abundance is highly correlated with geographic range size in many, if not most, bird assemblages (e.g., Gaston and Blackburn 2000; Blackburn, et al. 2006). Cassey

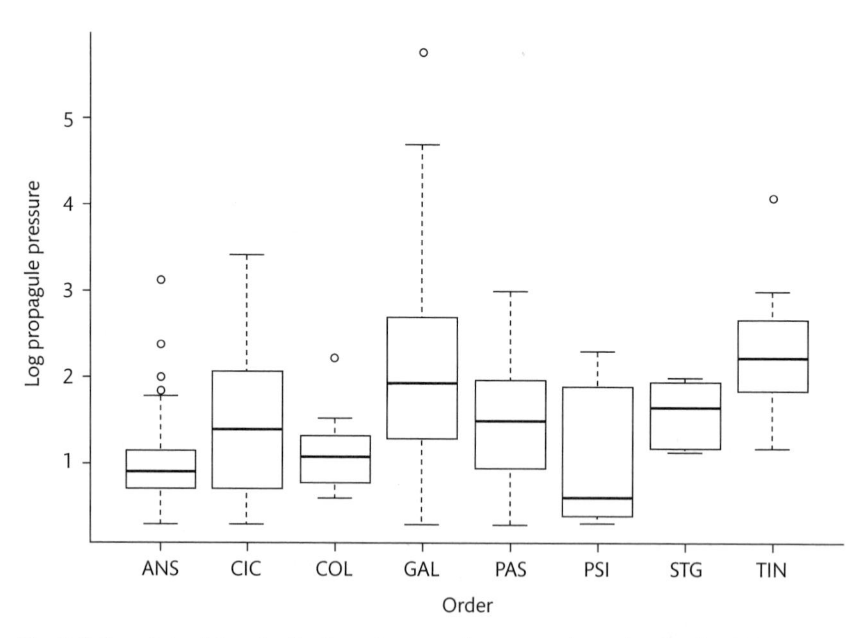

Fig. 3.6 Box plots of the distribution of propagule pressure by orders for the 542 introduction events with precise data on numbers of individuals released (orders with fewer than five estimates were excluded). Propagule pressure differs significantly across orders (Kruskal Wallis test: $\chi^2 = 98.0$, d.f. $= 7$, $P < 0.001$). ANS = Anseriformes, CIC = Ciconiiformes, COL = Columbiformes, GAL = Galliformes, PAS = Passeriformes, PSI = Psittaciformes, STG = Strigiformes, TIN = Tinamiformes. From data supplied by D. Sol.

and colleagues included bird taxonomic order as a random effect in a mixed model, to control for these differences in average effort across taxa. Bird species with larger geographic ranges also have higher propagule sizes (Cassey, et al. 2004c).

Cassey, et al. (2004c) additionally explored associations between propagule size and avian life-history characteristics, again controlling for taxonomic associations. They found that, within bird families, smaller-bodied species were introduced on average in larger numbers, as were more fecund species. The relationship to body mass was reversed if family membership was not accounted for: propagule size is generally higher for species in large-bodied families, but higher for smaller-bodied species within those families. Again, we can cite the example of the Galliformes. These are large-bodied relative to other birds (geometric mean mass of 450g vs 53g for all bird species: Dunning 1992; Blackburn and Gaston 1994) and tend to be introduced in larger numbers than other orders. However, smaller-bodied Galliformes are introduced in larger numbers, on average, than are larger-bodied Galliformes (Figure 3.7). The reverse situation to body mass pertains for migratory tendency. Cassey, et al. (2004c) found that migrants are introduced in lower numbers than non-migratory species, but only if family membership is ignored. Thus, migrants are introduced in lower numbers overall, but in larger numbers than non-migrants in the same family.

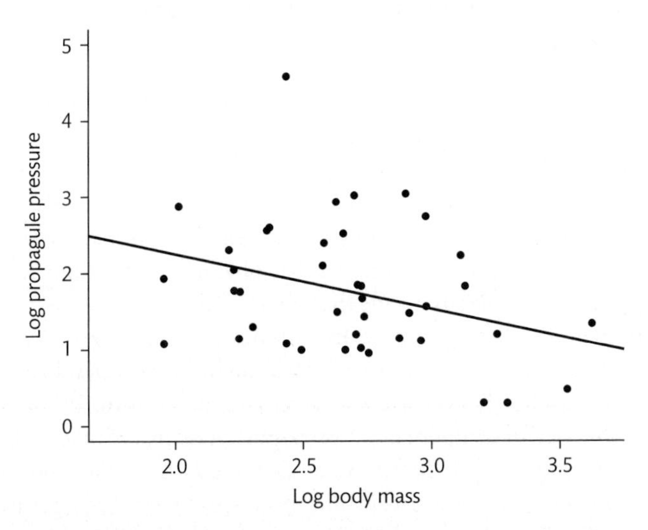

Fig. 3.7 The relationship between $\log_{10}$ median propagule pressure (number of individuals released) and $\log_{10}$ body mass (g) for species of Galliform for which precise estimates of propagule pressure were available (estimate $\pm$ standard error $= -0.714 \pm 0.0323$, $r^2 = 0.11$, $N = 41$, $P = 0.033$). From mass data in Dunning (1992) and propagule pressure data supplied by D. Sol.

Propagule size differs across bird taxa, but so too does mean success rate. For example, Cassey, et al. (2004c) found that 29% of introduction events resulted in establishment for Sylviidae, compared with 70% for Sturnidae. However, interestingly, there is no evidence that taxa with higher mean propagule size have concomitantly higher levels of establishment success: mean success and the rate of increase in success with numbers introduced are not significantly correlated ($r = 0.284$, $N = 10$, $P = 0.21$; Cassey, et al. 2004c). Yet again, the classic example is game birds, which have higher propagule sizes than species in other taxa (Figure 3.6), but also have relatively high failure rates. Presumably, there are species-level characteristics, shared by species in higher taxa due to their shared evolutionary history, which cause the relationship between propagule size and success to be elevated or lowered for these higher taxa. A review of evidence supporting the influence of such traits on success will occupy the next chapter, although the analyses above suggest body mass and migratory tendency as potential candidates. Here, we conclude by simply noting that the failure of propagule size to predict the ranking of bird taxa with respect to success, even though it explains variation in success within bird taxa (section 3.2.1), strongly suggests that there is more than simply propagule pressure at work.

3.3.2 Do Some Areas Consistently Receive Larger Propagules than Others?

Just as some species are introduced in larger numbers than others, so too is it possible that some areas receive larger numbers of individuals than others. This suggestion does indeed seem to be the case. Notably, Cassey, et al. (2004c) showed that islands receive smaller total numbers of exotic bird individuals, on average, than do mainland locations, and that this difference is significant whether or not taxon membership is controlled for. This has obvious implications for the paradigm that islands are easier to invade than mainland regions (see Section 5.4).

Colautti, et al. (2006) catalogued five studies that had assessed the effect of propagule pressure on the invasibility of a location by bird species. In all five cases, invasibility was positively associated with propagule pressure, although it is not clear in this case which components of propagule pressure are involved, or indeed which areas. Cassey, et al. (2004c) found that areas on similar lines of latitude to the location of origin of the exotic bird species also received larger numbers of introduced individuals. Presumably, this pattern reflects the fact that more individual birds are likely to survive the journey to an exotic location that is closer to the species' native range. Whatever the cause of the association, however, it is clear that propagule pressure has the potential to confound conclusions about both species- and location-level effects on establishment success.

3.3.3 Further Consequences of Propagule Pressure

Small numbers of individuals introduced have at least two consequences for exotic populations beyond their influence on the probability that populations will succumb to the perils of demographic or environmental stochasticity, or Allee effects. First, it is well known that small populations lose alleles, while inbreeding is also more likely in small populations. Both of these processes can have negative fitness consequences for individuals and thus the overall exotic population. We review the genetics of bird introductions in detail in Chapter 8.

Second, propagule pressure affects the probability that exotic parasites will be introduced along with their exotic host populations. Since not all individuals of a native population are infected with all the species of parasite indigenous to that host species, it follows that the probability that all parasite species are introduced with the host, and hence parasite species richness, should be positive functions of propagule pressure (Paterson, et al. 1999; Drake 2003). Indeed, some exotic populations are apparently entirely free of at least some types of parasites, such as the absence of blood parasites from the common myna *Acridotheres tristis* population introduced to the Cook Islands (Steadman, et al. 1990).

As far as we are aware, no study has yet related propagule pressure directly to the loss of native parasite species. The closest to date has been Paterson, et al.'s (1999) study of New Zealand birds and their associated parasitic louse species, to assess the probability that louse species have been lost due to 'missing the boat' during colonization. Paterson and colleagues compared the louse species richness of native and exotic populations of introduced bird species, original and colonizer populations of self-introduced bird species, and New Zealand native bird species with closely related bird species from elsewhere. They found that most bird species (34/48) have fewer louse species in New Zealand than in the non-New Zealand population (or species) with which they were compared. The loss was greatest (mean = 1.9 species) for exotic species, intermediate for native species (mean = 1.2 species), and smallest (and not significantly less than in the native population) for self-introductions (mean = 0.3 species). These differences were argued to reflect differences in the size of the founding population, because those of exotic introductions may often be smaller than those of self-introductions (cf. Clegg, et al. 2002). Paterson, et al. (1999) argued that their data fit the hypothesis that 'missing the boat' is an important factor driving patterns of host–parasite co-occurrence, while acknowledging that sampling effects or parasite extinction following arrival in New Zealand could also play a role.

If parasites 'miss the boat' when their hosts are introduced, this could influence subsequent host introduction success. The 'enemy release hypothesis' (Williamson 1996; Torchin, et al. 2001; Torchin, et al. 2003; Keane and Crawley 2002; Drake 2003; Mitchell and Power 2003) suggests that introduction success is, at least in part, due to a decrease in the regulation of the exotic population by natural enemies such as predators and parasites (see Chapter 5). If parasites are indeed important regulators of bird populations—and the evidence is that their effects are weak relative to those of weather and predation (Newton 1998)—escape from them may increase the likelihood that an exotic population will be able to grow from initial low numbers. However, any advantage to a small exotic bird population of losing parasites is likely to be outweighed by the concomitantly greater risk of demographic stochasticity or Allee effects (see also Drake 2003).

3.4 Conclusions

Three recent meta-analyses have assessed the evidence for different broad causes of establishment success. Cassey, et al. (2005c) used quantitative methods to assess effect sizes of event-level, species-level, and location-level variables on establishment success in studies of bird species introduced to island regions around the world. They found that event-level variables, and notably components of propagule pressure, were the most consistent predictors of success. There was some evidence for an effect of location on success, but none for a consistent effect of the characteristics of species. Colautti, et al. (2006) and Hayes and Barry (2008) used simpler (and concomitantly less powerful) vote-counting methods to summarize studies from a wider range of taxa, but with essentially the same message: propagule pressure is a key variable for success at the establishment stage. Colautti and colleagues also recapitulated the message of Cassey, et al. (2004c) that propagule pressure may be correlated with a range of other traits that may be important for establishment, and that it can influence the invasibility of areas (section 3.3.2) as well as the invasiveness of species (section 3.3.1). Colautti, et al. (2006) concluded that propagule pressure ought to be a null model for invasions, its influence needing to be removed first before the effects of other variables can be considered properly (see also section 10.2.2).

Propagule pressure's correlation with a range of traits that may be important for establishment is one reason why the establishment process often appears to be so idiosyncratic (Lockwood, et al. 2005; Colautti, et al. 2006), but there is a second reason. Cassey, et al. (2004c) used nested analysis of variance to explore the distribution of variance in propagule pressure with respect to the taxonomic hierarchy. They showed that most variation in propagule pressure (58%) is clustered among introduction events within species, with only 8% explained by differences among species within genera, 4% among genera within families, and 30% among families within orders. In other words, any given species is likely to have been introduced in widely varying propagule sizes across the locations to which it has been introduced. While we showed above that some locations or species tended to be associated with higher propagule pressures than others, it nevertheless follows that it will be difficult to identify characteristics of either that are consistently associated with establishment success when the primary determinant of that success is itself so idiosyncratic.

With this in mind, in the next two chapters we assess the roles of species traits and location in exotic bird establishment success. The reader would be forgiven, from our comments about the difficulties engendered by the idiosyncrasy of propagule pressure, for thinking that these are tasks we approach with trepidation. In fact, they are tasks about which we believe it is now possible to be optimistic. We think that our conclusions present an opportunity to move beyond old debates and on to a deeper understanding of invasions. Rather than recapitulating arguments about the primacy of the species or the location, or simply throwing our hands up in despair about the inherent complexity of the invasion process, it should now be possible to explore this complexity within a framework based around propagule pressure. Adopting a multi-level approach allows us to consider how variation in success that is unexplained by the role of numbers in the game of chance may relate to evolutionary or ecological differences between taxa, or geographical or biogeographical differences between locations. If propagule pressure is a major cause of the idiosyncrasy inherent in the invasion process, knowing that fact is a huge step forward.

4
The Role of Species Traits in Establishment Success

Frustrating many straightforward paths to invasion prediction is the combination of stochastic and deterministic processes that influence populations behaviours.

S. M. McMahon, M. W. Cadotte, and T. Fukami (2006)

4.1 Introduction

Avian introductions tend to concern the liberation of relatively few individuals, so resulting in exotic populations that are necessarily initially small, even by the standards of threatened species (Figure 3.1). It is risky for a population to be small, because this increases the chances that it will become extinct simply by bad luck. Demographic stochasticity, environmental stochasticity, Allee effects, or reductions in genetic variation can all lead to the extinction of a population that under other circumstances would be expected to persist and grow. For these reasons, there are clear advantages to an introduced population to be as large as possible, and indeed there is good evidence that propagule pressure or its components (propagule size and propagule number) are important factors in the establishment of viable populations in exotic birds (section 3.2).

That said, in the previous chapter we also identified several reasons why propagule pressure may not be the only determinant of establishment success, and that characteristics possessed by the species may also play a role in determining their successful introduction. First, the sigmoidal relationship between propagule pressure and establishment success in birds implies that, beyond a propagule pressure of about 100 individuals, success is no longer determined by numbers. Some introductions with large propagule pressures succeed while others fail, and there must be reasons for this.

Second, some introductions with small propagule pressures do succeed while many others fail, and there must be reasons for this also. Recent work on extinction probability has emphasized that, while population size matters, it is only under the most intense selection pressures or under the action of multiple stressors that the biological context becomes irrelevant (Brook, et al. 2008). Theory and data both suggest that the degree to which species are susceptible to Allee effects, demographic stochasticity, and environmental stochasticity depends on their life-history characteristics.

For example, models show that variation in reproductive rates affects the likelihood that demographic stochasticity will influence establishment probability (Grevstad 1999). A comparative analysis by Sæther, et al. (2004) found that interspecific variation in the magnitude of demographic stochasticity was related to where a species is placed on the 'fast–slow' continuum of avian life-history traits (Stearns 1983; Sæther 1987; Gaillard, et al. 1989; Read and Harvey 1989; but see Bielby, et al. 2007). Species at the 'slow' end, with lower reproductive rates and longer average lifespans, have lower demographic variance than species at the 'fast' end, presumably because variability in fitness is more constrained by life history in the slow species. Allee effects are more likely in species with widespread dispersal, and that aggregate as a mechanism of predator avoidance (Taylor and Hastings 2005). Genetic effects associated with small populations, such as loss of genetic variation or inbreeding depression, may also be moderated or exacerbated by specific traits, including mating system or heterozygosity in the native range.

The effects of environmental stochasticity on population dynamics are expressed through their effects on the population's vital rates. Thus, suboptimal conditions may increase mortality rates, and hence lower the intrinsic rate of population increase below zero. However, the extent to which species are susceptible to environmental stochasticity also depends on their life history. For example, Cawthorne and Marchant (1980) showed that populations of small-bodied British bird species showed the largest percentage reductions following harsh weather in the cold winter of 1978–9. Conversely, some models suggest that higher population growth rates improve the likelihood of establishment in the face of environmental stochasticity (Grevstad 1999). Either way, where species lie on the fast–slow continuum of life histories may not only influence their demographic variance (Sæther, et al. 2004), but also their ability to resist the vagaries of a new environment when at low population sizes.

Third, there is no relationship between mean propagule pressure and establishment success across bird families, while the slopes of the relationships between propagule pressure and establishment success within families also vary (Cassey, et al. 2004c). This suggests that species in each bird family respond differently to incremental increases in numbers released, as do the families themselves, presumably because they possess different traits.

There is thus a variety of good reasons to believe that the traits a species possesses will influence its establishment success, and that these traits should be those that integrate to overcome the problems faced by small populations. Indeed, the literature on avian introductions is replete with attempts to identify such traits (see recent reviews by Duncan, et al. 2003; Sol, et al. 2005a; Sol 2007). Yet, success in this endeavour has been rather low. The meta-analysis of studies of establishment by Cassey, et al. (2005c) found no evidence for consistent effects of species-level traits on success in bird introductions to islands. How can this apparent contradiction be reconciled? We can see two potential explanations.

First, while propagule pressure can increase the apparent influence of some traits, because of associations between those traits and the mean number of individuals released, it can also mask species-level relationships with success because of the high within-species variation in the numbers of individual birds released in any given event (Cassey, et al. 2004c). The way to assess such relationships is thus with propagule pressure as a covariate. Second, a wide range of species-level traits has been suggested to influence success. Some of these suggestions are contradictory. For example, high reproductive rates may allow an exotic species rapidly to escape the dangers of small population size. Conversely, investment in somatic growth rather than reproduction may allow species to ride the waves of environmental fluctuations. It is easy to imagine situations where one or other of these life-history strategies would be the more advantageous. These subtleties, however, would be very hard for analyses of historical introductions to detect. It is also possible that the key traits have not in general been identified, or have been identified, but their importance lost amongst irrelevant traits.

So, there are good reasons why species characteristics should affect establishment success, and good reasons why such effects may be difficult to detect in practice. There is little for it but to see what we can learn on this score from studies of avian introductions. That is the principal aim of this chapter. First, though, we explore a related line of evidence that species-level factors are likely to be important determinants of establishment success (Ehrlich 1989)—the suggestion that species show an 'all-or-none' pattern of exotic population establishment.

4.2 The 'All-or-None' Pattern: Does it Really Exist?

Regardless of the predictions from mathematical models, of inferences from analyses of propagule pressure versus introduction success, or of meta-analyses of the factors associated with success, there is one pattern of establishment outcomes that more or less guarantees that species-level traits are important. If a species always succeeds, or always fails, in establishing wherever it is introduced, then that implies that it possesses characteristics that determine its success regardless of the vagaries of location or the introduction process.

Simberloff and Boecklen (1991) argued that just such an 'all-or-none' pattern applied to the establishment success of bird species introduced to the Hawaiian

Islands. Of forty-one species introduced to more than one island in the chain, twenty-one always succeeded in establishing viable populations, while sixteen always failed. Only five species had mixed outcomes. Extending this analysis to introductions of these forty-one species to other locations around the globe, they found that success rate in Hawaii was highly rank correlated with success rate elsewhere. They concluded that species that were good colonists on one Hawaiian island were good colonists on them all, and indeed good colonists elsewhere. Hence, features of the species must be more important in determining success than features of the location or event. Simberloff (1992) argued that the same held for avian introductions to the Mascarene Islands, where nine of the eleven species introduced to more than one island show the all-or-none pattern.

The existence of an all-or-none pattern for avian introductions has been controversial since its inception. Moulton (1993) criticized Simberloff and Boecklen's (1991) analysis for being heavily influenced by phylogeny (pigeons in their data tended always to fail while passerines mostly succeeded), for misclassifying species with respect to their success or failure, for including events as introductions when in fact they resulted from colonization from other islands in the Hawaiian chain, and for failing to consider the number of islands to which species were introduced. This last point was important because species showing an all-or-none pattern were introduced to fewer islands than species showing mixed outcomes, and so simply had less opportunity to show mixed results (Moulton 1993). Moulton and Sanderson (1997) compiled data from nine oceanic islands to show that twenty-three out of the fifty-three species introduced to more than one island show mixed outcomes. Those with an all-or-none pattern of establishment had been introduced to significantly fewer islands, and the probability of a mixed outcome for a species increased with the number of introduction events. Our own assessment of introduction events also shows that mixed outcomes are a common feature of avian introductions (Figure 4.1): of the 207 species in Figure 4.1, 58% show mixed outcomes, 29% always failed, and 13% always succeeded. Species with mixed outcomes were introduced significantly more frequently (mean = 8.18 times) than those that always failed (2.84) or always succeeded (5.07; Kruskal-Wallis test, $\chi^2 = 25.95$, d.f. = 2, $P < 0.001$).

While a strict all-or-none pattern clearly does not apply to avian introductions (see also Moulton, et al. 2001a; Donze, et al. 2004), it is unlikely that we would expect it to. Even the very best invaders may sometimes be introduced in low enough numbers that they fall prey to the vagaries of chance (Chapter 3). A more sensible question to ask is perhaps how many mixed versus other outcomes would we expect to see, given the influence of chance events, if some species were genuinely good invaders and other species genuinely poor? To answer this question, Duncan and Young (1999) modelled the pattern of establishment outcomes that would be expected if some species succeeded with probability $p = 0.8$, while others succeeded with probability $p = 0.2$ (equivalent to failing with $1-p = 0.8$).

These can be considered good and poor establishers, respectively, or alternatively, species that would always succeed or always fail without the intervention of chance. Assuming that each introduction is an independent event, and that the probability of success is the same for a species at all locations, then the probability of a mixed outcome for a species π_s over n introduction events is $\pi_s = 1-[p^n + (1-p)^n]$, regardless of whether it is a good or a poor invader (Duncan and Young 1999). The expected number of mixed outcomes for a given set of species is then simply the sum of π_s over all species. The expected variance can be calculated as the sum of $\pi_s(1-\pi_s)$ over all species, and the 95% confidence intervals can be approximated as 1.96 times the square root of this variance.

The results from Duncan and Young's (1999) study showed that mixed outcomes in Moulton and Sanderson's (1997) study were almost exactly as frequent as expected given that some species were typically good invaders ($p = 0.8$) and others poor ($p = 0.2$). The expected number of mixed outcomes from the fifty-three species analysed by Moulton and Sanderson was 24 ± 7, compared to the observed value of twenty-three. Duncan and Young argued that while it was unlikely that introduced bird species could be neatly divided into good or poor invaders, twenty-three mixed outcomes was nevertheless a reasonable number to expect under a scenario where species show wide variation in intrinsic establishment probability. Interestingly, when we apply Duncan and Young's (1999) model to the data for global introductions in Figure 4.1, we more or less exactly replicate their results. Using the same probabilities of success and failure, we would

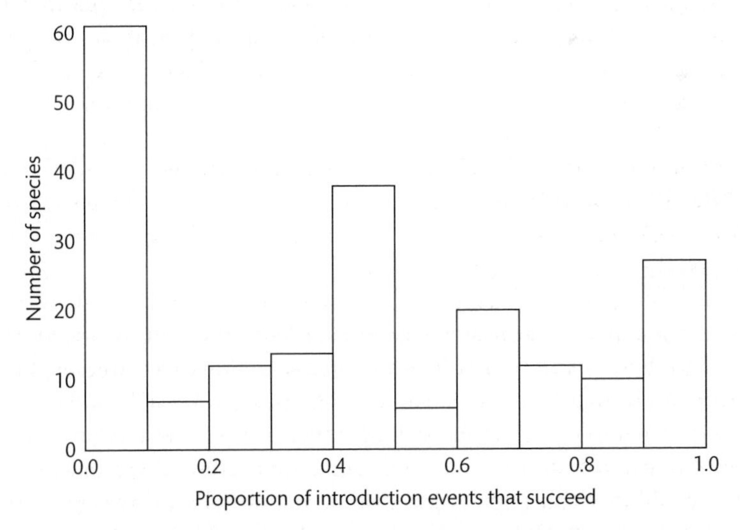

Fig. 4.1 The frequency distribution of the proportion of successful introductions for the 207 bird species with more than one introduction event in the data analysed by Blackburn and Duncan (2001a).

expect 122 ± 12 mixed outcomes from the 207 species modeled. We observe 119. Moulton and Sanderson (1999) responded that the predictive value of this model for any one species is poor, as a species introduced twice with a probability of success of 0.8 has only a 64% chance of establishing at both locations (or a 32% chance of a mixed outcome, with an additional 4% chance of failing at both locations). However, this is beside the point. The frequency of mixed outcomes in bird introductions is consistent with a model that posits species-level differences in establishment success.

There are, nevertheless, alternative explanations for this pattern. Differences in success amongst bird species could arise because some species tend to be introduced to locations that are more conducive to establishment, or because they tend to be introduced in larger numbers (although this in turn may be correlated to characteristics of the species). Nevertheless, there is further evidence for species-level differences in establishment success from Blackburn and Duncan (2001a) and Cassey (2002b). Blackburn and Duncan used mixed models to assess location-level and event-level correlates of success, while controlling for membership of different levels of the taxonomic hierarchy using random effects. They found highly significant unexplained variation in introduction success across species, once the variance explained by variables related to the location and the event had been removed (Blackburn and Duncan 2001a). Cassey (2002b) used randomization tests to show that successful introduction attempts are not distributed randomly with respect to species (or family) identity, at least for species introduced more than five times each. Overall, therefore, while there is little evidence for an all-or-none pattern of establishment success *sensu strictu* in birds, the evidence is consistent with a situation where species differ in their probability of establishment, which is likely, therefore, to be due to their intrinsic characteristics.

4.3 Intrinsic Characteristics as Determinants of Establishment Success

4.3.1 Population Growth Rates

Introduced populations in general have to grow from low numbers of individuals (Figure 3.1). Since small populations are more susceptible to extinction because of demographic and environmental accidents (section 3.1.1), and the longer a population remains at low levels the longer there is for these accidents to occur, it follows that exotic species with high population growth rates may be more likely to establish viable populations (Moulton and Pimm 1986a; Pimm, et al. 1988; Ehrlich 1989; Pimm 1989; 1991; Cassey 2001a). Alternatively, Sæther, et al. (2004) showed that bird species with slower population growth rates had lower demographic variance, while such species tended on average to decline less in the face of environmental stochasticity (Cawthorne and Marchant 1980; 1989; 1991). Either way, we would expect population growth rates to be related to establishment probability in birds.

It is difficult directly to estimate population growth rates: they are affected by density dependence (both positive and negative), while deterministic and stochastic components of the rates would also need to be decomposed. Thus, most tests of these ideas about population growth rates have concentrated on surrogate measures thought to be correlated with those rates. Bird species with slower population growth rates tend, amongst other things, to have longer life expectancies, older ages at first reproduction, smaller clutch sizes, lower rates of reproduction, and larger body sizes. The easiest of these life-history traits to measure for birds are body size and clutch size, and so it is for these that most analyses exist. However, there is also evidence that these traits are not neutral with respect to propagule pressure. Large-bodied species, and species with higher annual fecundity, tend to be introduced in larger numbers, although the relationship for body mass changes sign when controlling for phylogeny (Cassey, et al. 2004c).

Table 4.1 summarizes published relationships between establishment success and life-history variables related to where a species lies on the fast–slow continuum. Several generalities can be extracted from this table. First, the only life-history variables that enter minimum adequate models, and thus explain variation in establishment success independently of other variables in an analysis, are body size and clutch size. Thus, while Cassey (2002b) showed negative interspecific relationships between annual fecundity and incubation period and establishment success, and a positive interspecific relationship to breeding period, his minimum adequate model across species included none of these traits. These variables fail to enter minimum adequate models whether or not propagule pressure is controlled for.

Second, clutch size is significantly associated with establishment success only in Green's (1997) study of birds introduced to New Zealand: species with smaller clutches are more likely to establish, and this effect remains even when propagule pressure and other variables are controlled for in a multivariate analysis. However, the significance of clutch size in the minimum adequate model is marginal (Green 1997). Other analyses of avian introductions to New Zealand do not find an effect of clutch size (Veltman, et al. 1996; Cassey 2001b; Duncan, et al. 2006), suggesting that the result is dependent on the exact set of species analysed, which varies across studies (Duncan, et al. 2006). Moreover, the signs of the clutch size effects tend to be positive or negative in approximately equal proportion (Table 4.1), again suggesting that the effect is not consistent. Overall, and despite its presence in at least one multivariate model for establishment success (Table 4.1), there is little evidence that clutch size is a general predictor of such success in exotic birds.

Third, there is some evidence that larger-bodied bird species are more successful in establishing exotic populations. Nine of the twenty-eight relationships in Table 4.1 show significant positive effects of body size (mass) on establishment

Table 4.1 Relationships between establishment success and life history variables that may be surrogates for population growth rate. NZ = New Zealand. Effect: + = significantly positive, − = significantly negative, NS = not significant with the sign of the correlation, if reported; Model: type of statistical model reported (Univariate, Maximal, or minimum adequate model MAM); P.P.: whether or not propagule pressure was also included in the multivariate analysis; Phyl.: whether or not the analysis accounted for phylogenetic autocorrelation; †at the family level.

Source	Details	Effect	Model	P.P.	Phyl.
Body mass					
Veltman, et al. 1996	Introductions to NZ	NS	Maximal	Yes	No
	Introductions to NZ	NS	MAM	Yes	No
Green 1997	Introductions to NZ	NS	Univariate		No
	Introductions to NZ	+	MAM	Yes	No
Sorci, et al. 1998	Introductions to NZ	NS	MAM	Yes	No
Cassey 2001b	Introductions to NZ	NS	Univariate		No
	Introductions to NZ	+	Univariate		Yes
	Introductions to NZ	NS	MAM	Yes	No
Cassey 2001a	Land birds globally	−	Univariate		No
	Land birds globally	−	Univariate		No
	Land birds globally	+	Univariate		Yes†
	Land birds globally	−	Univariate		Yes†
	Land birds globally	−	Univariate		Yes†
Duncan, et al. 2001	Introductions to Australia	NS +	Univariate		No
	Introductions to Australia	+	Univariate		Yes
	Introductions to Australia	+	MAM	Yes	No
Cassey 2002b	Land birds globally	−	Univariate		Yes†
	Land birds globally	+	Univariate		Yes†
	Land birds globally	+	MAM	No	Yes†
	Land birds globally	+	MAM	No	Yes
	Land birds globally	NS	MAM	No	No
Sol, et al. 2002	Global introductions	NS	Univariate		Yes
	Global introductions	NS	MAM	Yes	Yes
Cassey, et al. 2004c	Global introductions	NS	MAM	Yes	Yes
Cassey, et al. 2004b	Parrot introductions	NS +	Univariate		No
	Parrot introductions	NS	MAM	No	No
Møller and Cassey 2004	Global introductions	NS +	Univariate		Yes
	Global introductions	NS	MAM	Yes	Yes
Allen 2006	Introductions to Florida	NS	Univariate		No

Table 4.1 *Cont.*

Source	Details	Effect	Model	P.P.	Phyl.
Duncan, et al. 2006	Introductions to NZ	NS –	Univariate		No
	Introductions to NZ	NS	MAM	Yes	No
Clutch size					
Newsome and Noble 1986	Introductions to Australia	NS +	Univariate		No
Veltman, et al. 1996	Introductions to NZ	NS	Univariate		No
	Introductions to NZ	NS	MAM	Yes	No
Green 1997	Introductions to NZ	–	Univariate		No
	Introductions to NZ	–	MAM	Yes	No
McLain, et al. 1999	Introductions to a range of islands	NS –	Univariate		Yes
Cassey 2001b	Introductions to NZ	NS	Univariate		No
	Introductions to NZ	NS +	Univariate		Yes
	Introductions to NZ	NS	MAM	Yes	No
Duncan, et al. 2001	Introductions to Australia	NS +	Univariate		No
	Introductions to Australia	NS +	Univariate		Yes
	Introductions to Australia	NS	MAM	Yes	No
Duncan, et al. 2006	Introductions to NZ	NS –	Univariate		No
	Introductions to NZ	NS –	Univariate		Yes
	Introductions to NZ	NS	MAM	Yes	No
Annual fecundity					
Cassey 2002b	Land birds globally	–	Univariate		Yes†
	Land birds globally	NS –	Univariate		Yes†
	Land birds globally	NS	MAM	No	Yes†
	Land birds globally	NS	MAM	No	Yes
	Land birds globally	NS	MAM	No	No
Sol, et al. 2002	Global introductions	NS	Univariate		Yes
	Global introductions	NS	MAM	Yes	Yes
Cassey, et al. 2004c	Global introductions	NS	MAM	Yes	Yes
Cassey, et al. 2004b	Parrot introductions	NS –	Univariate		No
	Parrot introductions	NS	MAM	No	No
Møller and Cassey 2004	Global introductions	NS +	Univariate		Yes
	Global introductions	NS	MAM	Yes	No
Number of broods per year					
Veltman, et al. 1996	Introductions to NZ	NS	Univariate		No
	Introductions to NZ	NS	MAM	Yes	No

Table 4.1 *Cont.*

Source	Details	Effect	Model	P.P.	Phyl.
Duncan, et al. 2001	Introductions to Australia	+	Univariate		No
	Introductions to Australia	NS +	Univariate		Yes
	Introductions to Australia	NS	MAM	Yes	Yes
Generation time					
Cassey 2001b	Introductions to NZ	+	Univariate		Yes
Incubation period					
Cassey 2002b	Land birds globally	–	Univariate		Yes†
	Land birds globally	NS +	Univariate		Yes†
	Land birds globally	NS	MAM	No	Yes†
	Land birds globally	NS	MAM	No	Yes
	Land birds globally	NS	MAM	No	No
Cassey, et al. 2004b	Parrot introductions	NS +	Univariate		No
	Parrot introductions	NS	MAM	No	No
Duncan, et al. 2006	Introductions to NZ	NS –	Univariate		No
	Introductions to NZ	NS –	Univariate		Yes
	Introductions to NZ	NS	MAM	Yes	No
Fledging period					
Cassey 2002b	Land birds globally	NS +	Univariate		Yes†
	Land birds globally	NS +	Univariate		Yes†
	Land birds globally	NS	MAM	No	Yes†
	Land birds globally	NS	MAM	No	Yes
	Land birds globally	NS	MAM	No	No
Cassey, et al. 2004b	Parrot introductions	+	Univariate		No
	Parrot introductions	NS	MAM	No	No
Breeding period					
Cassey 2002b	Land birds globally	+	Univariate		Yes†
	Land birds globally	NS +	Univariate		Yes†
	Land birds globally	NS	MAM	No	Yes†
	Land birds globally	NS	MAM	No	Yes
	Land birds globally	NS	MAM	No	No
Age at maturity					
Cassey, et al. 2004b	Parrot introductions	NS +	Univariate		No
	Parrot introductions	NS	MAM	No	No

success, whereas there are only three significant negative effects. Moreover, roughly three-quarters of all reported effects are positive. Green (1997) and Duncan, et al. (2001) found significant positive effects of body mass even when propagule pressure was controlled for in avian introductions to New Zealand and Australia, respectively, although Duncan, et al. (2006) did not in their analysis for New Zealand.

Finally, Table 4.1 suggests that if there is an effect on establishment success of where a species lies on the fast–slow continuum, it is species towards the slow end that do better. Positive effects predominate for fledging period, breeding period, age at maturity, generation time, and body size, and negative effects for annual fecundity. Clutch size is ambiguous, whereas the only significant effect for incubation time is negative. That said, the overall evidence that life-history correlates of population growth rate explain variation in establishment success is not strong. Many of the effects in Table 4.1 are not significant ($\alpha = 0.05$).

To explore further the effects of traits relating to population growth rate on establishment success, Blackburn, et al. (2009) performed a meta-analysis on those studies summarized in Table 4.1 for which quantitative information on effect sizes was available. They excluded clutch size from the analysis, because, unlike the other life-history traits in Table 4.1, it does not strongly align on the fast–slow continuum in birds (e.g., Sæther 1987; Bennett and Owens 2002). They also altered the sign of number of broods per year effects so that positive signs indicate that species lying towards the slow end of the fast–slow continuum of life-history traits have higher establishment success. Their analysis confirms the impression from Table 4.1 that establishment success tends to be higher for species with slower population growth rates.

This conclusion matches that of a study of extinction dynamics in native bird populations by Sæther, et al. (2005) using simulation models for populations of thirty-eight species for which long-term and highly detailed studies of population dynamics were available. Their simulations showed that environmental stochasticity had the most immediate effect on the risk of extinction, whereas the deterministic component of population growth rate had the strongest effect on long-term population persistence. In other words, the stochastic and deterministic components of population dynamics affect persistence differently. Yet, Sæther, et al. (2004) showed that these components are correlated in birds, raising the question of which of them more strongly affects extinction risk. To assess this, Sæther, et al. (2005) calculated the mean population size required for each species to produce an expected time to extinction of 1,000 years ($\log_{10}K_{T=1000}$), and compared this population size to clutch size and survival rate of the species. They found that $\log_{10}K_{T=1000}$ was higher for species with larger clutches and lower survival rates, and thus lying towards the fast end of the fast–slow continuum. This pattern arises because demographic and environmental stochasticity reduce the long-run population growth rate, but are also positively correlated to this growth rate in birds (Sæther, et al. 2004). What these results suggest is that extinction probability for a given population size should be higher for bird species with fast life histories, as seems generally to be the case for introduced bird populations.

Nevertheless, Blackburn, et al. (2009) also found significant heterogeneity in the effects of population growth-rate traits on establishment success. Thus, while large body mass, long fledging period, and late age at maturity all apparently increase establishment success, the effects for incubation period and number of broods per year are instead consistent with higher establishment success in species with fast population growth rates. This more complicated pattern of association perhaps suggests why previously there has been a general failure to identify life-history characteristics that were consistently related to establishment success in birds (Duncan, et al. 2003): different traits fit with opposing hypotheses about the effects of population growth rates.

Blackburn and colleagues' results suggest that establishment success would be most favoured by a combination of frequent and fast reproduction followed by longer post-hatching development to larger size. Presumably, this ideal combination of traits identifies a species with a relatively large deterministic component but a small stochastic component to population growth rates, which the simulations of Sæther, et al. (2005) suggest would maximize time to extinction. Such species would have both rapid population growth from low levels and an ability to ride out environmental stochasticity. Interestingly, these results find some parallels in a recent study showing that mammalian life-history traits appear to co-vary along two independent axes, rather than the single fast–slow continuum (Bielby, et al. 2007). One axis describes the timing of reproductive bouts, while the other describes the trade-off between offspring size and number. While the coincidence is not exact, it is interesting that Blackburn, et al. (2009) identify opposing effects on establishment success of life-history traits relating to timing versus reproductive output. An analysis for birds equivalent to that of Bielby, et al. (2007) for mammals would also be interesting in this regard.

4.3.2 Predisposition to Allee effects

Hypotheses about the potential importance of population growth rates to establishment success derive from the notion that life history determines the extent to which a species is buffered against the stochastic effects associated with a small population size. However, it may be that demographic and environmental stochasticity are less perilous, at least to bird introductions, than are Allee effects. The ability or otherwise of a population to grow away from the small numbers introduced may be irrelevant if that release initially finds itself at less than the threshold population size or density below which positive density dependence occurs. Theory suggests that Allee effects should vary with the degree of sexual selection, dispersal mode, and aggregation behaviour (Taylor and Hastings 2005), and hence that any or all of these may influence establishment success in birds.

Sexual selection has been considered to affect establishment success because it is argued to compromise the ability of males to adapt with respect to other components of fitness, such as interspecific competition, predation (e.g., Brook 2004), environmental change, and co-evolutionary responses of parasites. If so, sexually selected species may be more constrained in their responses to novel environments, and so less likely to establish exotic populations (McLain, et al. 1995; McLain, et al. 1999; Sorci, et al. 1998). Male reproductive success is also generally skewed in sexually selected species, because most females are mated by relatively few males. It follows that, for a given propagule size, effective population sizes may be smaller for sexually selected species than for others (Sorci, et al. 1998; but see Legendre, et al. 1999). If sexual selection influences the probability that an individual will find a mate, then this can lead to positive density dependence at low densities or population sizes.

Table 4.2 summarizes studies that have investigated relationships between establishment success and variables relating to the degree of sexual selection in birds. Most of these have considered the influence of plumage dichromatism. Dichromatism is thought to have evolved under sexual selection, and indeed Bennett and Owens (2002) presented phylogenetically controlled comparative analyses showing that the primary predictor of dichromatic plumage is a high frequency of extra-pair paternity. Whatever the underlying cause, several studies find that dichromatism significantly predicts establishment, and that dichromatic species tend to be more likely to fail. However, only for Sorci, et al.'s (1998) study of bird introductions to New Zealand is dichromatism significant in a model that also includes propagule pressure. In contrast, Duncan, et al. (2006) found no significant effect of dichromatism on success in this group, suggesting that the result is sensitive to the exact set of species analysed.

The strength of sexual selection is also thought to be greater in polygynous than monogamous species (Lack 1968; Kirkpatrick, et al. 1990; Bennett and Owens 2002). Legendre, et al. (1999) thus analysed the influence of mating system on the fate of bird introductions. They framed their study in the context of variation in mating opportunities on demographic stochasticity, though reduced mating opportunities in the face of low densities or population sizes may also be classed as an Allee effect (Courchamp, et al. 1999; Stephens and Sutherland 1999; Stephens, et al. 1999). Legendre and colleagues modelled extinction probability for a sexually reproducing population under different mating systems with demographic stochasticity. They found that extinction was more likely for monogamous than for polygynous species. This difference arises because when random fluctuations in sex ratio lead to an excess of females, these go unmated in monogamous species, but not in polygynous species. Thus, effective population sizes are lower under monogamy (cf. Sorci, et al. 1998). This effect is further heightened in models that assume that a proportion of monogamous females go unmated for reasons other than a shortage of males. Legendre and colleagues showed that the probability of extinction (failure) for passerine species introduced to New Zealand in different

Table 4.2 Relationships between establishment success and variables relating to the degree of sexual selection. NZ = New Zealand. Effect: + = significantly positive, − = significantly negative, NS = not significant with the sign of the correlation, if reported. Note that a negative effect means that dichromatic or polygynous species are less likely to establish than monochromatic or monogamous species; Model: type of statistical model reported (Univariate, multivariate, minimum adequate model MAM), or data tested against the predictions of an explicit mathematical model; P.P.: whether or not propagule pressure was also included in the multivariate analysis; Phyl.: whether or not the analysis accounted for phylogenetic autocorrelation; †at the family level.

Study	Details	Variable	Effect	Model	P.P.	Phyl.
McLain, et al. 1995	Oahu	Dichromatism	−	Univariate	No	No
	Tahiti	Dichromatism	NS −	Univariate	No	No
	Oahu and Tahiti	Dichromatism	−	Multivariate	No	Yes
Sorci, et al. 1998	Introductions to NZ	Dichromatism	−	MAM	Yes	No
Legendre, et al. 1999	Passerine introductions to NZ	Mating system	+	Fit to mathematical model	Yes	No
McLain, et al. 1999	Introductions to a range of islands	Dichromatism	−	Univariate		No
	Introductions to a range of islands	Paternal care	NS −	Univariate		No
	Introductions to a range of islands	Dichromatism	−	Multivariate	No	No
Duncan, et al. 2001	Introductions to Australia	Dichromatism	NS +	Univariate		No
	Introductions to Australia	Dichromatism	NS −	Univariate		Yes
	Introductions to Australia	Dichromatism	NS	MAM	Yes	No
Cassey 2002b	Land birds globally	Dichromatism	−	Univariate		Yes†
	Land birds globally	Dichromatism	−	Univariate		Yes†

	Land birds globally	Dichromatism	NS	MAM	No	Yes[†]
	Land birds globally	Dichromatism	–	MAM	No	Yes
	Land birds globally	Dichromatism	–	MAM	No	No
Sol, et al. 2002	Global introductions	Dichromatism	NS	Univariate		Yes
	Global introductions	Dichromatism	–	MAM	Yes	Yes
	Global introductions	Parental care	NS	Univariate		Yes
	Global introductions	Parental care	NS	MAM	Yes	Yes
Cassey, et al. 2004c	Global introductions	Dichromatism	NS	MAM	Yes	Yes
Cassey, et al. 2004b	Parrot introductions	Dichromatism	NS –	Univariate		No
	Parrot introductions	Dichromatism	NS	MAM	No	No
Donze, et al. 2004	Game bird introductions to islands	Dichromatism	NS +	Univariate		No
	Game bird introductions to islands	Dichromatism	NS +	Univariate		Yes
Møller and Cassey 2004	Global introductions	Dichromatism	NS –	Univariate		Yes
	Global introductions	Dichromatism	NS	MAM	Yes	Yes
Duncan, et al. 2006	Introductions to NZ	Dichromatism	NS –	Univariate		No
	Introductions to NZ	Dichromatism	NS –	Univariate		Yes
	Introductions to NZ	Dichromatism	NS	MAM	Yes	No

numbers well fitted a model assuming a monogamous mating system and that 5% of the females in the population go unmated for reasons other than demography. Legendre, et al.'s (1999) prediction that establishment success should be higher for polygynous than monogamous species appears to be falsified by the tendency for dichromatic species to fail more often (Table 4.2). However, the two sets of results may not be contradictory. Since the best predictor of dichromatism is not polygyny, but rather a high frequency of extra-pair paternity, dichromatism is often a feature of monogamous species (Bennett and Owens 2002). Alternatively, sexual selection may prevent establishment for reasons unrelated to demographic stochasticity or Allee effects.

Allee effects can be linked to dispersal in introduced species if, in spreading out from the point of release, individuals reduce the chance that they will encounter mates (Lewis and Kareiva 1993; Veit and Lewis 1996; Veltman, et al. 1996; Stephens and Sutherland 1999). All studies to date that have considered the effect of dispersal on establishment have considered whether or not an exotic bird species is a migrant in its native range (Table 4.3). Most of these studies find that success is lower for migrants than for sedentary species, albeit that there is some evidence that partial migrants have the highest levels of success (Duncan, et al. 2006). Nevertheless, the majority of effects are not statistically significant. Migration and dispersal are different processes (migration being characterized by its annual seasonality and directionality), but one may be a surrogate for the other (see also section 6.2.2). Paradis, et al. (1998) showed that, for British breeding birds, migrants tend to disperse further than residents. Thus, it is possible that an Allee effect contributes to the lower establishment success of migrant birds. However, it is also possible that the challenges of having to cope with new wintering grounds as well as new breeding grounds lower success, as indeed will the challenges of even finding a new wintering ground (Thomson 1922; Veltman, et al. 1996). Moreover, migrants have smaller brains and are less plastic in behaviour than are residents (Sol, et al. 2005c), and so may be less well equipped to deal with the novelty of a new environment (see section 4.3.3).

Finally, the likelihood that an exotic bird population will suffer from Allee effects will be influenced by its aggregation behaviour. Notably, species that gather together in flocks as a mechanism of defence against predators, or that breed colonially, may be disadvantaged in the early stages of establishment if the number or density of individuals is too low to effect such behaviours. Newsome and Noble (1986) compared the success and failure of bird species introduced in Australia with different typical flock sizes, distinguishing solitary species, species that tend to aggregate in small flocks (<10–20 individuals), and those that tend to aggregate in large flocks (>20 individuals). They found no effect of these different strategies on establishment success, either for introductions of Australian species or of foreign species. Duncan, et al. (2001) drew the same conclusion for Australian introductions from a binary comparison of species that did or did not

Table 4.3 Relationships between establishment success and variables relating to the degree of population movement. NZ = New Zealand. Effect: + = significantly positive, − = significantly negative, NS = not significant with the sign of the correlation, if reported. Note that a negative effect means that migratory/dispersive species are less likely to establish than sedentary species; Model: type of statistical model reported (Univariate, multivariate, minimum adequate model MAM); P.P.: whether or not propagule pressure was also included in the multivariate analysis; Phyl.: whether or not the analysis accounted for phylogenetic autocorrelation; †at the family level; § the success of partial migrants is greater than that of sedentary species, significantly so when controlling for phylogeny, but that of completely migratory species is less than that of sedentary species.

Source	Details	Effect	Model	P.P.	Phyl.
Veltman, et al. 1996	Introductions to NZ	−	Univariate		No
	Introductions to NZ	−	MAM	Yes	No
Sorci, et al. 1998	Introductions to NZ	NS	MAM	Yes	No
Sol and Lefebvre 2000	Introductions to NZ	−	MAM	Yes	No
Duncan, et al. 2001	Introductions to Australia	NS +	Univariate		No
	Introductions to Australia	NS−	Univariate		Yes
	Introductions to Australia	NS	MAM	Yes	No
Cassey 2002b	Land birds globally	−	Univariate		Yes†
	Land birds globally	−	Univariate		Yes†
	Land birds globally	−	MAM	No	Yes†
	Land birds globally	NS	MAM	No	Yes
	Land birds globally	−	MAM	No	No
Sol, et al. 2002	Global introductions	NS	Univariate		Yes
	Global introductions	NS	MAM	Yes	Yes
Cassey, et al. 2004c	Global introductions	NS	MAM	Yes	Yes
Cassey, et al. 2004b	Parrot introductions	−	Univariate		No
	Parrot introductions	−	MAM	No	No
Møller and Cassey 2004	Global introductions	NS−	Univariate		Yes
	Global introductions	NS	MAM	Yes	Yes
Duncan, et al. 2006	Introductions to NZ	NS +, −§	Univariate		No
	Introductions to NZ	+, NS −§	Univariate		Yes
	Introductions to NZ	NS	MAM	Yes	No

flock. Finally, Sol (2000b) found no evidence that colonial species were less likely to establish than non-colonial species.

In sum, there is evidence to suggest that species facing stronger sexual selection (Table 4.2), and migratory species (Table 4.3), tend to be more likely to fail at the establishment step on the invasion pathway. These conclusions are also supported by Blackburn, et al. (2009) in their meta-analysis of the effects of life-history traits on establishment success, based on data from the studies in Tables 4.2 and 4.3. The quantitative effects on establishment success of traits related to both sexual selection and population movement were consistent with the expectation that susceptibility to Allee effects should decrease success. It is also possible that the effects of life history on establishment success noted in the previous section could also be because a slow life history is advantageous in overcoming Allee effects. The longer life expectancies of slow species may help ameliorate some of the effects that lead to positive density dependence, such as by reducing the urgency of issues like finding a mate. If so, we might expect that the effects of life history on establishment success would be greatest at low propagule pressures.

4.3.3 Coping with Novel Environments

While characteristics that affect the ability of species to cope with stochastic or Allee effects on population growth rates seem to have some influence on establishment success, a more fundamental aspect governing success may be whether an introduced population can grow at all in the novel environment. Exotic species are confronted with sudden environmental changes to which they are unlikely to be fully adapted. These primarily involve dealing with novel environmental conditions, locating novel resources, or dealing with novel natural enemies. An exotic species that cannot overcome such difficulties will not establish. It follows that pre-adaptations that increase the probability that a species will cope well in a new environment should predict success. Three broad classes of traits have been proposed as potential pre-adaptations: whether a species has a broad or narrow niche, the flexibility of a species' behaviour, and immunocompetence.

4.3.3.1 The Breadth of the Niche

Characteristics that have frequently been suggested to typify successfully introduced species are those relating to the ability to tolerate a wide range of abiotic conditions, or to utilize a wide range of resources (e.g., Ehrlich 1989; Lodge 1993; Forys and Allen 1999; Dean 2000; Brooks 2001; Cassey 2001b; 2002b; Duncan, et al. 2003; Sol, et al. 2005a). It makes sense that catholic generalist species ought, on average, to find abiotic conditions or resources in the introduction location more to their liking than would more ecologically restricted specialists (Daehler and Strong 1993). The range of conditions and resources that a species can tolerate defines its (fundamental) niche, and so this hypothesis is sometimes framed in terms of the effect of niche breadth on establishment success (e.g., Forys and Allen 1999; Duncan, et al. 2003).

A significant problem with testing hypotheses about the tolerances of exotic species is in quantifying the breadth of those tolerances. Most studies use simple metrics of the number of different habitat types or the number of different types of food item used by the species. For example, Cassey (2002a; 2002b) quantified diet breadth for exotic bird species in terms of how many of seven different food types (grasses and herbs; seeds and grains; fruits and berries; pollen and nectar; vegetative material; invertebrate prey; vertebrate prey and carrion) they used, and habitat breadth similarly in terms of the number of major habitat types (mixed lowland forest; alpine scrub and forest; grassland; mixed scrub; marsh and wetland; cultivated and farm lands; urban environments) included in a species' range. Such metrics are basic, failing to convey the breadth of resource use (or its variation) within categories, the relative frequency of resource use across categories, and the extent to which the breadth of resource use by a species is determined by within- or between-individual preferences (Gaston 1994; 2003). Breadth of resource use may also be confounded with a species' rarity, either because species have been studied less (so that their full repertoire of behaviours is unknown), or because rarer species utilize narrow ranges of resources because they are rare (i.e., their realized and fundamental niches are confused) (Gaston 1994; Gaston, et al. 1997; Gregory and Gaston 2000).

Nevertheless, those analyses of establishment that have used such simple metrics of habitat or diet breadth often find that success is higher for more generalist species (Table 4.4). Notably, all reported correlations between habitat breadth and establishment success are positive, and most of these relationships are robust to the inclusion of propagule pressure as a covariate. Although the studies by Cassey, et al. (2004c), Møller and Cassey (2004), and Sol, et al. (2005b) all use very similar data sets, the current evidence suggests that bird species that utilize more habitats are indeed more successful at establishing exotic populations. Quantitative assessment of habitat breadth effects using the data in Table 4.4 (Blackburn, et al. 2009) confirm this conclusion: there is a strong positive relationship between habitat breadth and establishment success in birds. The results for diet breadth are similar: all reported correlations between diet breadth and establishment success are also positive, albeit that the strength of this association is weaker than that for habitat breadth.

A positive relationship between geographic range size and introduction success has also been argued to support the influence of niche breadth or generalism on establishment (Moulton and Pimm 1986a; Moulton and Scioli 1986; Williamson 1996; Blackburn and Duncan 2001a; Brooks 2001; Duncan, et al. 2001; Moulton, et al. 2001a). Brown (1984) suggested that species that are widespread and abundant in their native environment are so because they have broad niches. Brown argued that these relationships should arise because greater niche breadth allows species to occupy more locations, and hence have larger geographic ranges, but also to utilize more resources at any given location, and hence on average attain higher local population densities. Tests of these hypotheses are not in general supportive (e.g., Gaston, et al. 1997; Gregory and Gaston 2000; Hughes 2000;

Table 4.4 Relationships between establishment success and variables relating to the breadth of a species niche. NZ = New Zealand. Effect: + = significantly positive, − = significantly negative, NS = not significant with the sign of the correlation, if reported. Note that a positive effect means that more generalist or widespread species are more likely to establish than more specialist or range-restricted species; Model: type of statistical model reported (Univariate, multivariate, minimum adequate model MAM); P.P.: whether or not propagule pressure was also included in the multivariate analysis; Phyl.: whether or not the analysis accounted for phylogenetic autocorrelation; † at the family level

Source	Details	Effect	Model	P.P.	Phyl.
Habitat breadth					
Brooks 2001	Global introductions	+	Univariate		No
	Global introductions	+	Univariate		No
Cassey 2002b	Land birds globally	+	Univariate		Yes†
	Land birds globally	NS +	Univariate		Yes†
	Land birds globally	NS	MAM	No	Yes†
	Land birds globally	NS	MAM	No	Yes
	Land birds globally	+	MAM	No	No
Cassey, et al. 2004c	Global introductions	+	MAM	Yes	Yes
Møller and Cassey 2004	Global introductions	+	Univariate		Yes
	Global introductions	+	MAM	Yes	Yes
Sol, et al. 2005b	Global introductions	+	MAM	Yes	Yes
Diet breadth					
McLain, et al. 1999	Introductions to a range of islands	+	Univariate		No
	Introductions to a range of islands	+	Multivariate	No	No
Cassey 2001b	Introductions to NZ	+	Univariate		No
	Introductions to NZ	+	Univariate		Yes
	Introductions to NZ	NS	MAM	Yes	No
Cassey 2002b	Land birds globally	+	Univariate		Yes†
	Land birds globally	NS +	Univariate		Yes†
	Land birds globally	NS	MAM	No	Yes†
	Land birds globally	NS	MAM	No	Yes
	Land birds globally	NS	MAM	No	No
Cassey, et al. 2004c	Global introductions	NS	MAM	Yes	Yes
Cassey, et al. 2004b	Parrot introductions	+	Univariate		No
	Parrot introductions	+	MAM	No	No

Table 4.4 *Cont.*

Source	Details	Effect	Model	P.P.	Phyl.
Range size					
Moulton and Pimm 1986	Introductions to Hawaii	+	Univariate		No
Veltman, et al. 1996	Introductions to NZ	NS	Maximal	Yes	No
	Introductions to NZ	NS	MAM	Yes	No
Green 1997	Introductions to NZ	NS	Univariate		No
	Introductions to NZ	NS	MAM	Yes	No
Blackburn and Duncan 2001a	Global introductions	+	Full model	No	Yes
Brooks 2001	Introduced species with restricted range vs others	NS –	Univariate		No
	Introduced species endemic to a single EBA[1] vs others	NS –	Univariate		No
Duncan, et al. 2001	Introductions to Australia	+	Univariate		No
	Introductions to Australia	NS +	Univariate		Yes
	Introductions to Australia	NS	MAM	Yes	Yes
Moulton, et al. 2001a	Introductions to Hawaii	+	Univariate		No
Cassey 2002b	Land birds globally	NS +	Univariate		Yes†
	Land birds globally	NS –	Univariate		Yes†
	Land birds globally	NS	MAM	No	Yes†
	Land birds globally	NS	MAM	No	Yes
	Land birds globally	NS	MAM	No	No
Cassey, et al. 2004c	Global introductions	NS	MAM	Yes	Yes
Cassey, et al. 2004b	Parrot introductions	NS –	Univariate		No
	Parrot introductions	NS	MAM	No	No
	Parrot introductions	NS –[2]	Univariate		No
	Parrot introductions	NS[2]	MAM	No	No
Donze, et al. 2004	Game bird introductions to islands	NS	Univariate		No
Duncan, et al. 2006	Introductions to NZ	+[2]	Univariate		No
	Introductions to NZ	+[2]	Univariate		Yes
	Introductions to NZ	+[2]	MAM	Yes	No

[1] Endemic bird area: Stattersfield, et al. 1998.

[2] Latitudinal range rather than total range size.

Gaston 2003). Nevertheless, the idea that species with larger geographic ranges can tolerate more-varied environmental conditions in particular has persisted. It has also been appropriated to explain why bird species inhabiting higher latitudes tend to have larger ranges (section 7.4).

The literature reveals that measures of geographic range size are not consistent predictors of establishment success in exotic birds (Table 4.4). All significant relationships between these two variables are positive, but most reported relationships are not significant, and in several of these the association is negative. Nevertheless, a quantitative analysis of the data in Table 4.4 does find a general positive relationship between range size and establishment success (Blackburn, et al. 2009). Whether this reflects the ability of species with broad ranges to tolerate a wider range of conditions is unclear, however. Geographic range size may instead correlate with establishment success because species with larger ranges are more likely to be transported and released, and to be released in larger numbers (Chapter 2). The significance of range size in some analyses may thus arise as a consequence of this correlation (Cassey, et al. 2004c), and it is notable that in only one case (Duncan, et al. 2006) is the significance of a measure of geographic range size robust to the inclusion of propagule pressure as a covariate.

Intuitively, one would expect the range of environments a species can inhabit to be important in determining the probability that it encounters an exotic environment that it can tolerate. However, an exotic species will doubtless greatly increase its probability of establishment if it is introduced to an environment that it prefers. Thus, while the ability of a species to tolerate environments in general (niche breadth) may influence success, we would expect the ability of a species to tolerate the specific environment into which it is released ('niche position') also to do so. This is a subject that we will address in more detail in the next chapter (section 5.2).

4.3.3.2 Behavioural Flexibility

Another measure that has been argued to reflect the ability of bird species to utilize a wide range of resources is the frequency with which novel feeding behaviours are observed and reported on by ornithologists and other scientists. An ability to innovate with respect to diet or foraging techniques allows a species to exploit new ecological opportunities, and so respond to changes in its environment. The advantage to an exotic species of this sort of 'opportunistic generalism' (Sol, et al. 2002) should be obvious. The number of reported innovations can also be used directly as a metric of this behavioural flexibility (Lefebvre, et al. 1997; Lefebvre, et al. 1998; Nicolakakis and Lefebvre 2000), supported by evidence that the metric is not confounded by methodological biases such as greater research effort on certain taxa (Nicolakakis and Lefebvre 2000).

Martin and Fitzgerald (2005) presented an experimental test of the idea that behavioural flexibility is associated with bird invasiveness. They showed that house sparrows *Passer domesticus* from a recently established (twenty-eight years) and currently expanding population in the Republic of Panama were more likely to approach and consume novel food items than individuals from a long-established (150 years) population in the USA (Figure 4.2). Martin and Fitzgerald suggest that a predilection towards trying new foods may contribute to invasiveness in this species. Why this predilection then apparently declines over time remains unexplained, but could have to do with populations identifying and eventually specializing on certain food types at the exotic location. Of course, other explanations for this difference are possible, including differences in the range of diets available at tropical and temperate locations (Martin and Fitzgerald 2005), while a comparison of only two populations does not provide any power to distinguish between alternative hypotheses. Moreover, a comparison of established populations from a single species says nothing about why some species are more likely than others to establish in the first place (Figure 4.1).

The importance of behavioural flexibility for establishment success has mainly been championed in a series of more taxonomically extensive comparative studies by Dani Sol and colleagues (Sol and Lefebvre 2000; Sol, et al. 2002, Sol, et al. 2005b). The most recent of these is probably the most robust, spanning as it does the largest range of introduction events, species, and locations, and employing

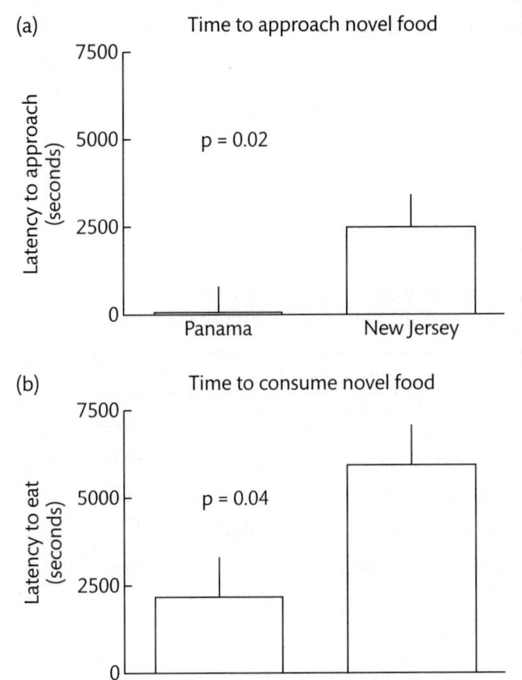

Fig. 4.2 Evidence that individuals in a more recently established exotic house sparrow *Passer domesticus* population (Panama) more readily approach and consume novel food items than do individuals in a long-established exotic population (New Jersey). Bars represent means ± 1 standard error. Reprinted with permission of the International Society of Behavioural Ecology from Martin and Fitzgerald (2005).

the kinds of statistical methods suggested in section 2.5 to address the issues of phylogenetic and spatial autocorrelation that different species and locations raise. Sol, et al. (2005b) investigated the interrelationship between exotic bird establishment success, innovation frequency (as defined above), and relative brain size (brain mass controlling for body mass) for 646 introduction events worldwide, for 196 species from thirty-five families. They included relative brain size in the analysis because this has been shown to correlate with innovation frequency in birds (Lefebvre, et al. 1997), and also to investigate the link between brain size, cognitive ability and establishment.

Sol, et al. (2005b) found that establishment success was higher for bird species with larger relative brain sizes. Moreover, this association was robust to the inclusion of a variety of potentially confounding variables in the analysis. The minimum adequate model for establishment success in their analysis included propagule size, habitat generalism, and relative brain size (Table 4.5), and so is not confounded by these different traits. Sol and colleagues next investigated the link between innovation frequency and establishment success. Innovation frequency is quantified at the family level, requiring that analyses compare data for families rather than species. Innovation frequency and establishment success are significantly positively associated at this level (Figure 4.3a), as indeed are relative brain size and establishment success (Figure 4.3b). Relative brain size and innovation frequency are also positively correlated ($r^2 = 0.39$, $N = 13$, $P = 0.016$), as might have been expected from the previous two relationships.

Table 4.5 Fixed and random effects in a minimum adequate generalized linear mixed model explaining variation in bird establishment success, while controlling for geographical region and taxonomic levels. The minimum adequate model was obtained by backward selection, removing non-significant fixed effects. The random effects show that there is significant unexplained variation in establishment success between related species, and between regions, due to unmeasured effects operating at these levels. s.e. = standard error of the model estimate. From Sol, et al. (2005).

Fixed effects	**Estimate**	**s.e.**	***F***	***P***
Propagule size	0.973	0.134	52.83	< 0.0001
Habitat generalism	0.625	0.146	18.28	< 0.0001
Relative brain size	0.526	0.185	8.08	0.0047
Random effects	**Estimate**	**s.e.**	**Z Value**	***P***
Orders	0.266	0.289	0.92	0.1785
Families within orders	0.000	.	.	.
Species within families	0.846	0.286	2.96	0.0015
Biogeographic region	0.739	0.442	1.67	0.0472

The relationship between relative brain size and establishment success could arise either because brain size enhances the ability of species to innovate, or because of other non-cognitive mechanisms. To assess these alternatives, Sol and colleagues performed a path analysis to assess the extent to which the link between brain size and establishment success is explained by the link between brain size and innovation frequency. Their path models were consistent with the idea that the direct effect of brain size on establishment is weak compared to the indirect effect associated with innovation propensity. Sol, et al. (2005b) concluded that large brains appear to help bird species respond to novel environments by enhancing their cognitive skills (i.e., their intelligence) rather than by

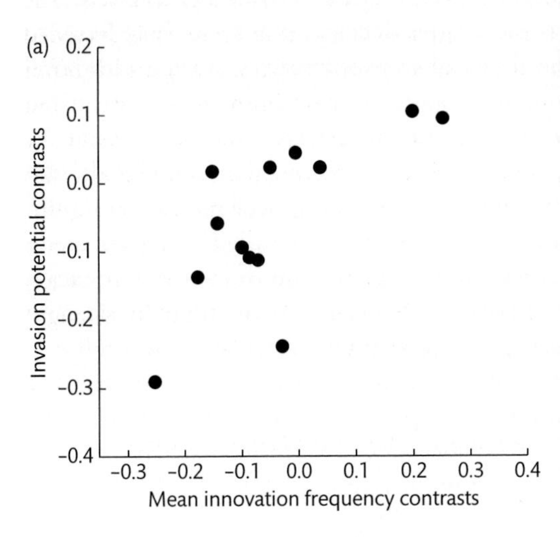

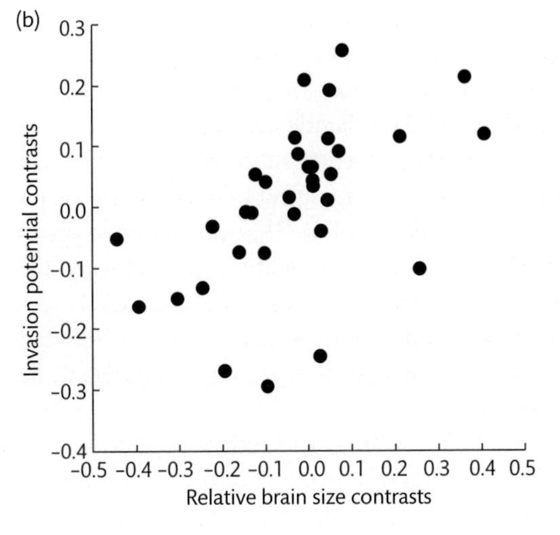

Fig. 4.3 Relationship between invasion potential (a family-level measure of establishment success) and (a) mean feeding innovation propensity ($r^2 = 0.55$, $N = 13$, $P = 0.002$), and (b) mean relative brain size ($r^2 = 0.26$, $N = 34$, $P = 0.001$), for avian families. These relationships control for phylogenetic effects using independent contrast analysis (Purvis and Rambaut 1995). Reprinted with permission from Sol, et al. (2005b). © 2005 National Academy of Sciences, USA.

other mechanisms, assuming that innovation frequency is a reasonable metric of cognitive ability. Establishment in exotic environments is thus aided by an ability to respond to novelty. They also noted that their results are in agreement with the wider hypothesis that large brains have evolved because they are advantageous in helping species respond to changes in the environment.

4.3.3.3 The immune system

There is an extensive literature on the effects of parasites on avian ecology and evolution, including the role of immune system response on natal dispersal distance (Møller, et al. 2003). Vertebrate immune systems employ an array of defences to protect individuals against parasites (Klein 1990; Wakelin 1996). Some of these defend against specific parasite strains, whereas others provide a general response against a range of parasites. It is these latter defences that seem most likely to be important in determining the ability of an exotic species to cope with novel infections. One such general immune response is the strength of T-cell mediated immunity (Klein 1990). If an individual has a stronger T-cell mediated immunity it may enjoy higher survivorship and reproductive success than individuals with low immunities. This all suggests that species with high T-cell mediated immunity may be more likely to establish exotic populations because (1) they are better able to defend against any newly encountered parasite in the non-native range, and (2) if they are subjected to parasitism, they can increase in population size when their abundance is low because they enjoy higher survival and fecundity.

The strength of T-cell mediated immunity formed the basis of a comparative test of the effect of variation in immunocompetence on avian establishment success by Møller and Cassey (2004). Phylogenetically informed analysis showed that bird species with stronger T-cell immune responses as nestlings had higher establishment success, but that there was no effect of adult immune response on success. The minimum adequate model for success in these data included only propagule size and habitat generalism as significant explanatory variables (cf. section 4.3.3.1). However, when habitat generalism was controlled for, there was an interaction between immunocompetence and propagule size: introduction events involving more than 100 individuals were more likely to succeed if the species had a stronger nestling T-cell response, but there was no effect of T-cell response on success for small propagules (Møller and Cassey 2004). This interaction suggests that immunocompetence has a negligible effect on population survival when the initial population is small, when stochastic factors and Allee effects are likely predominantly to influence success. However, it may make the difference between success and failure for a population introduced in high enough numbers to be relatively immune to the vagaries of chance.

These results are consistent with Drake's (2003) theoretical prediction that reduced establishment success due to parasites should only occur when propagule sizes are very large and when parasite loads are particularly high. Larger initial propagule sizes may ensure that a larger number of parasites from the species native range are also introduced simply through a sampling effect, increasing the probability that the exotic species will continue to be plagued by its usual set of native parasites. Greater immunocompetence will thus increase the likelihood of establishment in the face of these parasites. However, Møller and Cassey's (2004) results are also consistent with the alternative explanation that the population bottleneck imposed by small propagule sizes may compromise immunocompetence (cf. O'Brien, et al. 1985; Spielman, et al. 2004; Swinnerton, et al. 2005; Hale and Briskie 2007), so that the strength of the immune response for a species does not reflect the strength of the response in the exotic population.

The importance of immunocompetence to invasion potential also receives support from a study by Lee, et al. (2006), who experimentally compared the immune responses of house sparrows and tree sparrows *Passer montanus* from sympatric populations in the United States. They showed that there were no differences between control and immunologically challenged house sparrows in terms of metabolic rates, locomotor activities, or reproductive output. In contrast, tree sparrows showed decreases in metabolic rates, locomotor activities, and egg production in treatment versus control individuals. Lee and colleagues suggest that a relatively higher immune response in tree sparrows may be affecting population growth and spread, and so inhibiting invasion in this species relative to its more successful congener. This evidence suggests that the cost of a non-specific response in an environment housing novel parasites may be more important than the benefit of the response in fighting these infections, which appears to contradict Møller and Cassey's (2004) conclusions. However, whether these studies are indeed contradictory is unclear, because Lee, et al. (2006) do not compare the relative strengths of the immune responses in the two species. It is possible that house sparrows exhibit a stronger immune response but pay lower life-history costs. Clearly, the effect of the immune system on establishment success is a topic that deserves more attention.

4.4 Conclusions

As we noted at the end of the previous chapter (section 3.4), a quantitative meta-analysis of studies of avian introductions to islands by Cassey, et al. (2005c) found little evidence for a consistent effect of species-level traits on establishment success. The vote-counting approach adopted by Colautti, et al. (2006) reached similar conclusions for a wider range of taxa (within which birds were included): there was little evidence that body size, population growth rates, lifespan, reproductive output, or physiological tolerances were consistent predictors of establishment success. Nevertheless, the frequency of mixed outcomes of introduction events across species (Duncan and Young 1999; Figure 4.1) is consistent with a model whereby species are either generally good or generally bad at establishing (section 4.2). This evidence implies at the very least that there may be species differences in establishment probability that are attributable to differences in their characteristics. Further, the more recent quantitative meta-analysis by Blackburn, et al. (2009), focusing solely on the influence of species-level traits on establishment success, suggests that evidence for effects of these traits is starting to accrue. While there is obviously variation across studies in the extent to which most of the traits that have been considered to determine establishment success actually do so (Tables 4.1–4.4), this is unsurprising given the range of other factors (e.g., event-level and location-level) that have the potential to moderate species-level effects.

One trait for which evidence of an effect on establishment is firming up is habitat breadth. To date, studies that addressed this variable have used basic metrics, but its effect is reasonably consistent and, for an ecological relationship, relatively strong (Blackburn, et al. 2009). The relationship between the breadth of environmental tolerances and establishment success is one that would seem likely to repay more study. The same is true for the ability to respond behaviourally to new opportunities or challenges, as measured by both relative brain size and innovation frequency, which to date is supported by few studies. It will be interesting to see whether further study strengthens the case for their general effect on success, or, as is so often the case, explodes it (cf. Drake 2007; Sol, et al. 2008).

We also find some evidence to suggest that a fast life history reduces establishment success in birds. More nuanced studies may be particularly useful here, because of the various contradictory predictions about the advantages and disadvantages of slow life histories (section 4.3.1), and because different traits appear to influence establishment in different ways (Blackburn, et al. 2009). It seems likely to us that close attention will need to be paid to the specific context of the introduction event in expanding or eliminating a role for such traits, to assess when stochastic effects, Allee effects or deterministic components of population growth rate are of greatest importance: when is it better to live fast, and when is it better to live slow? Of course, the answers will need to be assessed within the framework of the overarching influence of propagule pressure.

Nevertheless, we predict that the effects of life-history traits on establishment success will never be more than weak. We base this prediction on the pattern of variation that establishment success shows across the avian phylogeny. Blackburn and Duncan (2001a) and Sol, et al. (2002) assessed how this variation is partitioned with respect to different nested levels of the avian taxonomic hierarchy. Both studies concluded that most variation in success is distributed at lower levels of this hierarchy. Thus, Sol, et al. (2002) showed that 72% of the variance in success clusters at the species level, relative to 20% at the order level. In other words, closely related species tend not to be similar in terms of their establishment success. This pattern suggests in turn that species-level traits associated with introduction success must therefore be evolutionarily labile, varying even among closely related species. Thus, there will be a limit to the capability of phylogenetically conserved traits, which are typically shared by related species, to explain variation in avian introduction success. This list includes traits like body mass, generation time, and population growth rate (Böhning-Gaese and Oberrath 1999; Owens, et al. 1999; Freckleton, et al. 2002). That said, the presence of some variation in establishment success at higher taxonomic levels (e.g., 20% across families within orders) does leave a role for such traits.

The corollary of this, of course, is that more promise in explaining establishment success is likely to come from traits that are phylogenetically labile. These tend to be characteristics of a species' ecology rather than its life history (Freckleton, et al. 2002). In that regard, one characteristic that is notably labile is how common is a species (Gaston and Blackburn 1997; Gaston and Blackburn 2000; Freckleton, et al. 2002; Webb and Gaston 2003; 2005): it is difficult to say much about how abundant or widespread a species is, given the abundance or distribution of a close taxonomic relative. Perhaps, the characteristics that make a species successful (common, widespread) in its native range are the same ones that influence its success in exotic locations? This observation may explain why the ability to tolerate a range of environments appears to be important, because it means that a species can find opportunities in most of the places it finds itself. However, in terms of explaining native abundance and distribution, more important than the range of a species' tolerances (i.e., niche breadth) is the match between its specific tolerances and the available conditions (i.e., niche position; Gregory and Gaston 2000; Gaston 2003). This suggests that more important than what a species does per se is what it does relative to where it is put. This is a topic we will consider in the next chapter.

5 The Role of Location in Establishment Success

We . . . as observers and students of our native birds should take particular pains to guard against any such calamity as the establishment within our limits of any foreign species . . . Of course in the great majority of cases the birds die harmlessly within a longer or shorter time on account of the radically new conditions of food and climate which they are physically unable to meet.

J. Grinnell (1906)

5.1 Introduction

We have introduced the idea that three categories of factors can determine whether or not a species becomes an invader—characteristics of the event, species, and location. Propagule pressure is a key variable that determines the probability that an introduction event succeeds, because larger releases are more likely to overcome the problems of environmental, demographic and genetic stochasticity, and Allee effects, that typically afflict small populations. However, the degree to which species are susceptible to small population problems depends on their life-history characteristics, suggesting that species-level traits should also affect establishment success. Nevertheless, the most important species-level traits appear not to be those related to susceptibility to small population problems, but rather those related to the ability of a species to cope with novel environments (Chapter 4). This association in turn suggests that the recipient location poses challenges to the establishment of exotic species, which behavioural flexibility helps species to overcome. Significant differences in exotic bird establishment success among global regions (Case 1996) also suggest the importance of characteristics of the recipient location. In this chapter, we consider what those characteristics might be.

Three broad aspects of a location are likely to influence establishment: the physical environment, availability of resources, and interspecific interactions (Shea and Chesson 2002). Logic suggests that establishment should be favoured at locations where there are fewer enemies and more mutualisms, more available resources, or a more suitable environment. Environments may be considered unsuitable for establishment because temperatures inimical to life (especially extremes of cold) are typical, because of the absence of available water, or because broad ranges of environmental conditions may be encountered on daily or annual timescales. However, the harshness of an environment can only be judged in the context of the tolerances of any given species, suggesting that characteristics of the physical environment will be less important for species in general than the match between the native and exotic environments for each species individually.

In this chapter, we review the evidence that location influences introduction success in exotic birds. We begin by considering the role that the local climate plays in establishment. The rest of the chapter is devoted to understanding the role of biotic interactions. This subject can be divided into two parts depending on whether the exotic population is viewed as one entering a community of strongly interacting species in the non-native range or as one escaping from strongly interacting species in their native range.

5.2 Environmental Matching

It is likely that some introduced bird populations failed to establish simply because they were introduced into environments to which they were completely maladapted. While it is difficult to assign individual cases of failure to specific causes, unsuccessful translocations such as those of the Kalij pheasant *Lophura leucomelanos* from south-east Asia to Alaska and the willow grouse *Lagopus lagopus* from northern Europe to Fiji (Long 1981) certainly cannot have been helped by the evident mismatches between the native and recipient environments.

Exotic bird populations may still fail to establish even if the climatic conditions are broadly favourable. Individuals introduced into a new environment may experience conditions that are benign enough to allow them to survive and reproduce, but harsh enough to prevent them from realizing peak survival and fecundity (e.g., Holt, et al. 2005), hence reducing the likelihood that they will overcome the small population problem (section 3.1.1). In these instances, we would expect that the difference in environmental conditions between the native and non-native ranges would be a significant predictor of establishment success, such that small differences result in higher probability of success and vice versa.

The results of several studies of bird introductions are consistent with a role for environmental differences in establishment success (see also Hayes and Barry 2008). For example, Blackburn and Duncan (2001a) found that establishment success was significantly greater when the difference between a species' latitude of origin and its latitude of introduction was small. Cassey (2001b; 2003) similarly found that the great circle distance between a species' native and recipient ranges was negatively correlated with the success of bird introductions, and that land bird introductions were more successful within than between biogeographic regions. Nevertheless, these studies may be confounded if translocations over longer distances fail because they impose higher levels of stress on the transported individuals, which subsequently render introduced individuals less capable of surviving in the recipient environment. To our knowledge no one has quantitatively estimated the stress levels of birds in transport, or evaluated the effect of stress on establishment success, so that this assertion is difficult to evaluate.

A more sophisticated investigation of the role of environmental matching in establishment comes from a study by Duncan, et al. (2001) of exotic bird species introduced to Australia. They explored the effects on establishment success of climatic matching (defined as the extent of climatically suitable habitat in Australia), life-history traits (e.g., population-growth rates), and historical factors (e.g., propagule pressure). Duncan, et al. (2001) used a climate-matching procedure that quantified variables such as mean annual temperature and mean annual rainfall within each species' native range and compared this to climate variables within Australia. Species were then categorized by the number of points in Australia that had a difference in climatic variables of 10%, 20%, 30%, etc, as compared to the native range values. Thus, species that had many points in Australia that evinced a low climate difference (10% to 50%) were expected to establish more readily than those with many points where the climate differences were large (up to 80% difference).

Exotic bird species that successfully established in Australia had significantly larger numbers of points in climate matching classes below all but the highest value (Figure 5.1). This effect was strongest when only the matching classes of

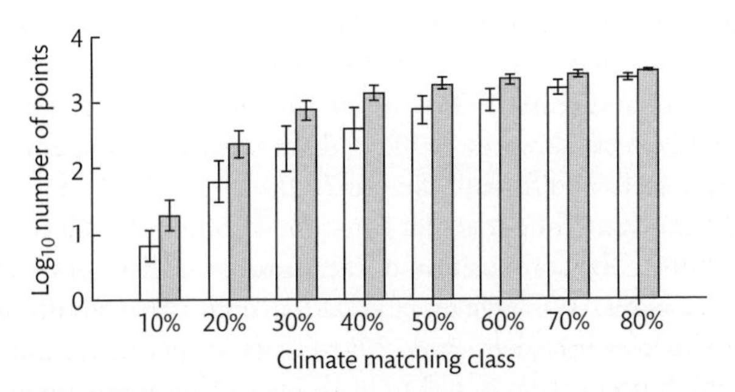

Fig. 5.1 The effect of climate matching on establishment success in bird introductions to Australia. The bars represent the mean (± 95% confidence limits) of the $\log_{10}$ total number of geographic points in climate matching classes below a progressively higher threshold of climatic difference, for successful (shaded) and unsuccessful (open) introductions. The plot shows that successful introductions relate to species for which more of the geographic points in Australia have smaller percentage climatic differences from points in the native range. Reprinted with permission of John Wiley & Sons Ltd from Duncan, et al. (2001).

50% or below were considered. The greater the extent of closely matched climatic conditions, the greater the probability of establishment (Duncan, et al. 2001). Indeed, none of the nine species that had twenty-five or fewer points in the lowest two climatic matching classes (i.e., points with high climatic matches between Australia and their native range) successfully established in Australia.

Duncan, et al. (2001) additionally incorporated data on species- and event-level traits to produce a minimum adequate model to explain observed variation in establishment success. Successful establishment was more likely for bird species that were introduced to more than two sites, that were not game birds, that had a greater area of climatically suitable habitat available in Australia, that had been introduced successfully elsewhere, and that had a larger body mass (Duncan, et al. 2001). Climate matching added explanatory power to the overall model, but by itself was not a good predictor of success as it could not sort species as either successful or failed as effectively as could propagule number. These results suggest that an introduced population will tend to fail if released at only a few locations even if it is climatically well matched to the recipient community. However, once sufficient release events have occurred, the degree to which the species is suited to the local climate will influence success.

5.3 Human Commensalisms

It is increasingly obvious that human actions have altered, and continue to alter, environmental conditions, and that they can do so at multiple spatial scales. A long-understood effect of human activities is the creation of urban landscapes. Urban habitats have several consistent aspects no matter where they are created, in large part because they are maintained to satisfy a narrow set of human needs (McKinney 2006a). Thus, urban areas tend to have a typical vegetation structure (e.g., urban savanna; Gobster 1994), sustain their own temporally stable climates (e.g., heat island effect; Kowarik 1990), and have abundant sources of particular food items (e.g., seeds and fruits; Marzluff 2001).

There is a clear trend for particular functional groups of birds to thrive in cities (Luniak 2004). The composition of these groups is distinct from those that thrive in native habitats (Clergeau, et al. 2001), and urban environments often favour exotic species (Garden, et al. 2006). Exotic species do not necessarily constitute a high proportion of urban bird species, but may numerically dominate urban bird assemblages. For example, a survey of resident bird species in urban Melbourne, Australia, recorded forty-three species, of which only nine (21%) were exotic. Yet, these few species represented over two-thirds (69%) of all bird sightings (Green 1984). In some cases, exotic bird species exhibit an apparently obligate association with human presence, to the extent that no exotic populations are known to live independently of humans (e.g., house crow

Corvus splendens: Nyári, et al. 2006). For example, during winter months in Chicago, 100% of observed monk parakeet *Myiopsitta monachus* feeding bouts were on birdseed provided at backyard feeding stations (South and Pruett-Jones 2000). Thus, humans may be artificially enhancing environmental matching by creating uniform urban habitats around the world in which a specific set of urban exploiter species can establish.

The creation of uniform urban habitats alone could account for the success of some commonly introduced bird species, but the success of these species may be further enhanced by the importance of availability to humans on the probability of transportation (section 2.2.4). Urban exploiter species may be more readily available to the avicultural trade (past and present), which provides birds for use as pets or to satisfy acclimatization societies or for prayer releases. The accidental or deliberate release of captive species is also obviously most likely to occur in urban areas. Thus, although the number of urban exploiting birds is a relatively small fraction of all birds—8% to 25% of native bird species in North America are typically classified as urban exploiters in the sense that they regularly and nearly exclusively exist in urban areas (Johnston 2001)—these are precisely the species that are most likely to be transported and released.

The likelihood that urban exploiter species may end up with exotic populations is further increased by an apparent tendency for such species to have a generalist life history (section 4.3.3). Bonier, et al. (2007) showed that urban bird species have larger geographic and elevational range sizes than non-urban congeners. They argued that species with larger ranges also have broader environmental tolerances, and hence that these broader tolerances have facilitated the invasion of urban environments by such species. However, there is currently no strong support for a relationship between range size and environmental tolerance, with most tests to date being confounded by sampling artifacts (Gaston 2003). Stronger evidence of a link between generalism and urbanization comes from a study by Devictor, et al. (2007), using an independent classification of generalism not based on range size. Using information from the French breeding birds surveys, Devictor and colleagues found that specialist species had a 54% higher chance of becoming locally extinct from a survey plot than generalist species, and that this proportion increased significantly as urbanization increased in some plots.

Overall, these trends suggest that there is a strong positive feedback loop between urbanization and the widespread successful establishment of a limited number of bird species. Urban-exploiting birds are more likely to be transported and released, are more likely to be released in other urban habitats that provide ideal conditions for their establishment, and possess a set of life-history characteristics that may further increase their probability of success once released. Finally, if exotic birds prey upon fruits and then disperse seeds that differ in composition or relative proportion to those eaten by native species they can, in fact, further alter the floristic composition of human-modified environments (Williams and Karl 1996). This interaction can make urban habitats even more attractive for these same species and perhaps lead to accelerated impacts on native ecosystems—instigating an invasional meltdown (Simberloff and Von Holle 1999).

5.4 Biotic Interactions

We have, so far, considered the effects of the physical environment on the success of exotic bird populations, but there is a substantial body of ecological theory that suggests that biotic interactions in the non-native range are also likely to have a strong influence on their success or failure. This theory produces conflicting expectations depending on whether an exotic species is viewed as entering a complex, strongly interacting community in the non-native range or escaping such a community within its native range. In the former case, characteristics of the recipient community have been viewed as key to establishment, and if they have particular traits (such as high species richness) we expect these communities to be resistant to invasion. In the latter case, it is the composition of the recipient community that has been considered to matter more, and we expect species to be more successful if the community lacks the type of interactions that limit fecundity or survival in the native range (competitors, predators, pathogens). These two sides of this ecological coin are not well integrated in the ecological literature, and this situation is reflected within research on exotic birds.

5.4.1 Interspecific Competition

The role of interspecific competition in structuring ecological communities has a long and controversial history (Keddy 2001). The foundations for claiming that the presence of one species at a site will determine whether or not another species will be able to persist there goes back at least to Gause (1932), if not further, and persists today in the form of niche theory (Chase and Leibold 2003) and determinants of species diversity (Hubbell 2001). The study of exotic birds has contributed to this broad literature through the study of two questions. First, what role, if any, does native species richness per se play in determining the successful establishment of exotic bird species? This question addresses whether or not high native species richness prevents or limits the establishment of exotic bird species. Second, what role, if any, do the existing species in a community play in preventing an introduced exotic bird from establishing via competition? This investigation is essentially a narrower statement of the first question, which looks for combinations of species that would prevent a newly arriving species from establishing due to intense competition. We review the evidence for each below.

5.4.1.1 The Role of Species Richness

If communities are a collection of tightly co-evolved species, we may expect them collectively to utilize all available resources, and to do so in an efficient manner. This view of communities is akin to the view of Clements (1916), and assumes that species will evolve in response to one another and their environment such that they attain maximum fitness. If such a situation predominates in nature, then we might suspect that intact native communities will be hard to invade since there would be little or no opportunity to capitalize on unused resources (indeed there should not be many unused resources to capitalize on).

As the extent to which communities use resources is difficult to quantify directly, studies of such effects have generally considered variables with which resource use is thought to be correlated. For example, oceanic islands typically harbour far fewer native species and functional groups than equivalently sized pieces of a mainland, because not all mainland taxonomic or functional groups are capable of colonizing sites so remote. Thus, we may expect that island communities do not utilize all available resources, simply because there are fewer native species there to exploit them. If this were true, then islands should be easier to invade than mainland areas (Elton 1958).

Certainly, many islands have more established exotic bird species than much larger neighbouring mainland regions. New Zealand (270,000 km^2) has thirty-four exotic bird species (Duncan, et al. 2006) compared with thirty-one in Australia (7 million km^2), and Hawaii (17,000 km^2) has forty-nine exotic bird species (Moulton, et al. 2001a) compared with the ten established species in the state of California (400,000 km^2). Several studies have addressed the question of whether avian introduction success is indeed higher on islands than mainlands. Newsome and Noble (1986) found that bird species introduced to Australia had a higher failure rate when introduced to the mainland than to offshore islands. However, a problem with this study was that different sets of species were introduced to the mainland and islands—the difference in establishment success could thus be a result of species-level characteristics. Sol (2000a) addressed this problem by comparing the success of the same species when introduced to islands or mainlands, in this case New Zealand versus Australia, and the Hawaiian Islands versus mainland USA. Thus, for forty-two species introduced to both Australia and New Zealand, eighteen species always succeeded in both locations, and sixteen always failed in both. Of the eight species with mixed success, five succeeded in New Zealand but not Australia, and three vice versa. Overall, Sol (2000a) found no evidence that islands were easier to invade. This conclusion has subsequently been generalized in three global analyses of bird introductions (Blackburn and Duncan 2001a; Cassey 2003; Cassey, et al. 2004c), the most recent of which also accounted for propagule pressure and a variety of other species- and location-level characteristics. Thus, the high numbers of exotic bird species found on islands appears to be a consequence of the many attempts to introduce birds to islands, rather than any inherent feature of islands that makes them easier to invade (Simberloff 1995).

A related test of the hypothesis that saturated communities will better resist invaders comes from broad-scale correlations between native and non-native species richness. Here, the assumption is that, all else being equal, communities with more native species are likely to be using more of the available resources. It follows that if native and exotic species richness are negatively correlated across communities, then the native species may in some way be preventing the exotic species from successfully establishing (Elton 1958).

Stohlgren, et al. (2006) evaluated the relationship between exotic and native bird species richness across the USA. They found that, for 3,074 counties across forty-nine states, native bird richness was positively related to exotic bird richness. When these numbers were scaled by area (US counties vary greatly in size, and increase in area on average as one moves from east to west), the relationship became even stronger (Figure 5.2). Although this result does not imply anything about how native species richness affects the probability that an exotic species will establish a viable population (because there is no information on failure rate), neither is it consistent with the idea that high native richness in some way limits the distribution of exotic birds. Rather, both native and exotic bird species appear to be responding in a very similar way to basic biophysical properties of the landscape. In particular, both groups attain higher species richness in low elevation areas close to the coastlines, which are sites with high precipitation and primary productivity (Stohlgren, et al. 2006).

Bird introductions also provide little support for the idea that establishment success is lower in assemblages with more native species. Cassey, et al. (2005b)

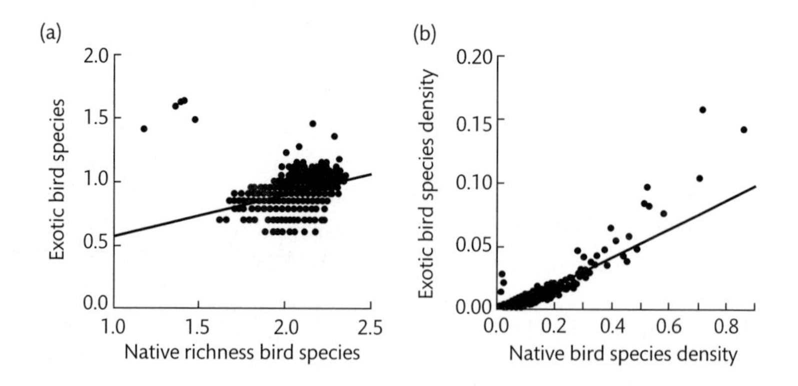

Fig. 5.2 Relationships between (a) exotic versus native bird species richness ($\log_{10} x + 1$ transformed), and (b) exotic versus native bird species density (richness scaled by area), across counties in the USA. From Stohlgren, et al. (2006).

found no significant relationship between the number of native bird species on a sample of islands around the world and the probability of exotic bird establishment. Similarly, Case (1996) found that, for a sample of island and mainland regions, introduction success did not decline significantly with the species richness of the native avifauna (after controlling for the effects of native extinctions and area). Interestingly, Case (1996) did find a negative relationship between the numbers of native and exotic bird species across a range of island and mainland habitats. He concluded that this was more likely to be due to environmental differences in native- versus human-modified habitats than the importance of species-rich avifaunas repelling exotic species. However, he conceded that the relative importance of these factors would be difficult to uncouple.

In fact, competition between native and exotic birds is probably unlikely to regulate exotic establishment because most exotics establish in highly modified habitats, such as farmland and urban areas, which are little used by native species (Diamond and Veitch 1981; Simberloff 1992; Smallwood 1994; Case 1996). Forys and Allen (1999) found that established exotic birds in Florida tended to have different diets and different niches than endangered species, which suggests that the former are neither driving extinctions in the latter nor moving into vacated niches.

5.4.1.2 Competition between Exotic Bird Species

While there is little evidence that native bird species richness negatively impacts upon exotic bird establishment success, it is still possible that introduced species that compete heavily with previously established exotic species in the recipient community will have a reduced probability of establishing. While Cassey, et al. (2005b) found no relationship between the richness of exotic bird species on a sample of islands around the world and the probability of exotic bird establishment, interactions between species that are ecologically very similar may be more important pieces to this ecological puzzle than exotic species richness per se. Thus, if we narrow our focus to include only these species, we may be more likely to detect patterns consistent with the competition hypothesis.

There are surprisingly few studies that document the fitness effects of competition between two or more exotic bird species. For the obvious reason that it is impossible to document the fitness of a species that is now locally extirpated, there are no studies of competition between species where one failed to establish once introduced. This absence of information means that the role of competition in establishment success has to be tested using a proxy measure of the degree of competition that can be readily measured for all species, including those that had the chance to establish but failed to do so. Two obvious candidate proxies for birds are taxonomic affiliation and morphology (Ricklefs and Travis 1980; Moulton and Pimm 1986b). The assumption is that two species in the same taxonomic group (e.g., genus) are more likely to utilize similar resources and are thus more likely to compete strongly with each other. Likewise, species that share similar morphologies, such as bill length and body size, are assumed to compete more strongly with each other for available resources than they will with species that have very different morphologies.

Using these surrogate indices of the presumed strength of competitive interaction, a number of authors have looked for patterns in establishment success of exotic birds that are consistent with the notion that interspecific competition plays a dominant role. The Hawaiian Islands are an obvious location for studies on the ecology and evolution of exotic birds because a large number of bird species have been introduced there, and there are good records on when these species arrived, whether they established breeding populations, and if and when they became locally extinct. This record has been used most notably by Michael Moulton and Stuart Pimm (Moulton and Pimm 1983; 1986a; 1986b; 1987; Moulton 1985) to show that interspecific competition may have influenced establishment success.

First, Moulton and Pimm showed that extinction rates (measured as the number of exotic species that went locally extinct per ten-year period) were correlated with the number of exotic species present (i.e., richness, measured as the number of exotic species present at the end of each decade) in a quadratic fashion. Thus, as the richness of exotic birds increased through time (1860–1960), the number of exotic birds that failed to establish increased at a greater rate (Figure 5.3).

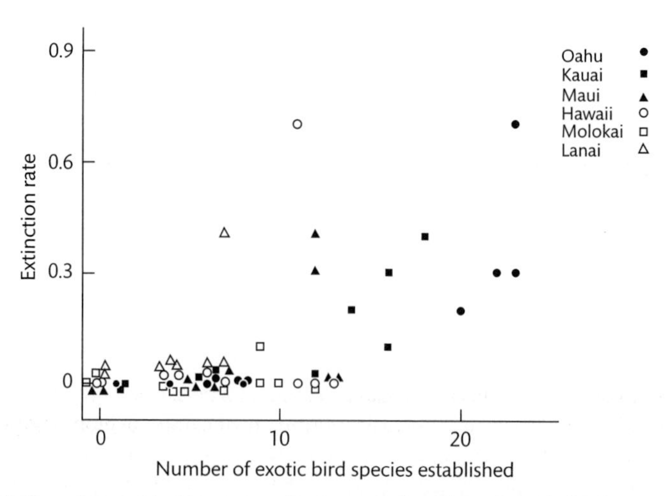

Fig. 5.3 The relationship between extinction rate (number of exotic bird extinctions per year, averaged over ten-year periods from 1860 to 1960) and the number of exotic bird species established, for six islands in the Hawaiian archipelago. The data are better fitted by a curvilinear than a linear statistical model, indicating that the per-species extinction rate increases as the number of species increases. Reprinted with permission from Moulton and Pimm (1983). © 1983 University of Chicago Press.

This result is consistent with competition amongst exotic species influencing their establishment success.

Second, when congeneric pairs of exotic passeriform and columbiform species were introduced to the same island, the differences in their bill lengths were greater when both species established than when one of the pair failed. Moulton (1985) interpreted this result as evidence that the failure of one of the similar species was due to competition.

Third, for three of the six main islands (Oahu, Kauai, and Hawaii), the exotic passerines that established in forested habitats showed a pattern of morphological over-dispersion relative to random subsets of the same number of species drawn from the list of all passerines that were introduced (Figure 5.4). Morphological over-dispersion is thought to arise when interspecific competition predominates across all species in the community such that only those species that are particularly different from one another in their morphology will establish, while species that are too similar to an already established species will fail.

These results were subsequently challenged by Simberloff and Boecklen (1991), who noted that most could be overturned by a slightly different interpretation of the underlying information on which species were introduced, which established, which failed, and when. They used the Hawaiian exotic bird data to propose the 'all-or-none' pattern of establishment success (section 4.2), whereby some exotic species always succeeded when introduced to these islands, while some always

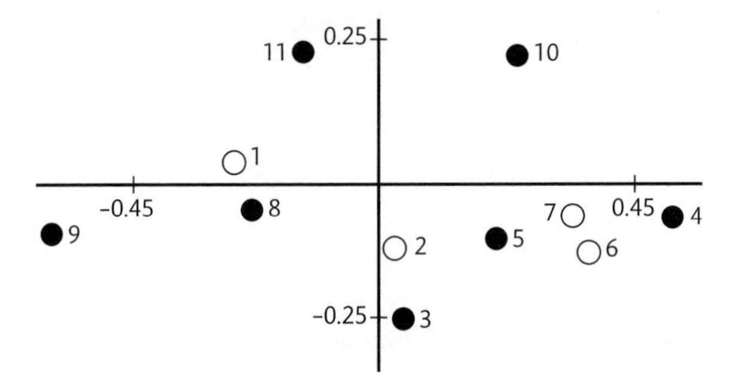

Fig. 5.4 Morphological overdispersion in exotic passerines introduced to forested habitats on Kauai in the Hawaiian archipelago. The axes are measures of morphological distance calculated from six morphological variables using principal components analysis. Established exotic species (filled circles) are more dispersed in morphological space than expected by chance. Open circles represent the positions of failed introductions. Species are: (1) Varied tit *Parus varius*, (2) Oriental magpie-robin *Copsychus saularis*, (3) White-rumped shama *C. malabaricus*, (4) Greater necklaced laughing thrush *Garrulax pectoralis*, (5) Hwamei *G. canorus*, (6) White-throated laughing thrush *G. albogularis*, (7) *Black-throated Laughing thrush G. chinensis*, (8) Red-billed leiothrix *Leiothrix lutea*, (9) Japanese white-eye *Zosterops japonicus*, (10) Northern cardinal *Cardinalis cardinalis*, (11) House finch *Carpodacus mexicanus*. From Moulton and Pimm (1987).

failed. This hypothesis would probably have more to do with the intrinsic traits of the species than with competition (Simberloff and Boecklen 1991). As we noted in section 4.2, there is little evidence for an all-or-none pattern of establishment success *sensu strictu* in birds, but existing evidence is consistent with a situation where species differ in their probability of establishment, with some being generally good at establishing exotic populations, and some being generally poor. Subsequent debate about the specifics of the Hawaii dataset (e.g., Simberloff and Boecklen 1991; Moulton 1993; Keitt and Marquet 1996; Duncan and Young 1999; Moulton and Sanderson 1999) has failed to produce consensus about the role of competition in this exotic avifauna.

Several other attempts to explain establishment success using interspecific competition theory have nevertheless built on the initial work by Moulton and Pimm. For example, Lockwood and colleagues (Lockwood, et al. 1993; Lockwood and Moulton 1994) showed clear patterns of morphological over-dispersion amongst the exotic passerines of Tahiti and Bermuda, and reaffirmed the pattern amongst the exotic birds of Oahu using updated statistical tools and species information. Brooke, et al. (1995) showed a 'priority effect', whereby exotic passerines that were introduced earlier, when few if any other exotic birds were present on St Helena, were more likely to succeed than those introduced later, when more species persisted. Moulton, et al. (1996) and Simberloff (1992) found evidence for a priority effect and morphological over-dispersion amongst the exotic passerines of La Reunion but not for Mauritius. The ubiquity of the results consistent with the view that interspecific competition plays a role in establishment success suggests that there may be some merit to the notion after all. Nevertheless, arguments persist over this topic, though of late they have centred on the role of competition in determining the success of exotic bird species introduced to New Zealand.

As we noted in Chapter 3, Veltman, et al. (1996), Duncan (1997), Green (1997), and Cassey (2001b) all showed that the main determinant of establishment success for the exotic birds in New Zealand was propagule pressure. However, Duncan's (1997) analyses of exotic passerines also bear heavily on the debate on the role of competition in determining establishment success. His results showed support for a priority effect, and for the role of propagule pressure in establishment success within each of four New Zealand acclimatization districts. Duncan went on to show, however, that propagule pressure is confounded with the timing of release for each species such that species released earlier were released in larger numbers or more times. Thus, those species released when fewer other species were present could be more successful than those species released later when more species were present (the priority effect) because they were released in larger numbers, and not because they enjoyed a less-competitive environment. While an influence of competition could not be ruled out, Duncan concluded that the role of competition in structuring exotic bird communities might have been overstated in earlier studies.

Duncan (1997) also proposed a mechanism underlying his result: New Zealand acclimatization societies had a predilection for introducing species native to Britain that were readily available for transport and release. These are species that are common and abundant in Britain, and thus were either highly desired by residents of New Zealand, or were simply easy to trap and export in large numbers. Either way, these species were more likely to be introduced early, in larger numbers, and more times. They were thus also the species that quickly established exotic populations. In contrast, later introductions were presumably relatively less desirable or available, were consequently released in lower numbers or less frequently, and thus tended to fail to establish. It is also possible that the acclimatization societies simply began to wane in their efforts as time wore on and thus the later introductions involved releases of far fewer individuals than their earlier counterparts.

Finally, Duncan (1997) suggested that the pattern of morphological over-dispersion could also be confounded with propagule pressure if more effort was invested introducing species that were very different from one another. Written records from New Zealand indicating that acclimatization societies endeavoured to fill empty niches on the islands support this idea. However, a firm test of this supposition would require a positive relationship to be demonstrated between morphological differences measured in multi-dimensional space and some metric of propagule pressure. That is, established species that are very different from other established species in an exotic bird community would have to have been introduced in larger numbers, or more times, or some combination of these.

The only test of this mechanism published to date is for exotic galliforms introduced to New Zealand. Duncan and Blackburn (2002) calculated two measures of morphological similarity using the fifteen species introduced, based on four morphological variables originally reported by Moulton, et al. (2001b). The first measure was MDI, which is the minimum difference between each species and all others introduced up to the date of introduction for the focal species. This metric quantifies the likelihood that a new introduction represents a novel body size as compared to the species already introduced. One would expect that as more species are introduced, the probability that later introductions represented something novel would go down simply because of a sampling effect: there are only so many different sizes of game birds in the world to release. The second measure was MDA, which is the minimum difference of the principal component score of the focal species and the score of any other species introduced. If species that successfully established show higher MDA scores than those that failed, we could assume that there is a pattern of morphological over-dispersion.

The key to this analysis, however, was the availability of propagule size data to accompany the morphological difference measures. If there was a tendency for acclimatization societies to release morphologically different species in higher numbers, then (1) MDI scores would be positively correlated to introduction effort and (2) propagule size should explain establishment success once MDA is controlled for, statistically, but not vice versa. Duncan and Blackburn (2002) found that MDI did decrease through time, such that later introductions were more likely to encounter similarly sized species than earlier ones. They also found a positive correlation between MDI and propagule size, which suggests that early releases involved more individuals and encountered fewer putative competitors. Game birds that were morphologically distinct from all other species released (measured as either MDI or MDA) where more likely to succeed. However, species released in larger numbers were also more likely to succeed and this result held when MDA was statistically controlled for. There was no evidence for the reverse result whereby MDA was correlated with success once propagule size was controlled for.

In summary, we again see evidence that propagule pressure explains a large amount of variation in the establishment success of exotic birds. More interestingly, there can be a relationship between the timing of introduction and the effort expended (i.e., through propagule pressure) to ensure establishment success that emanates from the peculiarities of human-derived introduction processes. This effect can confound attempts to understand the influence of several purported mechanisms of establishment success (Cassey, et al. 2004c; Lockwood, et al. 2005), including competition (Duncan 1997; Duncan and Blackburn 2002). It does appear that competition is not the overwhelming force determining introduction success that early theory suggested should be the case. Yet, evidence that interspecific competition may sometimes determine establishment persists. Patterns consistent with the competition hypotheses are found across several localities, each with its own peculiar historical contexts for species introductions. Even some critics of the competition hypothesis find evidence in their own analyses that competition could account for some variation in establishment success (e.g., Moulton, et al. 1996; Duncan 1997; Duncan and Blackburn 2002). So where do we proceed from here?

The answer may be a clear consideration of the relative roles of stochastic and deterministic forces in determining establishment success. Thus, even a species that is competitively dominant, is matched perfectly to the non-native climate it is entering, and has a life history that is tailor-made for rapid population growth, will not establish successfully if it is introduced in low enough numbers, or is unlucky enough to encounter adverse conditions in the non-native range. Essentially, this is to advocate integrating event-level, species-level, and location-level factors into population models of establishment success.

The first step towards such integration has come from a study by Duncan and Forsyth (2006). They constructed a two-species stochastic birth–death model to explore how interactions between propagule pressure, the abundance of a competitor species, and the strength of competition determined the fate of introduced species. The model showed that higher propagule pressure (measured as the initial population size in which a species was entering a community) substantially increased establishment probability. However, establishment probability decreased as both the abundance of a competitor and the strength of competition increased. Most importantly, these three factors interact such that competition has little effect on establishment probability when the competitor trying to invade is present in low initial abundance. This pattern holds regardless of the strength of the competitive interaction. The influence of competition on success increases as the initial abundance of the invader increases, and as the abundance of the competitor increases (Figure 5.5).

These results led Duncan and Forsyth (2006) to make two predictions. First, the abundance of a competitor and the per capita strength of competition should be important in determining the outcome of an introduction event. Second, there should be an interaction between these factors, such that the strength

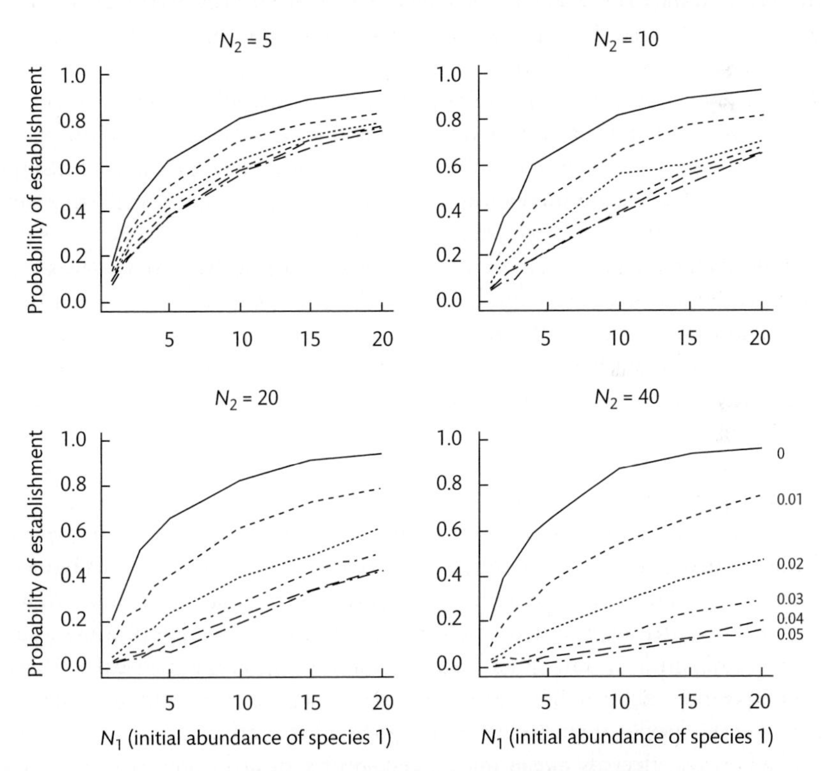

Fig. 5.5 Results of a stochastic birth–death model showing the probability of establishment as a function of the initial abundance of an introduced species (N_1), given a resident competitor at four different abundances (N_2; the four different panels), and with the per capita strength of interspecific competition taking six values from 0 to 0.05 (shown by different lines). Reprinted with kind permission of Springer Science and Business Media from Duncan and Forsyth (2006).

of competition should depend on the relative abundances of the competitor and the species trying to invade. Hence, competition should be of much less importance in determining establishment when a species is introduced in small numbers, in which case stochastic effects will dominate, or when the resident competitor is present in low numbers. We should see clear evidence for competition when both the exotic species and its resident competitor are present at high abundances.

Duncan and Forsyth (2006) used the information on exotic passerines introduced to New Zealand originally published in Duncan (1997) to test these predictions. Following previous work, they used the degree of morphological similarity as a proxy for intensity of competition between two species. They used the number of years since initial release, based on written accounts of population expansion rates from Thomson (1922), as a proxy measure of competitor abundance. In particular, they used the time delay (in years) between the introduction of a focal species and the introduction date of its closest competitor as a measure of that competitors' abundance. They found that statistical models that included propagule pressure, per capita strength of competition (as indexed by similarity in morphology), and relative abundance of the closest competitor, garnered substantial support. As expected, establishment probability increased with higher propagule pressure, and decreased when morphological similarity and competitor abundances were higher.

Duncan and Forsyth (2006) found that the relationships observed in the actual data from New Zealand fit remarkably well with the predictions generated from their simple mathematical model (Figure 5.6). When the morphological similarity or abundance of the competitor is low, the probability of establishment for a newly arriving species is predicted almost solely by propagule pressure. In contrast, when both the abundance of the competitor and the newly released species are high, there is a pronounced effect of the strength of competition as indexed by morphological similarity.

It seems that the persistent debate on whether competition determines establishment success is partly a product of a profound and largely unrecognized interaction between propagule pressure and the strength of competition. When historical information is used to test various hypotheses about establishment success in exotic birds (or any other taxon), sometimes there will be clear evidence for competition, but usually there is not. The likelihood of finding that evidence goes up if the species are introduced in large numbers, if there are strong per capita competitive effects between introduced species, or if the abundance of competitors is relatively high. These conditions must vary greatly between exotic bird assemblages depending on the historical circumstances surrounding the release of birds. In addition, these conditions may vary to a great extent between different species introduced to the same location. This possibility suggests that some species will fail due to competition and some will fail due to problems of small population

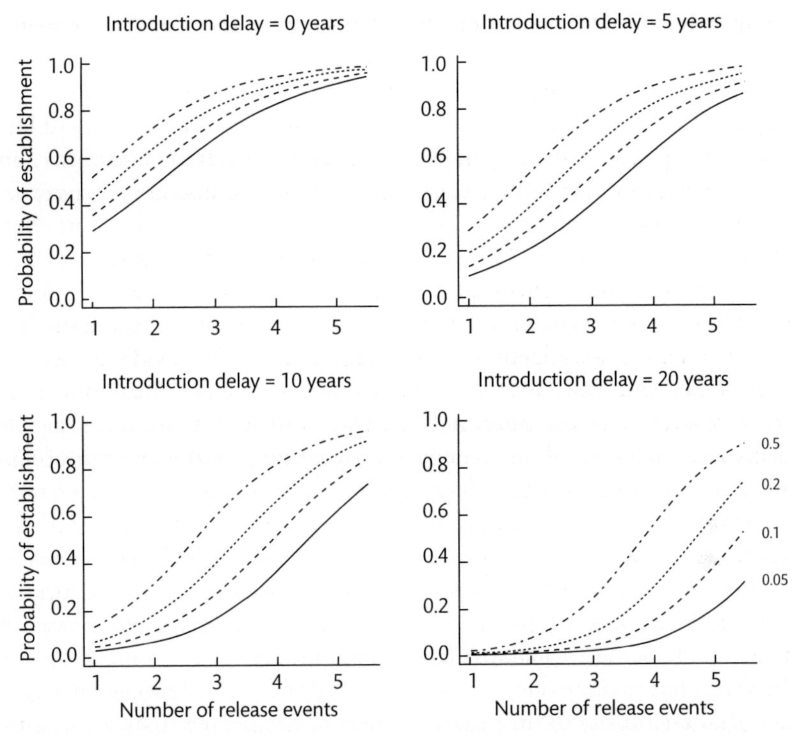

Fig. 5.6 The relationship between probability of establishment and number of release events (a measure of propagule pressure) for four values of introduction delay (the number of years after the introduction of its closest competitor that a target species is introduced, as a measure of competitor abundance; the four different panels) and for four values of morphological distance between the target species and its closest competitor (a measure of the strength of competition), for passerine birds introduced to New Zealand. Reprinted with kind permission of Springer Science and Business Media from Duncan and Forsyth (2006).

size, and it will be very difficult to tell the difference without detailed information on each species' history of population growth.

5.4.2 Predation

The prolonged debate over the role of interspecific competition in determining the successful establishment of exotic bird populations has perhaps distracted invasion biologists from a more important process. If we look at the broader literature on invasive species, the one biological interaction that seems to have swift and profound effects on community membership is predation, especially by an invasive species that represents a novel source of mortality for its prey (Blackburn and Gaston 2005; Lockwood, et al. 2007). Invasive predators have caused the extinction of entire suites of species (e.g., Fritts and Rodda 1998; Hadfield, et al. 1993), most notably in the context of this book, of island birds (Blackburn, et al. 2004; 2005a; Blackburn, et al. 2005b; Blackburn and Gaston 2005; Steadman 2006). Given these results, we should perhaps expect that exotic bird populations are also very susceptible to the deleterious effects of predators.

Cassey, et al. (2005b) evaluated the ability of a variety of factors, including predation pressure, to explain variation in number of established non-native bird species across forty-one islands or archipelagoes around the world. As we have now come to expect, they found that propagule pressure explained most of the variance in avian establishment success between islands. There was also some evidence that island area, human population size, number of native bird species, number of exotic mammalian herbivore species, and number of exotic mammalian predator species each explained some fraction of the variation across islands. When Cassey, et al. (2005b) considered the effect of all variables together in a single model, however, the only variables that explained significant and independent variation in bird introduction success across islands were propagule pressure and the number of non-native mammalian predator species.

These results provide evidence that it is not only native bird species on oceanic islands that are subject to the negative effects of exotic mammalian predators: exotic bird species can feel them too. It may be that mammalian predators prevent exotic birds from establishing only on oceanic islands. Mammalian predator abundance and density can be very high on islands as typically there are no natural checks on their populations in the form of larger predators (see Rayner, et al. 2008) or disease. Nevertheless, it is also possible that there is a general effect of mammalian predators on the probability of exotic bird establishment. Predator communities are hard to quantify for continental regions, even in terms of simple species numbers, and so this has yet to be addressed empirically.

Sinclair, et al. (1998) studied the effects of predation on the conservation of endangered prey, but their conclusions are also relevant to exotic populations post-introduction. The impact of predators on prey populations depends on how the per capita mortality rate changes with prey density. When the effects of predation are inversely density dependent, and where a prey species is not the primary food source of the predator, a predator can cause the extinction of a prey species. In these situations, there may be a boundary density above which it is unlikely that the predator drives the prey population to extinction. The likelihood that a newly arriving exotic bird population persists in the face of such predation may thus be improved by greater propagule pressure, suggesting another reason for the general effect of propagule pressure on establishment success (Chapter 3).

Either way, the impact of predation on establishment success seems to us to be a subject that would repay further investigation.

5.4.3 Enemy Release

The other side of the ecological coin to the idea that biotic interactions with the recipient community reduce the success of introduced species at an exotic location is that exotics instead benefit from 'release' from their natural enemies (predators, parasites, or pathogens), as these are left behind in the native range (Keane and Crawley 2002). This idea has become formalized as the enemy release hypothesis (ERH), which posits that exotic species typically suffer from fewer enemies in the exotic range compared to the native range. The reduction in enemy impacts on reproduction and survival allows exotic populations rapidly to increase in numbers, escaping the stochastic effects that afflict small populations (Chapter 3). Enemy release is also posited to allow exotic species to spread quickly over a large geographic area and to have a high impact on coexisting native species (Chapter 6).

Results from non-avian assemblages suggest that more specialist parasites are lost most easily, with the proportion of generalist parasite species increasing in exotic communities (Kennedy and Bush 1994). Torchin, et al. (2003) showed that exotic species tended to have lost about half of their parasite species after establishment in the exotic range. These authors suggested that this drop in parasite numbers was due largely to the nature of the introduction process. Exotic populations are by definition founded by only a few individuals from the species' native range. Fewer parasites should be present within the exotic propagule if fewer individuals were released, or if the geographic extent of origin of these individuals was relatively narrow.

Colautti, et al. (2004) reviewed the literature purporting to test the ERH and found that nearly 60% of the twenty-five studies they considered reported some evidence in its favour. They divided studies into two types: biogeographical studies compared native and introduced populations of a given host, while community studies compared native and exotic species living within the same community. In general, biogeographical studies were consistent with the ERH in that exotic species tended to have fewer enemies in the exotic versus the native range, although there were only a few instances where the release from enemies was shown to translate into real increases in demographic rates. The community studies, on the other hand, provided more equivocal results, with native and exotic species that co-occurred in the same community generally showing no significant difference in the number of enemy species affecting them.

Exotic birds have played a minor role in the literature on the ERH, and as we noted in Chapter 3, the evidence to date that exotic populations of birds support fewer parasite species is equivocal (cf. Steadman, et al. 1990; Paterson, et al. 1999;

Torchin, et al. 2003; Ishtiaq, et al. 2006). The one study to date explicitly to test the enemy release hypothesis in birds (Ishtiaq, et al. 2006) found little evidence that exotic common myna *Acridotheres tristis* populations had left behind their blood parasites, although they could not determine whether these exotic populations had carried all the parasite species with them or acquired new species at the exotic location. Either way, enemy release was not implicated in the success of these populations.

One of the species that Torchin, et al. (2003) used to show general support for the ERH was the European starling *Sturnus vulgaris*. These authors reported forty-four parasite species as being present in European populations of this species. We know that around 100 European starlings founded the exotic North American population. Through a simple random sampling procedure, Torchin, et al. (2003) showed that we would expect a population of this size to bring with it about twenty-eight of the forty-four native parasite species. They found nine native parasite species in the North American population. They suggested that the lower than expected number of parasite species could have resulted from a European starling population bottleneck after initial release, thus reducing the effective population size, or because some parasites could not survive in the new environment because their complex life cycles are not supported there (e.g., no intermediate hosts were present).

Colautti, et al. (2004) challenged this result by questioning the validity of the comparison. Torchin, et al. (2003) compared the number of parasite species found in North American starling populations with the number found across the native range in Europe. However, the individuals released in North America were not a random sample of all European populations, but were instead taken from one or a few locations, most likely in southern Britain (Colautti, et al. 2004). If there is substantial spatial heterogeneity in the distribution of starling parasite species across Europe, we should not expect European starlings in North America to have had a chance to bring all of these species with them. Colautti, et al. (2004) showed that the proper comparison with parasite species present in the source population of southern Britain provides little support for a reduction in parasite species richness due to translocation to North America. In addition, some avian parasites native to North America switched to using European starlings as hosts once they arrived on the continent. When these parasites are included in the calculations, the number of parasite species hosted by European starlings in the exotic range is actually higher than in the native range (Colautti, et al. 2004). The loss or otherwise of native parasite species may thus be less important than the intrinsic ability of a species to deal with exotic parasites via its immune system (section 4.3.3.3). This intrinsic ability may be doubly important given that extreme environmental conditions (which, relative to the tolerances of the exotic species, may more likely be encountered at the recipient location) can increase disease susceptibility (Hoffman and Parsons 1991; Hale and Briskie 2007).

No study has yet evaluated the relative reduction in predation pressure between native and introduced populations of exotic birds. The results of Cassey, et al. (2005b) concerning the negative influence of exotic predatory mammals on establishment success of exotic birds suggest that there is at least no reduction in the per capita effects of predators in the introduced range of many exotic bird species. Nevertheless, even with the widespread introduction of exotic mammal species, it is clear that some exotic bird populations (especially those on oceanic islands) are subjected to a lower absolute richness of predators as compared to populations in their native range. This observation raises the question of why exotic bird populations still seem to fail in the teeth of exotic mammal predation, and, more generally, should we expect that exotic bird populations are more likely to establish if they have fewer enemies in their introduced range?

Two factors complicate the search for an answer to the question of whether the escape from enemies, including the effects of a reduction in predators, enhances establishment success. First, not all enemies have the same effect, and thus a reduction in the number of enemy species may not translate into a release from enemy effects if the abundance of the few enemy species is quite large (cf. section 5.4.1). There are two contrasting possibilities for how hosts or prey may respond to enemies. On the one hand, enemies may elicit a strong defence response (behavioural or immunological) on the part of the prey or host. Alternatively, the host may not be well protected against enemies, and thus suffer pronounced reductions of fecundity or survival in the presence of the enemy. In the latter case, we would expect that the loss of enemies would result in a quick and dramatic increase in population growth rate as the exotic species is functionally released from a strong regulatory agent (regulatory release). In the former case, we may not expect much change at all in population growth rate since the lost enemies had no appreciable effect to begin with (compensatory release). In this case, the exotic population may evolve towards the loss of enemy defences if these defences are costly, but the evidence for this is decidedly mixed across all taxa and non-existent in the case of exotic birds (although oceanic island bird species do lose anti-predator responses over evolutionary timescales; Milberg and Tyrberg 1993; Massaro, et al. 2008). In terms of the effects of enemy release on exotic bird establishment, we would expect an enhancement of success only in the case of regulatory release.

Second, exotic bird populations may lose enemies from their native range but gain enemies that are already present in their exotic range (native and previously established non-natives). In many cases, this could result in no net loss of enemies, or even an increase, as we noted above for European starlings introduced from Britain to North America, and as is also true for exotic mallard *Anas platyrhynchos* populations (Torchin, et al. 2003). Exotic mallard populations harbour thirty-one parasite species, twenty-one of which do not occur in the mallard's native range. The mean prevalence of the novel parasites is also slightly higher than that of the parasites introduced along with the host (12% vs 10%). If exotic species are encountering pathogens (and predators) for the first time, these enemies may have particularly deleterious effects. These effects could be exaggerated because the loss of genetic diversity that sometimes accompanies the founding of an exotic bird population may also result in increased susceptibility to commonly encountered enemies (see section 8.4.1).

Perhaps the question to ask is not whether exotic bird populations have fewer enemies in the introduced range as compared to the source populations, but rather do they suffer fewer impacts from the enemies they encounter in the introduced range? In the case of introduced exotic predatory mammals, the evidence suggests that the impacts are at least as strong as in the native range, and could be much worse owing to either increased densities of these predators or prey naivety. However, this suggestion is informed speculation. The field needs research on which enemies exotic bird populations encounter in their non-native ranges, and what effects these enemies might have.

5.5 Conclusions

The experiment in nature represented by the introduction of exotic bird populations has not been well designed: the introduction of multiple populations of multiple species with diverse characteristics to multiple locations using a range of different protocols means that variation in establishment success is likely to have been driven by a wide variety of uncontrolled factors. Yet, in dealing with the complexity of this natural experiment, the tendency in the past has been to attempt to assess the effects of individual processes. The result has been some heated debates about the importance or otherwise of particular drivers, especially competition, but little in the way of consensus about the mechanism by which establishment occurs.

It seems to us that the advances that have been made in the last decade or so in terms of understanding the establishment process have derived from approaches that have increasingly attempted to model the complexity of this experiment in nature. These studies recognize that factors acting at different levels of organization are important, that variation within and across these levels is non-random, and that this autocorrelation needs to be controlled to allow robust conclusions. One of the key developments has been the recognition that effects need to be partitioned into those that are deterministic and those that have a strong element of stochasticity, and that we are more likely to see effects of the former once we have accounted for the effects of the latter. In particular, we need to remove the uncertainty in establishment success that accompanies low propagule pressures to reveal a role for other factors (Cassey, et al. 2004c; Lockwood, et al. 2005; Colautti, et al. 2006). Happily, fewer studies now consider establishment success without attempting to incorporate metrics of propagule pressure.

Going beyond the effect of propagule pressure has unveiled the importance of the ability of introduced species to cope with the novel environment. Climate matching and behavioural flexibility are probably the most consistently supported location- and species-level determinants, respectively, of establishment success. These two characteristics are probably related, as the novel environment is more likely to fall within the environmental tolerances of a generalist species. They may also be related through the tendency for urban-dwelling bird species to be generalist, more likely to be introduced into a novel location, and more likely to find that location to be another urban habitat.

If reasonable numbers of individuals have been introduced to an amenable environment, biotic interactions then come into play. Evidence for their importance overall is more equivocal, perhaps because much of it relates to older studies, which, lacking the necessary controls on stochastic factors and autocorrelation, are easier to criticize. Nevertheless, significant influences of competition, predation, and positive density dependence (Allee effects) on establishment success have been revealed by recent analyses. Further progress seems to us likely to come from direct measures of these effects rather than relying on surrogate indices, sometimes many times removed. Approaches that explicitly model interactions between propagule pressure and other, deterministic factors (e.g., Duncan and Forsyth 2006), to identify and test situations in which those other factors should and should not be important, will be equally profitable. There is certainly much scope for such models, given the range of deterministic factors for which, as we have shown over the last two chapters, a role in establishment success receives some support in the literature on avian invasions.

6 Geographic Range Expansion of Exotic Birds

> The task of tracing the spread of the starling proves to be a difficult one indeed. Many accidental movements have occurred through unusual winter wanderings and by wind transportation as well as by unknown causes.
>
> L. Wing (1943)

6.1 Introduction

Once it has been ascertained which set of introduced populations (and species) have successfully established as exotics, it becomes pertinent to address what behaviours they exhibit in their novel geographic range. One of the most obvious behaviours is to increase the size of the exotic distribution, through combinations of population growth and dispersal. In fact, birds possess some of the greatest dispersal abilities in the terrestrial environment and individuals of even relatively sedentary species can disperse over tens or hundreds of kilometres. Other species undertake deliberate and predictable annual migrations that cover thousands of kilometres in a matter of days. The dispersal and establishment powers of birds are such that even the most remote oceanic islands (e.g., those in the subantarctic region and the Hawaiian chain) have native breeding bird species.

Given that their natural powers of dispersal and establishment are so well developed, we might expect that birds, once established as exotic populations, would easily pass through the final stage on the invasion pathway and become invasive and widespread in their recipient environment. Yet, in the majority of cases, exotic distributions tend to be of relatively small magnitude (Figure 6.1), and indeed the median exotic range size of the species in Figure 6.1 is <8% of the median for native bird geographic ranges (Orme, et al. 2006). Thus, despite being able to maintain a self-sustaining population at the recipient location, in most cases exotic bird species have not (yet) spread far from their point of release. This observation is consistent with Williamson's (1996) suggestion that the 'tens rule' applies to all stages of invasion, such that only approximately 10% (in the range 5% to 20%)

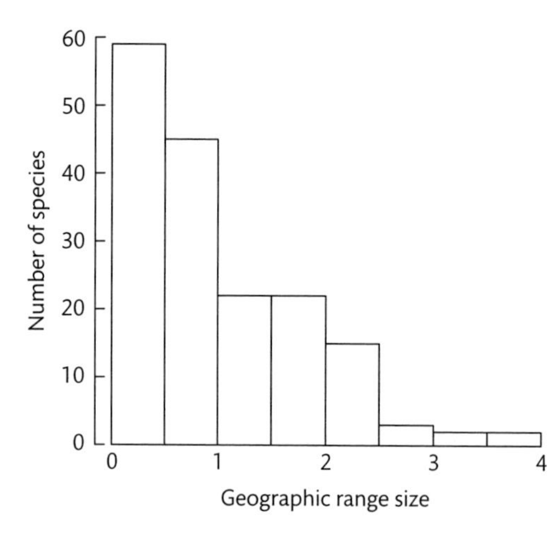

Fig. 6.1 The frequency distribution of geographic range sizes ($\log_{10}$ number of 96.486 × 96.486 km grid cells occupied) of the successfully established exotic bird species listed in Long (1981). These range sizes are likely in many cases to have increased in the thirty years or so since Long's publication. The data were compiled by M. Parnell using methods described in Orme, et al. (2006).

of established species proceed to spread from their location of establishment. An important question then is why some species have spread out from the original introduction bridgehead to invade the exotic location, but most have not. Indeed, more generally, why is it that species differ in the extent of their distribution at any given location?

In this chapter, we summarize how studies of range expansion in exotic bird populations have informed questions regarding changes in invasive species distributions. We begin by reviewing what evidence there is for patterns in the rate of range expansion of exotic bird populations, and for determinants of variation in range extent. The two basic requirements for the range expansion of an already established exotic species are population growth and dispersal. Therefore, mechanistic understanding of how spread occurs requires information on population dynamics, and especially on how far and frequently individuals move away from their location of birth (or introduction). The key approach to understanding spread in exotic species has been spatio-temporal modelling (Lockwood, et al. 2007). Hence, we go on to consider the application of models of range expansion to exotic bird populations. We conclude by suggesting how future studies of exotic birds could be useful in helping to understand the spread of invasive species.

6.2 Patterns of Spread

Geographic range expansion (hereafter, 'spread') captured substantial attention in the early ecological literature regarding exotic species invasions (e.g., Elton 1958; Hengeveld 1989). Yet, relative to the establishment stage on the invasion pathway, comparative studies on patterns of spread among exotic birds are relatively rare, as indeed are case studies of spread for individual bird species (Duncan, et al. 2003; Sol, et al. 2005a). One obvious reason for this is that the species and populations that fail at each step on the pathway cause a reduction in the sample size available for analysis at the subsequent stage. By the time the spread stage is reached, few species are left on which to perform robust statistical analyses regarding the causes or correlates of variation. This effect is compounded by the fact that many exotic bird populations do not spread far from their original site of release after establishment: there is little obvious benefit to be gained from studying a phenomenon that is not seen to be occurring!

However, time is starting to mend this gap in the invasion literature. Continuing introductions, and the eventual spread of an ever-greater number of exotic bird species, is prompting an increasing number of studies of the rate and extent of spread. The availability of global databases of species' natural distributions (e.g., Orme, et al. 2005), of the environment over which those distributions are laid (e.g., University of Delaware Global Climate Resource Pages: <http://climate.geog.udel.edu/~climate/html_pages/download.html>, and tools for the analysis of such data in spatial and phylogenetic contexts (e.g., Rangel, et al. 2006; Dormann, et al. 2007; Diniz-Filho and Bini 2008), will eventually allow the biotic and abiotic contexts within which spread occurs to be quantified, and their influence on population distributions to be assessed. In the meantime, in this section we summarize what can be learnt about patterns in the spread of exotic bird species from the information that is currently available.

6.2.1 Rate of Spread

Although it is not absolutely necessary, it is highly likely that for spread to occur, a population must be growing. Indeed the observation that population growth is likely to result in spread follows from one of the most general relationships in ecology: as the size of an animal population increases so does its geographic extent, such that abundant populations (or species) tend to be widespread while rare populations (or species) have restricted distributions (Gaston 1996; 2003; Gaston, et al. 2000; Blackburn, et al. 2006).

Among the exotic bird species that are regarded as invasive, population growth is often close to exponential. A good example is provided by the monk parakeet *Myiopsitta monachus* (Figure 6.2), which is the most widely distributed and abundant exotic parrot species in North America (van Bael and Pruett-Jones 1996; Pruett-Jones, et al. 2007). Following an attempted control programme in the early 1970s, both the abundance of monk parakeets and their geographic range increased exponentially. The average annual rate of population growth in North America up to 1995 was 14.6%, yielding a population doubling time of 4.8 years (van Bael and Pruett-Jones 1996). Similarly, Domènech, et al. (2003) estimated an exponential growth of 8% per year for monk parakeets in the Spanish city of

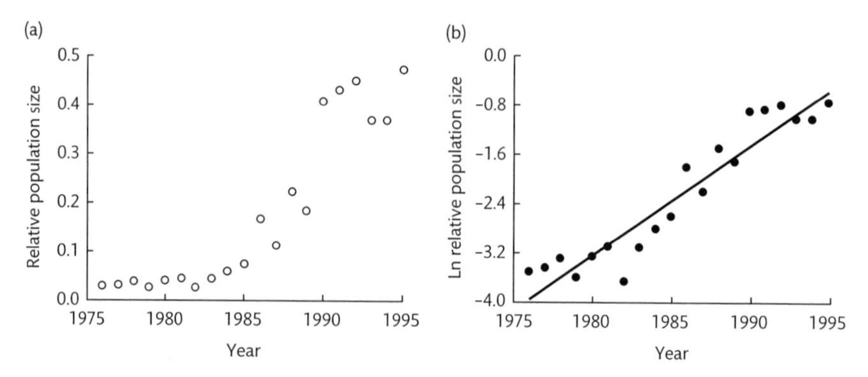

Fig. 6.2 (a) The relative population size (number of birds counted per party hour) of monk parakeets *Myiopsitta monachus* on Christmas Bird Counts in the contiguous USA each year between 1976 and 1995. Reprinted with permission of the Wilson Ornithological Society from van Bael and Pruett-Jones (1996). (b) The natural logarithm of the count data is approximately linear across years (r^2 = 0.91), which is suggestive of exponential population growth.

Barcelona based on an initial release date of 1974, and a population doubling time of approximately nine years. Lim, et al. (2003) determined the population sizes and habitat-abundance relationships of the three most successful invasive bird species in Singapore, including the house crow *Corvus splendens*. The house crow population grew more than thirtyfold in the sixteen years preceding the study, despite culling of the crows by the government since 1973: the number destroyed was several hundred a year in the early control period but increased to nearly 25,000 in the year 2000 (Lim, et al. 2003). Often the reported rates of population increase for exotic birds are high relative to the range for naturally increasing populations (e.g., Sæther and Engen 2002). However, it seems likely that such studies will be biased towards exotic species that are expanding reasonably rapidly, as these are the species most likely to reward such studies.

The exact mechanism by which population size and range extent are linked remains unclear (Gaston and Blackburn 2000; Gaston 2003), but a common feature of these relationships is that increases in population size tend to exceed increases in geographic range size. Species with larger ranges thus have higher population densities as well as population sizes, so that a doubling of a species' population size generally leads to an increase in a species' range extent of less than a factor of two. Nevertheless, growing populations will tend also to be spreading populations. One of the first studies to map the physical spread of exotic bird populations was Wing (1943), for the European starling *Sturnus vulgaris* and house sparrow *Passer domesticus* successfully introduced to continental North America (Figure 6.3). Subsequently, a growing body of studies has documented the rate at which the range sizes of exotic bird populations have increased (summarized in

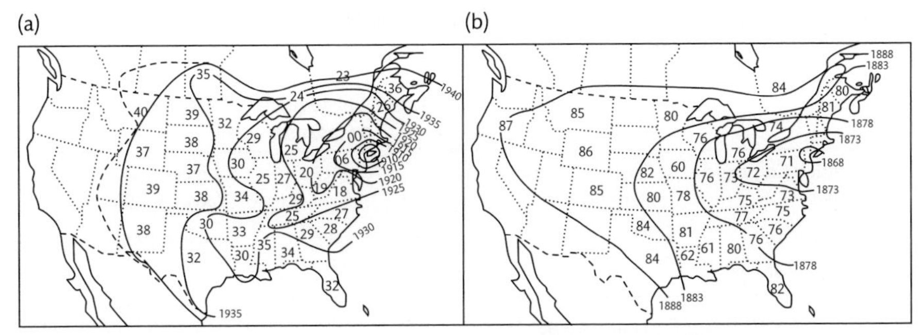

Fig. 6.3 (a) Isopleths of European starling *Sturnus vulgaris* spread as indicated by its appearance in the Bird-Lore Christmas censuses. The broken line indicates its known advance (in 1943) not at that time yet revealed by the censuses. (b) Isopleths of house sparrow *Passer domesticus* spread. Reprinted with permission of the American Ornithologists Union from Wing (1943).

Table 6.1), although many more studies have estimated changes in population size than have calculated rates of spread.

The simplest metric of the rate of spread can be calculated (following Shigesada and Kawasaki 1997) by converting change in the extent of the area occupied by a species into a linear measure that quantifies how the radius of the range area would have changed were it a perfect circle. Formally, this measures the annual change in

$$\frac{\sqrt{A}}{\sqrt{\pi}} \qquad \textbf{(Equation 6.1)}$$

where A is the area occupied (e.g., in km^2). For a constant rate of range expansion v, the range area at time t, $A(t) = \pi(vt)^2$. Solving for v gives

$$v = \frac{1}{t}\frac{\sqrt{A(t)}}{\sqrt{\pi}} \qquad \textbf{(Equation 6.2)}$$

Thus, if $\sqrt{A}$ is plotted against t, the slope of the line equals v (e.g., Figure 6.4).

It seems likely that the list in Table 6.1 will be biased towards species that are expanding their ranges reasonably rapidly, but the initial rates of spread of some of these species none the less appear relatively low. For example, the radius of the exotic range of the common waxbill *Estrilda astrild* in Portugal initially increased by only around 400 metres per year in the first eight years over which spread was measured (Figure 6.5). Nevertheless, the species with the fastest rate of spread (Table 6.1) in the early phase of invasion, the European starling in North America, increased the radius of its range by 11.2 km yr^{-1} over the first twenty-five years or so following its introduction (Shigesada and Kawasaki 1997).

Table 6.1 Examples of rates of spread of some exotic bird species showing, where estimated, the differences between the initial and subsequent rates of spread.

Species	Exotic location	Initial rate km/yr	Subsequent rate km/yr	Reference
House finch *Carpodacus mexicanus*	Eastern North America	3.5	20.7	Shigesada and Kawasaki 1997
Egyptian goose *Alopochen aegyptiacus*	The Hague Netherlands	1.16	4.59	Lensink 1998
	Drenthe Netherlands	2.83		Lensink 1998
Common waxbill *Estrilda astrild*	Portugal	0.42	7.43	Silva, et al. 2002
European starling *Sturnus vulgaris*	North America	11.2	51.2	Shigesada and Kawasaki 1997
	North America		91.2	van den Bosch, et al. 1992
Kalij pheasant *Lophura leucomelanos*	Hawaii		8	Lewin and Lewin 1984
Red-whiskered bulbul *Pycnonotus jocosus*	La Réunion	6.2	14.7	Clergeau and Mandon-Dalger 2001
	Oahu	1	5	Clergeau and Mandon-Dalger 2001
	Mauritius		30	Clergeau and Mandon-Dalger 2001

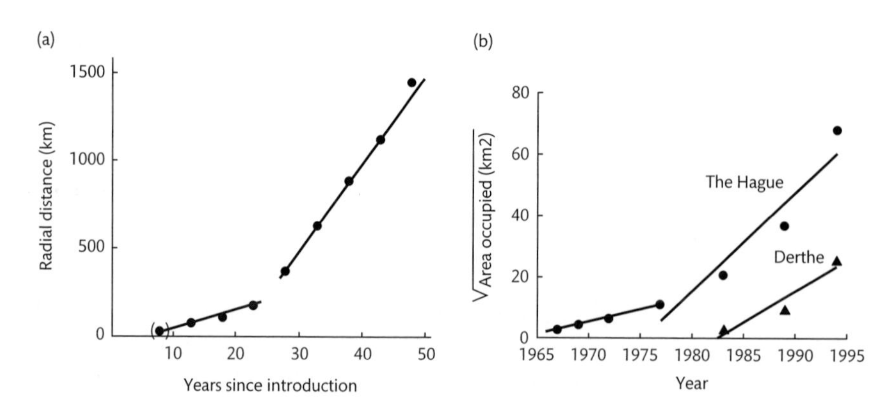

Fig. 6.4 (a) The radial distance of the European starling's *Sturnus vulgaris* North American range versus time since introduction. Reprinted with permission of Oxford University Press from Shigesada and Kawasaki (1997). (b) The square root of the area of the range of the Egyptian goose *Alopochen aegyptiacus* at two localities in the Netherlands. Reprinted with permission of John Wiley & Sons Ltd from Lensink (1998).

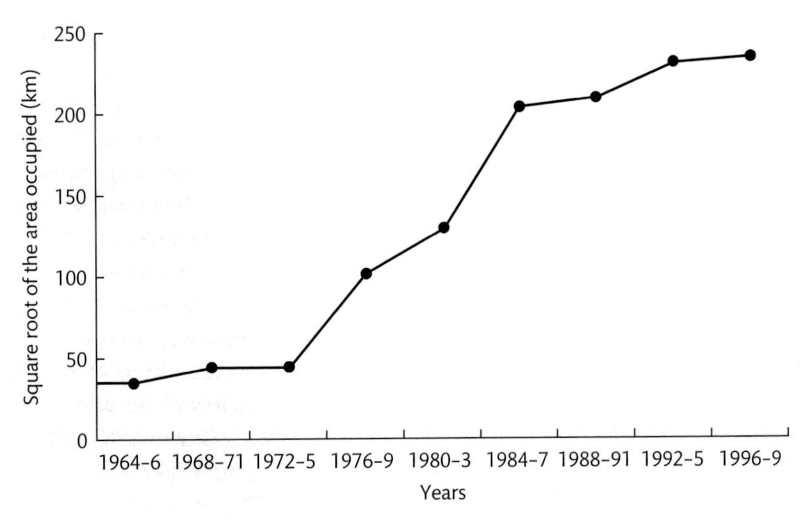

Fig. 6.5 Temporal changes in the square root of the range area occupied by the exotic Portuguese population of the common waxbill *Estrilda astrild*. Reprinted with permission of John Wiley & Sons Ltd from Silva, et al. (2002).

In some instances it is difficult to distinguish when a species has first spread to a new location from when it has increased in numbers to the point that we can actually detect it. This confusion may cause rates of spread to be underestimated. Wing (1943) noted that in later years the increase in distribution of the starling in North America was likely to be confounded by the increased mobility of census-takers using automobiles and good roads. Woolnough, et al. (2006) found that when they radio tagged European starlings in south-western Australia, the geographical range area occupied was more than double (225 km^2) the area previously known to be used by the species (103 km^2). More generally, Eraud, et al. (2007) found that when they mapped the distribution of an invading bird species, the lowest detection probabilities occurred in the regions with the lowest occupancy. Thus, as the occupancy of the species increased the more likely it was to be detected.

Despite sampling problems, many exotic bird species exhibit a distinct and apparently real step-up in the rate of spread as their numbers increase. Quantitative plots of spread frequently show a bimodal pattern in rate over time (Figure 6.4). For example, sixty-seven Kalij pheasant *Lophura leucomelanos* were introduced in 1962 at Puu Waawaa Ranch on the island of Hawaii. For the first five years following their release they were confined to a small breeding population in an exotic silk oak *Grevillea robusta* plantation. Their subsequent dispersal was so rapid that within a decade they were widespread and abundant enough to be declared a legal game bird (Lewin and Lewin 1984). Their linear rate of dispersal after the initial five-year lag was estimated at 8km yr^{-1}, and by 1984 they occupied an area of approximately 3,500 km^2. At a larger scale, the radial increase in the North American range of the European starling also increased markedly during the later phase of its establishment, although estimates of the rate of spread vary somewhat between authors. Shigesada and Kawasaki (1997) calculated a radial increase of 51.2 km yr^{-1} after about the thirtieth year post-introduction. Van den Bosch, et al. (1992) calculated that up to 1943 (*c.*50 years after introduction) the species had travelled 91.6 km yr^{-1}, based on maps of spread in Wing (1943). In 1978, the European starling was reported to be established in Fairbanks, Alaska (Kessel 1979)—a straight-line distance of 5,250 kilometres from the original release point in New York. The first starling was recorded at Fairbanks on 4 May 1960, which produces an estimated rate of spread of 87.5 km yr^{-1}. Regardless of the exact figure, the later rate of spread of the European starling has been precipitous, and within a century of establishment in New York it was virtually ubiquitous across North America. If exotic species with slow observed rates of spread are possibly in an early phase of invasion it may be prudent for conservation organizations to establish a good early warning and rapid response system rather than taking the more relaxed 'do nothing' view of currently range-restricted exotic species (see also Simberloff and Gibbons 2004).

At present there are insufficient data on rates of spread in exotic bird populations to speculate about the likely causes of observed variation in spread rates. In particular, it is not obvious how rates of spread are related to taxonomy or location in the examples so far published, while parameters of exotic bird spread can vary greatly among different populations of the same species. For example, Clergeau and Mandon-Dalger (2001) compared the initial rates of spread of exotic populations of the red-whiskered bulbul *Pycnonotus jocosus* on Reunion with rates reported from Mauritius, Oahu, Florida, and New South Wales. Results were presented as radial expansion and were based only on the first few years following introduction (Figure 6.6). Data from continental introductions indicated low spread rates of ~ 3 km in ten years, compared with much faster rates on the islands (between 5 km on Oahu and 30 km on Mauritius after ten years). Clergeau and Mandon-Dalger (2001) suggested that differences in the rates at which red-whiskered bulbuls spread at different locations could possibly be due to specific features of the recipient environment, such as habitat quality or the importance of biotic interactions (e.g., competition and predation). Wing (1943) suggested that birds tend to 'wander' more in areas of marginal habitat than in optimum areas, which he used to explain the faster spread of the European starling across the less suitable North American prairie habitat.

Geographic spread may in theory be slowed or even stopped altogether when there exists a threshold population density below which a species cannot persist (Keitt, et al. 2001; Tobin, et al. 2007). Therefore, Allee effects may prevent a species

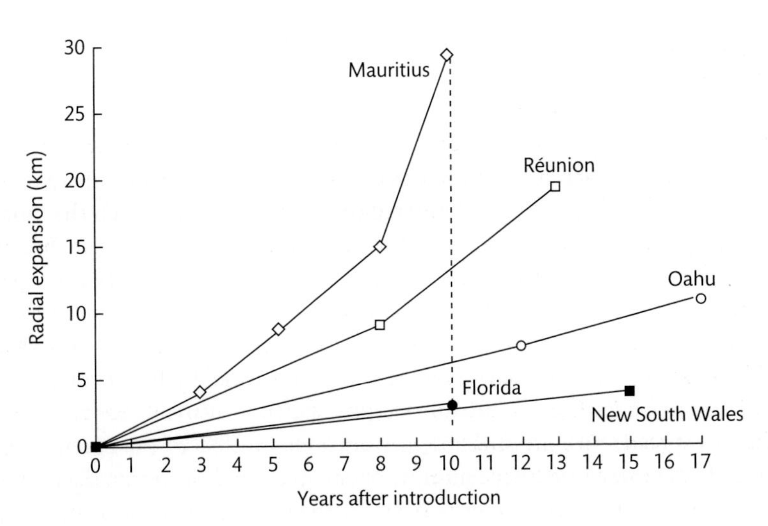

Fig. 6.6 Square roots of the range areas occupied by exotic populations of red-whiskered bulbuls *Pycnonotus jocosus* during the first few years after introduction at five different sites around the world. Reprinted with permission of John Wiley & Sons Ltd from Clergeau and Mandon-Dalger (2001).

from invading landscapes where suitable habitat is patchily distributed. This happens because the species cannot maintain positive growth rates in areas beyond a break in the distribution of suitable habitat, even if it can successfully disperse across the break. Keitt, et al. (2001) termed this 'invasion pinning'. Such effects mean that, in the absence of experimental evidence, one should be cautious in concluding that the failure of a species to spread is due to interspecific interactions or a decline in habitat quality at the immobile edge of an exotic species' range. Unfortunately, we currently lack robust quantitative studies that have explicitly tested the varying roles of different mechanisms in limiting the rate of spread.

6.2.2 Extent of Spread

No species can spread indefinitely. The finite nature of spread can be observed in decreases in spread rate observed in some exotic bird species over time. For example, Figure 6.5 shows how the rate of spread changed over the course of the expansion of the exotic Portuguese population of the common waxbill. This species shows the classic pattern of initial slow population growth rate in the early years of spread, followed by an increase in spread rate (cf. Figures 6.4, 6.6). However, the rate of spread subsequently declines towards the end of the census period. This pattern mimics the curve for logistic population growth to carrying capacity, and indeed Silva, et al. (2002) note that the slowdown in spread coincides with the occupation of the majority of available habitat. Presumably, other exotic populations with bimodal patterns of spread will show similar declines in spread rate as all available sites become occupied.

While all exotic populations must stop spreading at some point, there is none the less substantial variation in the size of the exotic range that results. Few species extend across the entirety of any region they occupy, and even the 'ubiquitous' European starling was absent from >10% of North American Breeding Bird Survey (BBS) routes in 2007 (Sauer, et al. 2008). In fact, most exotic species fall well short of ubiquity, even when the regions they occupy are not nearly so extensive as North America. For example, six of the ten most widespread land bird species on mainland New Zealand were introduced by European colonists over a century ago (Cassey 2001b). Despite the time available for these exotic species to spread, they are still absent from 11% to 28% of the country (Bull, et al. 1985), as measured by the number of occupied grid squares in the first New Zealand bird atlas scheme (Figure 6.7). Of the other exotic bird species established in New Zealand, seventeen occupy fewer than 10% of grid squares, and nine occupy fewer than 1%.

Given the considerable variation in exotic range sizes, and in the proportion of a region occupied, an obvious question is whether it is possible to identify factors associated with that variation? To date, relatively few studies have considered this question in birds (and studies in other taxa are no more common). As with previous stages in the invasion pathway, we can identify three broad categories of factors that may determine the extent of exotic ranges: features of the individual release event, of the species introduced, or of the location of introduction.

For a given locality, the different lengths of time since establishment can span thousands of years. Species established for a longer time have had longer

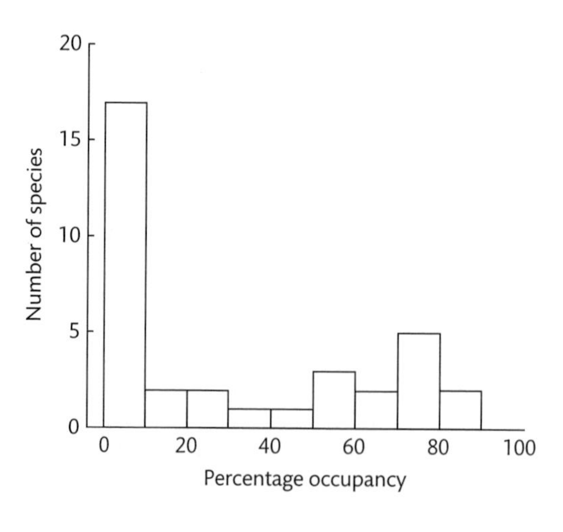

Fig. 6.7 Frequency distribution of the extent to which exotic bird species occupy all possible grid cells in the first New Zealand bird atlas scheme. This scheme plotted the distributions of bird species across 3,675 possible grid cells. From data in Bull, et al. (1985).

to adapt to (if necessary) and spread across the recipient environment, and so might be expected to have larger exotic distributions, all else being equal (Wilson, et al. 2007). At present, however, aside from the obvious fact that range sizes will tend to be small for all species in the period immediately following introduction, evidence for an event-level effect of time on exotic bird range size is mixed. Duncan, et al. (1999) found no significant relationship between date of introduction and exotic range size in New Zealand for the twenty-six established exotic bird species for which the year of first introduction was known. None of these species was introduced within the last hundred years, meaning that all have had ample time to spread from the site of initial establishment. However, there *is* a negative relationship between year of introduction and range size for species first introduced after 1850 (twenty-four out of the twenty-six species: Figure 6.8).

A negative relationship between year of introduction and range size may arise because later introductions in this period involved smaller propagules (Spearman's $r = -0.43$, $N = 24$, $P = 0.036$). Duncan, et al. (1999) found a strong effect of propagule pressure on exotic bird range sizes: species with larger exotic ranges in New Zealand tended to have been introduced more often and in larger numbers (Figure 6.9). Propagule pressure has an intuitive and strong positive effect on establishment success (Chapter 3), but its relationship to extent of exotic range size is unexpected. Duncan and colleagues speculated that species with large founding populations may have initially been able to capture a greater proportion of any shared resources from species with smaller founding populations. If this initial advantage compounded itself, those larger propagules could have led to faster population growth and spread rates, and further resource pre-emption at newly colonized sites as ranges expanded. Duncan, et al. (1999) predicted that this effect would be most pronounced among closely related species that would compete for

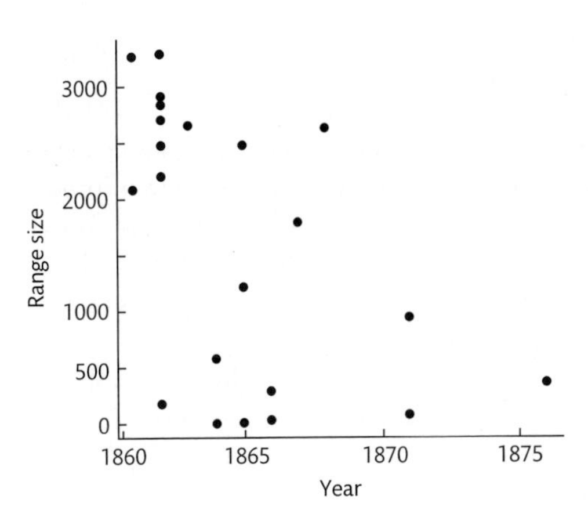

Fig. 6.8 The relationship between the first date on which a species was introduced to New Zealand (Year) and the size of the species' New Zealand range (Range size; $\log_{10}$ number of occupied cells on the New Zealand grid) for twenty- four exotic bird species introduced after 1850 (Spearman rank correlation $= -0.58$, $N = 24$, $P = 0.003$). For clarity, the point relating to little owl *Athene noctua* (introduced in 1906, range size = 152) has been omitted from the figure. Data supplied by R. P. Duncan.

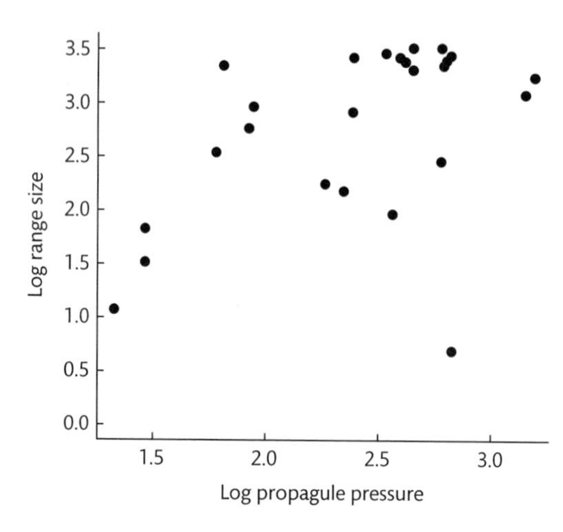

Fig. 6.9 The relationship between propagule pressure (log_{10} number of individuals introduced to New Zealand) and the size of the species range (log range size; log_{10} number of occupied cells on the New Zealand grid) for twenty-six exotic bird species introduced to New Zealand. The outlier is the grey partridge *Perdix perdix*, which has failed to spread despite relatively high propagule pressure. Data supplied by R. P. Duncan.

similar resources. Interestingly, within-taxon correlations between geographical range size and propagule pressure were indeed stronger than across-species correlations. This led the authors to infer that competitive interactions can play a role in limiting range sizes, and that the outcome of those interactions depends on the initial introduction effort. This result suggests that the same competitive priority effect that Duncan and Forsyth (2006) argued influences establishment success in New Zealand exotic birds (section 5.4.1.2) may also affect the extent of spread in a species that establishes despite competition.

Nevertheless, alternative explanations for an effect of propagule pressure on exotic range size can also be posited. Larger propagules may encompass greater genetic variation for selection to act upon and thus the population may be better able to adapt to a wide variety of local conditions and realize a broad geographic range. Larger propagules may also tend to be associated with widespread, abundant native species that have broader habitat or dietary tolerances, although the effect of propagule pressure on exotic range size in New Zealand was independent of native range size for species introduced from the United Kingdom. Whether the effect of propagule pressure on exotic range size is specific to New Zealand would benefit from further investigation, given that there is no effect of propagule pressure on the range sizes of exotic birds in Australia (Duncan, et al. 2001). A widespread relationship between propagule pressure and extent of exotic distribution would have significant ramifications in predicting which species remain localized, and which become widespread invaders.

Given that the two basic requirements for the spread of an already established exotic species are population growth and dispersal, we might expect species-level traits relating to these two requirements to be important in determining the ability to spread, and hence distributional extent. There is indeed some evidence for this. In New Zealand, exotic bird species with life-history traits associated with high population growth rates (small body size, rapid development, high fecundity) had larger geographical range sizes (Cassey 2001b), as did species that are partial migrants in their native range (Duncan, et al. 1999). A multivariate model for the ranges of exotic birds in New Zealand included positive effects of migratory tendency, number of broods per season, and propagule pressure, suggesting that metrics of population growth and dispersal are each important predictors of spread (Duncan, et al. 1999).

In Australia, migratory behaviour and traits associated with high population growth rates were also positively related to exotic bird geographic range sizes (Duncan, et al. 2001). However, an effect of migratory behaviour was only observed for phylogenetically controlled univariate analysis, and was not retained in a multivariate model of exotic range size. The only life-history trait in this multivariate model was body size: smaller-bodied exotic birds had larger range sizes in Australia. The effect of body size is interesting because it runs counter to the general trend observed in native species distributions for large-bodied species to have larger geographic ranges (Gaston 2003). It also runs counter to the trend noted in Chapter 4 for higher establishment success in larger-bodied bird species. It is nevertheless difficult to draw general conclusions about the effects of species-level traits on exotic geographic range sizes from just two studies.

Regardless of the relative importance of population growth and dispersal, spread is not possible into areas where environmental tolerances do not permit the exotic species to persist. There is evidence that climate, environmental matching, and biotic interactions are important determinants of exotic bird range size. Kawakami and Yamaguchi (2004) observed that the expansion of the melodious laughing thrush *Garrulax canorus* across four prefectures in Japan was most likely limited by elevational conditions and snowfall. Similarly, the strongest predictor of Australian range size in exotic birds in Duncan, et al.'s (2001) analysis was the amount of continental area with a close climatic match to a species' native range, mirroring the positive effect of a close climate match on establishment success (section 5.2). Species for which the Australian climate is more like that of their native range have greater exotic distributions there (see also Hayes and Barry 2008).

Further evidence for an effect of location-level characteristics on spread can be garnered from geographical variation in the extent of exotic ranges. MacArthur (1972) suggested that the southern range limits of many northern hemisphere species are determined by biotic interactions, with the corollary that many northern limits are determined by abiotic conditions. Sax (2001) showed that the highest latitudes in the native and exotic ranges tend to be correlated, but that species often exceed their natural high-latitude range limits in their exotic range. In contrast, bird species only rarely exist at lower latitudes in their exotic range than they do in their native range. Sax argued that the latter result was indicative of the influence of biotic resistance on southern range limits, as MacArthur (1972) postulated. We explore these ideas in more detail in the next chapter.

6.2.3 'Boom and Bust'

While most exotic populations remain relatively restricted in their distribution (Figure 6.1), and some smaller proportion go on to spread widely in the recipient environment, there is a subset of the latter for which their exotic range expansion is a temporary state. These exotic populations expand after establishment, but then dramatically decline in either numbers or extent, in some cases even to extinction. Williamson (1996) suggested that such 'boom and bust' dynamics might be a relatively common feature of biological invasions in general. Simberloff and Gibbons (2004) endeavoured to collate examples of boom and bust for a quantitative analysis of common features of population crashes among exotic species. Interestingly, of the seventeen crashes for which the authors presented detailed information, nearly one-third (five) were for exotic bird species (Table 6.2).

One of the most dramatic examples of boom and bust dynamics in exotic birds relates to the budgerigar *Melopsittacus undulatus* (Pranty 2001). An estimated 240,000 individuals of this species were imported to the USA between 1925 and 1940, mainly entering via California and Florida. A massive cottage industry of aviculturists rearing budgerigars developed in Florida, where a combination of accidental escapes and deliberate releases led to a large exotic population developing. At its peak in the late 1970s, this population may have numbered over 20,000 individuals, and spanned 160 km along the central Gulf coast of Florida from Pasco to Sarasota counties. However, a decline in the population was evident by the early 1980s, and by the 1990s they had almost completely vanished from Florida (Figure 6.10). Pranty (2001) estimated that the remaining population numbered only around 200 individuals, representing a 99% decline from its peak just twenty years earlier.

It is evidently not unusual for population crashes (or simple declines) in exotic birds to occur after long residence times in the recipient location. For example, exotic populations of Java sparrow *Padda oryzivora,* red avadavat *Amandava amandava*, and Cape canary *Serinus canicollis* on Mauritius all declined to extinction after a century or more on the island (Simberloff 1992; Cheke and Hume 2008). In the continental USA, the most widespread exotic bird species, the European starling, has been declining since 1980, according to trend data from the North American Breeding Bird Survey (Sauer, et al. 2008).

Cheke and Hume (2008) invoke competition with subsequently introduced exotic species as the likely driver of such declines in the Mascarenes—for example, suggesting that the extinction of the Cape canary on Mauritius followed establishment of the house sparrow there. Another possible example of competition between two exotic species concerns mynas introduced to Samoa. The jungle

Table 6.2 Examples of bird species showing 'boom and bust' population dynamics. From data in Pranty (2001), Simberloff and Gibbons (2004), and Cheke and Hume (2008).

Species	Location	Introduction date	Length of tenure	Peak abundance	Current status
Red-billed leiothrix *Leiothrix lutea*	Kauai	1918	*c.*50 years		Extinct
Cape canary *Serinus canicollis*	Mauritius	*c.*1820s	>100 years	'widespread and common'	Extinct
Java sparrow *Padda oryzivora*	Mauritius	1890s	133 years		Extinct
	La Réunion	Late eighteenth century	>50 years		Extinct
Grey-headed lovebird *Agapornis canus*	Mauritius	*c.*1735	*c.*200 years	Unknown, but common enough to become 'a cereal pest'	Extinct
Budgerigar *Melopsittacus undulatus*	Florida, USA	1950s	*c.*50 years	>20,000	99% decline
Spot-breasted oriole *Icterus pectoralis*	Florida, USA	Late 1970s/ early 1980s	*c.*50 years	'Established but apparently declining'	<100 pairs
Canary-winged parakeet *Brotogeris versicolurus*	Florida, USA	Late 1970s/ early 1980s	*c.*35 years	'Several hundred pairs'	<100 pairs
	California, USA	Early 1970s	*c.*45 years		
Crested myna *Acridotheres cristatellus*	Vancouver, Canada	1987	106 years	c. 20,000 birds	Extinct

myna *Acridotheres fuscus* was introduced to the capital city Apia in the 1960s, and is well distributed around human-inhabited sites on both main islands (Figure 6.11a). Records of the second species, the common myna *A. tristis*, date from 1988, again in Apia. This more recent arrival has, in less than twenty years, spread to form a continuous population west and east of the capital city, as well as establishing additional outlying populations (Figure 6.11b). In 1992, Gill, et al. (1993) found that 15% of mynas in Samoa were the more recently arrived common myna.

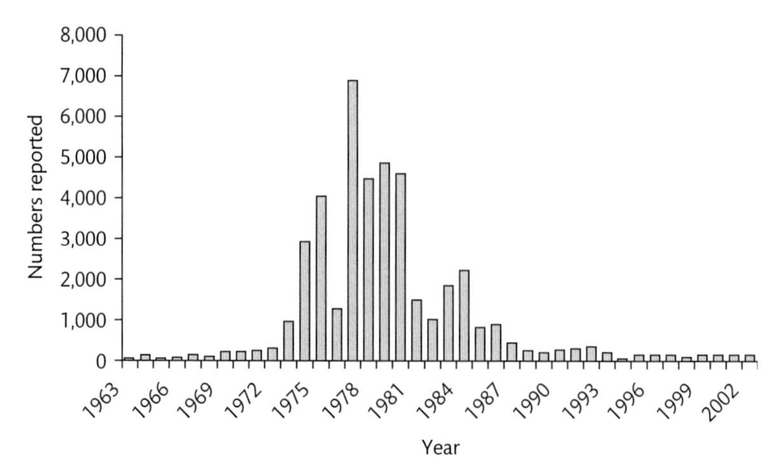

Fig. 6.10 The numbers of budgerigars *Melopsittacus undulatus* recorded on Christmas Bird Counts in Florida between 1963 and 2002. From Butler (2005).

By 1998, the common myna outnumbered the jungle myna in Apia by as much as three to one (Gill 1999). By November 2004, McAllan and Hobcroft (2005) were unable to locate any jungle mynas in the urban centre. Nevertheless, all examples to date of competition as a driver of boom and bust cycles in exotics are anecdotal, and Simberloff and Gibbons (2004) could identify no strong evidence as to cause for any of the exotic bird populations that have declined to extinction, or nearly so (Table 6.2). Clearly, however, understanding the causes of these declines (and future ones) will be of considerable theoretical and practical interest.

6.3 Models of Spread

There is a variety of different ways in which population growth and dispersal can proceed, and we would expect the extent and rate of spread to depend on exactly how they do. Over the course of any invasion, population growth and dispersal are difficult to observe and measure directly. Thus, insight into the mechanism of spread has been most frequently derived from theoretical models of the process (Shigesada and Kawasaki 1997). These models produce functions that represent different scenarios of population growth and dispersal from which the resulting patterns of spread can be directly compared to observations from the field.

A wide range of modelling approaches has been applied to the study of range spread, incorporating ever more realistic (and hence complicated) descriptions of population growth, dispersal, and the spatial and temporal contexts within which these processes occur. The full range of complications that we might expect models for spread to incorporate is beyond the scope of this chapter, and recently

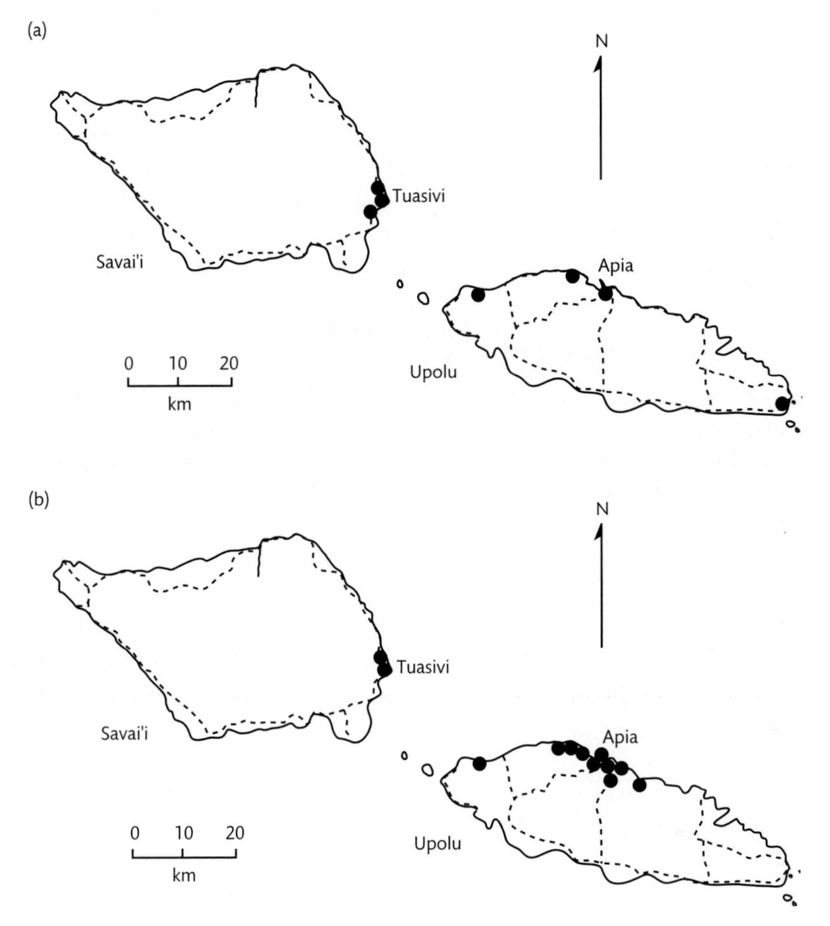

Fig. 6.11 Range distribution maps of (a) jungle myna *Acridotheres fuscus* and (b) common myna *A.tristis* in Samoa in 2004. The former was introduced in the 1960s, while the latter was first recorded in 1988. Reprinted with permission of the Ornithological Society of New Zealand from McAllan and Hobroft (2005).

entire books have been devoted to the topic (e.g., Shigesada and Kawasaki 1997; Petrovskii and Li 2005). Rather, our aim here is to summarize the development of models of spread, focusing on those examples where data on exotic bird populations have particularly informed these models.

6.3.1 Classical Models

Skellam (1951) was the first formally to investigate the example of a spreading organism. He used a diffusion process, originally derived by Fisher (1937) in a population genetic context, to model the spread of individuals via partial differential equations. Skellam showed that if the invading population is growing exponentially, the landscape is uniform, and the diffusion of individuals across the landscape follows a random walk, then the invasion front advances at a constant rate of spread. If spread proceeded in this manner, points in a plot of radial distance versus time (e.g., Figure 6.4) would fall on a straight line. Although the model introduced by Skellam assumes exponential growth, similar predictions are obtained if logistic growth is assumed. This congruence is because diffusing populations grow by expansion at the range edge, where they will not have reached carrying capacity, and so probably are growing more or less exponentially (Shigesada and Kawasaki 1997). Using Skellam's models, the constant rate of spread can be calculated using information about net reproduction, generation time, and average distance an organism disperses away from its previous location. Skellam's model is a useful start, as many invading populations (including birds) do tend to follow a pattern of constant range expansion (Shigesada and Kawasaki 1997).

The most commonly used models for invasive spread since the pioneering work of Skellam have been diffusion-reaction equations (DRE) and integrodifference equations (IDE). Other less well-known classes of models include exactly solvable models (Petrovskii and Li 2005), discrete time–space models (Hastings 1996), and hierarchical Bayesian spatio-temporal models (Hooten and Wikle 2008). Diffusion-reaction equation models, in general, are based on partial differential equations of the form

$$\frac{\partial N}{\partial t} = rN + D\left[\frac{\partial^2 N}{\partial x^2} + \frac{\partial^2 N}{\partial y^2}\right] \qquad \textbf{(Equation 6.3)}$$

where N is the population density at time t at a particular point (x,y) on a two-dimensional landscape, r is the intrinsic per capita population growth rate, and D is the diffusion coefficient (the rate of random movement across the landscape). The first part of the right hand side of this equation, rN, represents population growth rates. The second half represents dispersal from the point of origin, which for invasive species is the location of their initial release. Sometimes this section of the equation is called the 'dispersal kernel', the exact form of which is key to our understanding of geographic spread rates (see below). From this model it can be calculated that the rate of spread, v, with which a newly established population expands over space is given by

$$v = 2\sqrt{rD} \qquad \textbf{(Equation 6.4)}$$

The DRE model (Eqn 6.3) assumes that individuals reproduce continuously and that they disperse via simple diffusion. Implicit in the diffusion approximation is the assumption that the distance that an individual disperses in a fixed length of time is normally distributed. Van den Bosch, et al. (1992) showed that greater variation in dispersal distances increases the proportion of the population that moves further and thus increases DRE model estimates of the rate of spread at the invasion front. In addition, Kot, et al. (1996) observed that the rate of spread can be extremely sensitive to the precise shape of the dispersal kernel (or probability density function of dispersal distances), and in particular to the tail of the distribution. If the shape of the tail of the distribution declines as quickly as an exponential function, the spread will be a 'travelling wave'—as described by Skellam (1951). However, if the tail declines more slowly than an exponential function, the resulting wave will instead be a 'dispersive wave'. Importantly, the rate of spread of a dispersive wave continues to increase whereas the rate of a travelling wave approaches an asymptotic maximum (Kot, et al. 1996).

Simple DRE models may underestimate the rate of spread of an expanding exotic population in comparison with models that incorporate more realistic distributions of dispersal distance. Moreover, it is difficult to distinguish between different types of spread when the intrinsic rate of increase of a population is low (Frantzen and van den Bosch 2000), as is likely to be the case in most bird species (e.g., van den Bosch, et al. 1992). Empirical dispersal data for a wide range of exotic bird species typically show leptokurtic or 'fat-tailed' distributions, in which rare long-distance dispersal events occur more often than expected from the assumption of normality (Kot, et al. 1996). Clark (1998) provides an excellent discussion of how normal dispersal kernels, leptokurtic dispersal, and fat-tailed (or extremely leptokurtic) dispersal kernels can change estimates of spatial spread.

The leptokurtic dispersal kernels typically seen in data for bird dispersal mean that long-distance dispersal events are relatively more frequent than would be expected under a normal probability density function. One approach to modelling spread has thus been to assume 'stratified diffusion'. Here, the exotic bird's geographic range initially expands through random diffusion at the wave front, as in Skellam's classic model, but occasionally the population produces long-distance dispersers that go on to found new colonies, which then also start to spread. When these outlier colonies meet the expanding front of the core range, they are absorbed into it. The absorption of outlier colonies results in an increase in the rate at which the core range expands, which is maintained as subsequent colonies are encountered and absorbed (Shigesada and Kawasaki 1997). When the rate at which long-distance dispersers is produced is proportional to the radius (or circumference) of the core range (rather than its area), this 'coalescing colony' model generates a biphasic pattern of range expansion over time. This is consistent with the observation that many exotic bird populations produce two distinct phases of expansion during their invasion (Figure 6.4).

Detailed dispersal data can be directly incorporated within integrodifference equation (IDE) models, which separate population growth and dispersal into different stages. The IDE model is composed of two parts: a difference equation that describes population growth at each point on the landscape (here a one-dimensional transect) and an integral operator that accounts for the dispersal of individuals in space (i.e., the dispersal kernel). Discrete time integrodifference equation models thus have the general form

$$N_{t+1}(x_1) = \int_{-\infty}^{\infty} k(x_1, x_2) f[N_t(x_2)] dx_2 \qquad \textbf{(Equation 6.5)}$$

where $N_{t+1}(x_1)$ is the population density at some destination point x_1, which is a function of the population growth at each source point $x_2(f[N_t(x_2)])$ and the movement of individuals from x_2 to x_1 according to the shape of the dispersal kernel, k. For the interested reader, Lewis (1997) gives an excellent example of an age-structured IDE model using data from the spread of house finches *Carpodacus mexicanus* throughout their exotic range in the eastern USA.

Van den Bosch, et al. (1992) considered the range expansion of five populations of bird species, three of which have expanded their ranges naturally (collared dove *Streptopelia decaocto* and house sparrow in Europe, and cattle egret *Bubulcus ibis* in North America), and two of which were introduced through human agency (European starling and house sparrow in North America). They used independent field data from published sources to estimate age-specific survivorship and fertility, and dispersal distances approximated from mark-recapture studies. Models that lack stage-structured dispersal will always overestimate the rate of spread because not all life stages are equally likely to disperse (e.g., Neubert and Caswell 2000). In birds, natal dispersal distances are commonly greater than breeding dispersal distances (Paradis, et al. 1998). Van den Bosch, et al. (1992) compared the rate of population expansion (km yr^{-1}) calculated using a simple Fisher–Skellam DRE model with an IDE model that includes demographic and dispersal characteristics of an individual (van den Bosch, et al. 1990). The demographic characteristics were the probability that an individual is alive at a given age, and the rate of offspring produced (number yr^{-1}) by that individual. The dispersal characteristic was the conditional probability that an individual born at a given time was living at a particular locality given that it was still alive. The authors found that the square root of the area occupied increased linearly with time for all the species considered. For each of the exotic populations, the reproduction-and-dispersal IDE model gave less-biased estimates of the observed yearly rate of population spread than the Fisher–Skellam DRE model.

Lensink (1997) concluded that the more a species displayed long-distance dispersal, the faster its range would expand. He used the model of van den Bosch, et al. (1990) to estimate the range expansion of native raptor populations in Britain and the Netherlands following reductions in the use of organochlorine compounds from around 1970. Using published data on reproduction, survival, and dispersal of breeding birds, Lensink (1997) found that these three parameters were sufficient to obtain a reliable fit between the observed range expansion velocities and those predicted by the model. Interestingly, Neubert and Caswell (2000) found for plant species invasions that when dispersal contains long- and short-distance components, it is the long-distance component that governs the rate of spread in the expanding population, even when long-distance dispersal events are rare.

The shape of the dispersal kernel has often been assumed to be more important than demographic parameters in influencing the rate of spread of an invading population (van den Bosch, et al. 1992). Caswell, et al. (2003) compared the relative importance of stage-specific vital rates (survival probability and mean reproductive output) and population growth rates, versus dispersal distance. They compared the spread of the exotic European starling in North America, with naturally colonizing populations of pied flycatcher *Ficedula hypoleuca* in the Netherlands and the European sparrowhawk *Accipiter nisus* in both England and the Netherlands. Although dispersal was more important in determining rate of spread for the starling and pied flycatcher, vital rates and population growth were critical to all three. In fact for the sparrowhawk in England and the Netherlands, population growth and vital rates explained 86% of the observed spread rate (Caswell, et al. 2003). However, the authors also found that the 10% of individuals that moved the furthest had the largest single effect on the spread rate of all three species, regardless of the relative importance of either population growth or dispersal.

Diffusion models are obviously an oversimplification of bird movement, and one of their most unlikely assumptions is that spread proceeds at an infinite rate along infinitely random paths. In contrast, a model of telegraph dispersal assumes species move at finite velocities whose changes in direction from one step to the next are correlated (Okubo 1980). Holmes (1993) estimated the spatial spread of six populations, including the exotic European starling in North America ($rD = 552.2\ \text{km}^2/\text{yr}^2$), comparing a model of classical diffusion with a model of telegraph dispersal (*sensu* Goldstein 1951). Holmes (1993) substituted a logistic growth function into the general reaction-telegraph equation, and showed

$$\frac{\partial^2 N}{\partial^2 t}+2\delta\frac{\partial N}{\partial t}=\gamma^2\frac{\partial^2 N}{\partial x^2}+\frac{\partial}{\partial t}+\left[rN(1-N/K)\right]+2\delta rN(1-N/K)$$

(Equation 6.6)

where K is the 'carrying capacity' or equilibrium population size, γ is the population's finite velocity and δ is the rate at which individuals in the population change direction. Note, that the larger δ is, the less inertia the population has (Holmes 1993). Without a term for reproduction, the telegraph and diffusion models gave almost identical patterns of spread (~1% difference in the predicted rates of range expansion for the European starling), after an initial transition (lag) period. With a reproduction term, both models showed organisms moving into the new environment as an advancing travelling wave. However, the diffusion model predicted that the rate of range expansion would increase without bound as the reproductive rate of the exotic bird population increased, while the telegraph model predicted that, as the reproductive rate increased, the spread wave would increase to a maximum rate and would not increase beyond that upper limit. Holmes (1993) concluded that while significant errors can in theory arise when using the simple diffusion model to describe range expansion rates, these errors are small when the species' reproductive rate is small relative to the rate at which individuals change direction.

6.3.2 Time Delays

One potential shortcoming of the model analysed by Holmes (1993) is that she did not account for the time lag in dispersal and reproduction between generations. Typically in bird populations, there is a delay between reproduction in one generation and dispersal and reproduction by the offspring that result. Time delays are widely incorporated in population dynamic models, such as the classic discrete time versions of the exponential and logistic growth models. Time delays are also an integral part of IDE models, where reproduction and dispersal are separated, and models that lack time delays will always overestimate the rate of spread.

Méndez and Camacho (1997) and Méndez, et al. (1999) showed that the asymptotic rate of spread for the reaction-telegraph model is given by

$$v = \frac{2\sqrt{rD}}{1 + \frac{T}{2}r} \qquad \textbf{(Equation 6.7)}$$

where T is a measure of the time elapsed between successive dispersal events (e.g., times between generations) and conditions $v^2 < 2D/T$ and $rT/2 < 1$ must be satisfied (Méndez, et al. 1999). In the limit $T \to 0$, Equation 6.7 becomes Fisher's result—namely, Equation 6.4.

Ortega-Cejas, et al. (2004) extended the work of Holmes (1993) to include time delays in a telegraph dispersal model. In their model the diffusion steps can have different distances, and the delay time is due to the assumption of a characteristic time interval T during which young birds do not disperse. Ortega-Cejas, et al. (2004) used this approach to model spread over eastern North America by the exotic house finch population. In order to estimate the intrinsic growth rate r, they used the observed initial exponential stage of population growth from Breeding Bird Survey data, calculated as the slope of the linear regression of total population size versus time ($r = 0.020 \pm 0.003$ yr^{-1}). They estimated the parameter T as the mean interval elapsed between two successive dispersal events, which they calculated as the time (age) needed for a young bird to grow into a dispersing (reproducing) individual. The typical value for T was estimated to be 1.75 years (i.e., between 1.5 and 2 years old; from Hochachka and Dhondt 2000). The diffusion coefficient D was calculated as

$$D \equiv \frac{\overline{\Delta^2}}{4T} \qquad \text{(Equation 6.8)}$$

where $\overline{\Delta^2}$ is the mean-squared displacement per dispersal event. Ortega-Cejas and colleagues estimated $\overline{\Delta^2}$ from the frequency of dispersal distances presented by Veit and Lewis (1996) as

$$\overline{\Delta^2} = \sum_i \Delta_i^2 f_i \qquad \text{(Equation 6.9)}$$

where Δ_i are the observed dispersal distances and f_i are their respective observed frequencies ($\Sigma f_i = 1$). This resulted in a diffusion coefficient of $D = 10.1$ ($\pm$ 2.4) $\times$ 10^3 km^2 yr^{-1}.

The time-delayed model with these parameters yielded a predicted rate of spread for the house finch invasion front of $v = 28 \pm 4$ km^2 yr^{-1} (Ortega-Cejas, et al. 2004). The observed rate of spread had been previously reported as 28 $\pm$ 1 km^2 yr^{-1} (Okubo 1988), and thus coincides extremely well with the predicted value. Although the value predicted by the time-delayed model differs by only about 2% from that predicted by the classical DRE model (Eqn 6.3), Ortega-Cejas, et al. (2004) claimed that the DRE model performs much more poorly for populations in which the product of the delay time T and the initial growth rate r is much higher. As an example, they provided the estimate of r for the spread of the exotic collared dove population in North America (0.29 $\pm$ 0.02 yr^{-1}). Here, the time-delayed model gave a range for the rate of spread of the dispersing front of $v = 60 \pm 19$ km^2 yr^{-1}, versus $v = 76 \pm 19$ km^2 yr^{-1} for the classical DRE model. Only the confidence intervals of the former included the observed $v = 44 \pm$ 3 km^2 yr^{-1} (from van den Bosch, et al. 1992).

6.3.3 The Effect of Population Density on Spread Rates

In most models of the spread of an exotic bird population, population growth is assumed to be density independent, showing exponential (or geometric) rates of increase. However, the growth of populations, and in consequence their spread, may also be limited at low densities because of Allee effects. A common example of an Allee effect is that it is harder for individuals of a sexually reproducing species to find mates at low density (see section 3.1.1). The inclusion of an Allee effect may decrease the overall rate of spread. If the effect is strong enough, it may also introduce a critical density threshold that the population must overcome in order to spread (Lewis and Kareiva 1993; cf. Chapter 3).

In populations for which Allee effects are important, rates of spread can accelerate suddenly when the critical density threshold is overcome. Allee effects can slow down or reverse travelling wave solutions (rate of spread) of reaction-diffusion equations (Lewis and Kareiva 1993; Lewis and van den Driessche 1993), and can turn accelerating invasions into constant rate invasions in integrodifference equations (Kot, et al. 1996). Wang and colleagues provide both general solutions for DRE and IDE equations with an Allee effect (Wang and Kot 2001; Wang, et al. 2002) and numerical results suggesting that even weak Allee effects may, on occasion, have the same effect. In general, however, the stronger the Allee effect, the slower the rate of spread of the invasion. This observation suggests that the initial size of the propagule could influence the relative rate of spread of exotic bird species. This connection provides another potential explanation for Duncan, et al.'s (1999) finding (section 6.2.2) that the current geographic ranges of exotic bird species in New Zealand are positively related to the number of individuals released (Figure 6.9).

Keitt, et al. (2001) showed theoretically that given fine-scale habitat heterogeneity, the geographic range of a population for which Allee effects are strong may be both stable against contraction and prevented from expansion. This effect may provide an explanation for the observation that a 'lag time' between initial establishment and the onset of rapid population growth and range expansion is a common feature of invasions (Sakai, et al. 2001).

Allee effects also have the potential to explain the distinct bimodal pattern in rate of spread often observed for exotic bird species. This possibility was demonstrated by Veit and Lewis (1996) in their model for the spread of house finches in the eastern USA. The observed spread of house finches was strongly correlated with the rate of population growth near the centre of their range. Veit and Lewis's (1996) model includes distinct (separate) phases for the growth and spread of the population, and includes density-dependent dispersal and lowered reproduction at low densities due to an Allee effect. The model produces bimodal spread, where the initial low rate is due to the inclusion of the Allee effect at the front of the invasion wave. Numerical simulations for a variety of parameter values revealed that even a very small proportion of dispersing birds not finding mates dramatically slows the population spread, especially early on in the invasion, and that density dependence for the juvenile and adult dispersal fractions causes an increase in the length of time taken before the invasion achieves a constant rate of spread.

6.3.4 Variable Environments

All of the above models that have examined the spread of invasive bird species are based on the assumption that the factors affecting the population dynamics (population growth, life-history rates, dispersal) are homogenous in space. In particular, these models do not consider population growth and spread during the invasion process as a function of the type of landscape (With 2002). Yet, heterogeneous landscapes have the potential to affect the progress of spread across them, and it is currently unknown how different spatial distributions of habitat and resources will affect rate of spread. Muller-Landau, et al. (2003) proposed that long-distance dispersal events will be most advantageous in landscapes in which large areas of suitable habitat are consistently available at long distances from established populations. Van Kirk and Lewis (1997) used IDE models to study the theoretical persistence of dispersing populations in fragmented habitats of finite and patchy spatial domains. They showed that large dispersal distances are detrimental to population persistence when suitable habitats are either highly heterogeneous or occur as single isolated patches. However, if a species is capable of long-distance dispersal across unsuitable habitat, the amount of suitable habitat available may be more important than the spatial arrangement of that habitat on the landscape (With 2004).

Van den Bosch, et al. (1994) adjusted their earlier model of spread (van den Bosch, et al. 1990) by adding new parameters for the amount of suitable habitat available, the rate of settlement, and the fraction of long- and short-distance dispersal events. Van den Bosch and colleagues also modelled the proportion of the year in which breeding occurs, because reproduction can be seasonal. The rate of settlement accounts for the fact that the number of fledglings need not be the same as the number of reproducing birds in the next generation: some of the birds do not settle and breed, although they can stay alive, thus forming a 'shadow' population. A restricted number of available habitats as well as a given survival rate of the settlers reduces the dispersal rate. Depending on the reproductive rate, the fraction of short-distance and long-distance dispersal may be important in determining the rate of invasion. Unfortunately, comparisons to assess the fit of this model have not been attempted, given its substantially greater number of parameters.

Wikle (2003) used a hierarchical Bayesian model to revisit Veit and Lewis's (1996) analysis of data from the North American Breeding Bird Survey on the spread of the exotic population of house finch. His approach was stochastic, and key to his formulation was the specification that traditional DRE diffusion rates could vary through both space and time. Wikle (2003) showed that the diffusion rate for the house finch was indeed heterogeneous and provided estimates of the uncertainty associated with the different spatio-temporal rates. The variable that accounted for the change in mean population size with time decreased in the early phase of spread, consistent with Veit and Lewis's (1996) supposition that an Allee effect is important at this stage. However, Wikle (2003) also showed that there was sufficient uncertainty in the population estimates, given the BBS data, to question this previous interpretation.

Hooten, et al. (2007) and Hooten and Wikle (2008) extended the framework of Wikle (2003) to analyse the ongoing mid-invasion spread of Eurasian collared doves in North America. They found that collared dove populations are growing as well as dispersing, and that both dispersal and carrying capacity are heterogeneous and vary significantly through space and time. In particular, they showed that assuming homogenous dispersal can lead to both inflated and deflated population size estimates in space and time. Some locations reveal an initial decrease in the estimated rate of population growth (Hooten and Wikle 2008), and these decreases could be attributable to an Allee effect slowing down the parameters of invasion. The authors used estimated posterior distributions for population growth and dispersal parameters to predict the future spread of the collared dove beyond the current data range (1986–2003). They concluded that by the year 2016 the collared dove population will be at or near carrying capacity for the majority of the USA (Hooten, et al. 2007). This prediction is consistent with an empirical observation made by Romagosa and Labisky (2000) that, given the strong dispersal capabilities of this species, rapid colonization of North America was highly probable.

6.4 Conclusions

Andow, et al. (1990) proposed four basic questions for researchers trying to address the rate of geographic spread in an invading population: (i) Does range expansion occur primarily as the sum of many short steps, or does it reflect a few great leaps? (ii) Do expanding populations achieve an asymptotic (i.e., constant) rate of spread; and if so, how rapidly? (iii) Can the asymptotic rate of spread of a population, which is observable only over a large geographic area, be related to locally measured demographic and behavioural parameters? (iv) How sensitive is spread to variation in habitat? These questions remain particularly pertinent to the study of spread in exotic birds, which has been greatly hindered by the lack of quantitative studies of dispersal during the invasion process. Here, we suggest how future studies could be further directed to help understand the spread of exotic birds, and of invasive species in general.

Foremost, researchers studying the invasion of exotic bird populations need to collect more empirical data. These data should cover a variety of spatial scales in order to offer a better understanding of the co-varying processes that promote range expansion in complex habitats (Stohlgren, et al. 2006). The way that range size is measured can determine the rate of range size increase (e.g., linearly or sigmoidally) and this is particularly important given that most studies to date have relied on very coarse clinal maps of distribution (Hastings, et al. 2005). Given such coarse data, most studies find or assume that either the area of invasion is increasing linearly or the radial expansion rate is constant. However, if spread is more complex, the range expansion of an invasive species cannot simply be described in terms of a single rate. Although it is possible to model invasions in relatively simple terms using parameter-sparse models, this approach does not offer insight into the individual processes influencing population expansion. Currently we lack tests of the relative influences of either environmental- (spatial heterogeneity of resources and interactions) or population-level (variance in reproductive success and dispersal distances) processes that are predicted to affect the fine-scale invasion process of exotic birds.

For British breeding birds, dispersal distances among related species are lower for more abundant species and for species with larger geographic ranges (Paradis, et al. 1998). It would be interesting to know whether these are the same species that disperse less in their exotic range. Or, do dispersal capabilities differ during various stages of establishment and spread? More likely, dispersal capabilities are related to processes of population regulation and are density dependent. For example, Duckworth (2008) showed that dispersive phenotypes were maintained in a western bluebird *Sialia mexicana* population as an adaptation for the continual re-colonization of new habitat. Her results suggest that pre-existing dispersal strategies may explain rapid changes in dispersal phenotypes during invasions. In addition, results from non-avian vertebrate taxa suggest that dispersal kernels can be evolutionarily dynamic and respond rapidly to geographic spread (Phillips, et al. 2008). How dispersal capabilities vary among exotic bird species, and their relative importance in different recipient environments, is almost completely unknown.

Care needs to be taken before inferring mechanisms underlying exotic spread from patterns of change in spread rate. Starrfelt and Kokko (2008) use simulation modelling to show that a negative relationship between change in per capita rate of spread and per capita rate of spread itself is observed in models with no population regulation, and so is not necessarily indicative of a slowdown in population spread due to a regulating factor. The apparent slowdown in the rate of spread in systems without regulation is in part a sampling effect caused by the fact that spread is assessed by the occupancy of different sampling units (e.g., grid squares, counties), and hence that a population can continue to grow and spread in an unregulated manner without necessarily occupying new units. The association between patterns of spread under different models of population growth and regulation, and how this relates to the form of the near ubiquitous abundance-occupancy relationship amongst birds (Freckleton, et al. 2005; Freckleton, et al. 2006b; Blackburn, et al. 2006), would greatly benefit from further study.

One of the main sources of confusion in measuring the spread of invasive species across landscapes comes from the differences in lag times between when an exotic population becomes established and when it begins rapidly to expand its range (Crooks 2005). An explicit understanding of the mechanisms that produce these lag times is critical for developing our ability to compare differences in rates of spread. In particular, it is vital that we are able to test empirical data against ever-more sophisticated alternative model explanations for initially low invasion rates, followed by acceleration to higher steady state invasion speeds (e.g., Filin, et al. 2008).

We have reviewed the anecdotal evidence that community dynamics can influence population expansion. On Hawaii and Samoa the differences in spread of congeneric bulbuls and mynas have both been linked to the possible role of competition (Williams and Giddings 1984; McAllan and Hobcroft 2005), as has the extinction of some exotic bird species on Mauritius (Cheke and Hume 2008). Evidence for the role of disease in causing population declines among a spreading exotic species is best provided by Hochachka and Dhondt (2000), who demonstrated a causal relationship between high prevalence of a novel strain of a widespread poultry pathogen *Mycoplasma gallisepticum* and declines in house finch abundance through their exotic range in the eastern half of North America (see section 7.5.4). Presumably predators could also equally affect rates of spread. More data would greatly help elicit the relationships between spread and the effects of community dynamics.

It has been suggested that independent estimates of dispersal are required for mechanistic models of invasion, which preferably should not use data from the wave itself in order to prevent circular reasoning (Hengeveld and van den Bosch 1991). However, if parameters from the native range of the species are very different from those in the exotic range this can cause additional errors in estimating and predicting the spread of the population. For example, the expected rate of spread for the house sparrow in North America derived from European data was half the observed rate (Hengeveld 1994). The expected rate of spread was greatly improved by using American data on the house sparrow's reproduction within the initial phase of population expansion. During establishment, exotic populations in North America (and New Zealand) reproduced continually, the young birds before fledging brooding the eggs of the next clutch. Hengeveld (1994) observed that populations in Europe (and more recently in North America) do not exhibit this behaviour, which thus may further support the idea that density-dependent processes influence behaviours relevant to rates of spread.

Hastings, et al. (2005) noted that, in a large number of cases, empirically measured rates of dispersal combined with a model (e.g., Eqn 6.3) and a solution (e.g., Eqn 6.4) do not accurately predict the rates of range expansion in invasion. Although diffusion models do not consistently over- or underestimate the rates of spread for terrestrial species (Hengeveld 1994; Grosholz 1996), the discrepancies are usually explained by invoking the probability of rare long-distance dispersal events. In the absence of reliable quantitative estimates of long-distance dispersal, stochastic models that use extreme value distributions derived from fitted dispersal kernels to determine the rate of spread of the furthest-forward individual in an expanding population may provide a solution (Clark, et al. 2003). However, if long-distance dispersal events can have the single greatest effect on differences in the rate of spread of an invasion (Caswell, et al. 2003), then the explicit study of the frequency and distribution of these low probability events in different environments and across a range of exotic birds is likely to be vital in advancing the study of avian invasion events.

In non-avian taxa, it has been suggested that the mechanisms of long-distance dispersal may qualitatively differ from those of ordinary dispersal, requiring different formulation of the costs (Higgins, et al. 2003). For example, it has been proposed that decreased survivorship of dispersing individuals may be a reason why population density generally decreases near the boundary of a species' geographic range (Brown 1984). In addition, there is evidence that dispersal kernels can be evolutionarily dynamic, such that rates of spread can increase partly through increased dispersal on the expanding range front (Phillips, et al. 2008). Further empirical research is needed to identify the traits affecting long-distance dispersal in birds (native or exotic) and to investigate selective pressures upon them (Muller-Landau, et al. 2003).

Williamson and Fitter (1996) questioned whether population dynamics contribute more to the understanding of invasions than invasions contribute to the understanding of population dynamics. Their original data suggested that the number of established exotic species attaining high population density fitted the 'tens rule' of invasion biology. The question therefore remains whether exotic bird populations—in general—are attaining abundances and geographic ranges that are any different from an expected distribution of community abundance-range size relationships. In New Zealand, the strong positive correlation between range sizes of exotic birds and their native ranges in Britain (Duncan, et al. 1999) is certainly suggestive that the native community has simply reconstituted itself. More examples of this sort would be highly profitable.

In some cases, attempts to reduce the spread and abundance of invasive bird species may provide the best examples of the importance of characteristics related to population increase. Yap, et al. (2002) studied the factors affecting the roost characteristics of invasive white-vented *Acridotheres javanicus* and common mynas in urban Singapore. They found that while experimentally thinning tree canopy density and reducing refuse at food centres both reduced the abundance of roosting mynas, canopy density reduction had the greater effect. Availability of nest sites has been previously linked to the distribution, population increase, and spread of invasive bird species (Sol, et al. 1997). Alternatively, models of population growth rates offer a useful way of explicitly testing different management strategies for reducing population growth and spread. Ellis and Elphick (2007) used a stochastic stage-based population matrix model to test when control is most likely to be effective in reducing the exotic population of mute swans *Cygnus olor* in North America. They recommended that substantially to reduce a population, removal of individuals was much more effective than an approach which reduced reproductive rates without changing survival rates (Ellis and Elphick 2007). Pruett-Jones, et al. (2007) showed that it would be necessary to reduce the exotic monk parakeet population in North America by 20% of the adult population or to destroy 50% of the nests each year to effect reductions in the population size.

Finally, there are cases where the study of population expansion of exotic birds may be informative for detecting the differences in parameters that influence the decline of these same species in their native ranges. In Britain, house sparrow populations have markedly declined over the last thirty years (Robinson, et al. 2005). Robinson and colleagues concluded that declines in sparrow populations are in part related to increasing agricultural intensification, as is the case for many other British farmland birds. Substantial variability in population declines and increases also exist in the house sparrow's exotic range in North America. Thus, it would likely be rewarding to investigate whether variables related to declines in their native ranges are the same variables that structure changes in their population sizes and rates of spread in the exotic range.

7
The Ecology of Exotic Birds in Novel Locations

> It is, of course, known that it is impossible to maintain our bird fauna at anything like its original balance, whether new varieties are introduced or not, because of man's operation over the face of nature.
>
> J. C. Phillips (1928)

7.1 Introduction

So far, we have considered where exotic bird species come from, what influences whether or not they successfully establish in their novel recipient environment, and if establishment does indeed occur how and through what circumstances exotic species spread across novel environments. The ultimate outcome of these successive stages in the invasion pathway is that ecological assemblages will contain some set of exotic bird species within them. The obvious next step in terms of understanding exotic bird invasions is to ask about the novel ecological relationships these birds express.

Once an exotic bird species has established in a recipient location, it interacts with a suite of species and habitats, many or most of which will be different from those with which it evolved. Exotic birds comprise a conspicuous element of the global avian biota, especially in urban, agricultural, and other human-dominated landscapes (Luniak 2004). In a growing number of regions, exotic birds are now more abundant and more widely distributed than many or all of the native species (see also section 5.3). This circumstance is especially apparent within island assemblages. The exotic common myna *Acridotheres tristis*, for example, is reported to be the commonest breeding bird on Diego Garcia (Hutson 1975), while the two most widespread species in New Zealand (chaffinch *Fringilla coelebs* and blackbird *Turdus merula*) are both exotics (Cassey 2001b). Novel associations will develop between the exotic and native birds, and between different exotic species, while it is absolutely certain that the exotic birds will have some sort of impact on the recipient environment that would not have occurred in their absence.

Brown (1981) argued that a general equilibrium theory of diversity needed to embrace two broad sets of constructs, which he termed capacity rules and allocation rules. The former define the physical characteristics of environments that determine their capacity to support life. The latter define how that capacity is allocated amongst organisms, which determines assemblage structure and composition. These allocation rules may in turn be broken down into three sets of processes: entry rules, exit rules, and transformations (Kunin 1997; Gaston and Blackburn 2003; Gaston 2006), which determine, respectively, which species join an assemblage (through speciation or immigration), which species leave an assemblage (through extinction or emigration), and how species change when they are a member of the assemblage (Gaston 2006). The relative influence of these rules, and thus of breaking them, is likely to vary markedly in determining different aspects of the distribution of biodiversity (Gaston and Blackburn 2003). Clearly, however, exotic species are indicative of changes to the action of the rules, and indeed can change the rules themselves (Lockwood 2005).

In the first instance, entry rules are different for exotic species. For example, natural barriers to dispersal will exclude native populations from regions that they could perfectly well inhabit. Human intervention can effectively lift these barriers for any species that we choose to transport, anywhere. Exotic species also change exit rules, notably contributing to the elevated extinction rates currently being experienced in many habitats around the world (e.g., Blackburn, et al. 2004; Clavero and García-Berthou 2005). Finally, exotic species represent a new process influencing transformations. They may affect community composition by changing the abundance and distribution of native organisms, and may even drive evolutionary change in those natives (Vellend, et al. 2007). Exotic species thus provide an unprecedented opportunity to assess the effects of changing allocation rules (Lockwood 2005). This may provide information on how natural assemblages are structured, and the responses of exotic species may in turn provide information on the process of biological invasions (Sax and Gaines 2006).

The aim of this chapter is to review the information that has accumulated on the ecology of exotic bird species in their novel locations. We first consider the expression of classical patterns of species richness in exotic birds. We then assess how exotic birds have influenced patterns of spatial turnover in species richness, which in this context is studied under the name of biotic homogenization. The information that allows us to explore these patterns also allows consideration of how the extent of exotic distributions varies across space, and how exotic ranges relate to exotic bird abundances. We proceed to review the evidence for interactions between exotic and native species and conclude by assessing what this information tells us about the rules by which natural assemblages are structured, and about the invasion process itself.

7.2 Exotic Bird Species Richness

Perhaps the most general pattern in ecology is that the number of species inhabiting an area increases with the area's size (Arrhenius 1921; 1923; Gleason 1922; 1925): the 'species-area relationship' (SPAR). The precise form the relationship takes has been a matter of considerable debate (see reviews in Connor and McCoy 1979; Williamson 1988; Rosenzweig 1995; Gaston and Blackburn 2000), but it is generally considered to be a linear relationship on log-log axes for a broad range of area values. The slope, z, of the double-logarithmic relationship varies roughly in the range 0.1 to 1, and has been suggested to increase in magnitude with spatial grain size (Rosenzweig 1995; but see Williamson 1988). Plots that compare the faunas or floras inhabiting different areas of a mainland region typically exhibit values of $z \sim 0.1–0.2$, plots that compare the faunas or floras of different islands typically show values of $z \sim 0.25–0.35$, and those that compare the richness of different biogeographic regions have values of $z \sim 0.5–1$.

All published examples of SPARs for exotic birds relate to introductions to islands, all are positive, and all are approximately linear on log-log axes (Case 1996; Chown, et al. 1998; Sax, et al. 2002; McKinney 2006b; Blackburn, et al. 2008). Unfortunately, only three published studies report z-values for these relationships. Chown, et al. (1998) show $z = 0.107$ ($P > 0.05$) for exotic bird assemblages on twenty-five Southern Ocean islands, which is less than the z-value typical for islands in general, and less than that for native birds on the same islands ($z = 0.249$). Sax, et al. (2002) report $z = 0.17$ ($P < 0.05$) for exotic bird species on a sample of twenty-three islands/archipelagos around the world. Blackburn, et al. (2008) show z ($\pm$ standard error) $= 0.18 \pm 0.07$ for the exotic bird assemblages of thirty-five of the islands and archipelagos analysed by Cassey, et al. (2004b), as compared to $z = 0.25 \pm 0.03$ for the extant native species of these same islands. To these studies we can add the exotic bird species-area relationship for 186 of the oceanic islands studied by Blackburn, et al. (2004). For these data, $z = 0.23 \pm 0.02$ (Figure 7.1). Many of these 186 islands have no exotic bird species established on them, but when we exclude these islands, the form of the relationship changes only slightly ($z = 0.19 \pm 0.03$, $N = 120$). Although there is considerable variation between studies, overall these numbers suggest that SPARs for exotic bird species on islands conform reasonably well to expectations derived from calculations using native island bird assemblages (Rosenzweig 1995; see also Sax and Gaines 2006).

A range of plausible hypotheses exist for positive SPARs in natural assemblages (e.g., Rosenzweig 1995), of which, thanks to their different entry rules, only a subset can apply to exotic bird assemblages. However, before exotic bird SPARs can be considered as a probe into the causes of natural SPARs, it is first

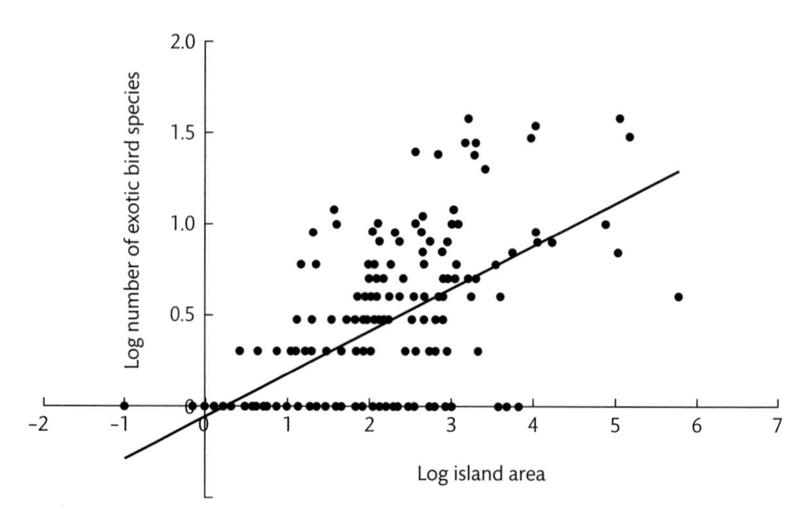

Fig. 7.1 Species–area relationship for exotic birds established on 186 islands around the world. Log_{10} (1+number of exotic species) = 0.23 x log island area – 0.05. $r^2 = 0.36$, $P < 0.001$. From data in Blackburn, et al. (2004).

necessary to consider a more prosaic explanation for their form: larger islands have more exotic bird species established on them because they have simply had more exotic bird species introduced. Blackburn, et al. (2008) found that $z = 0.20 \pm 0.07$ for the relationship between log area and log number of species introduced (i.e. both successful and failed) onto the thirty-five islands in their study. At first sight, the similar z-values for native and exotic SPARs (0.25 vs 0.18) for these islands might suggest that similar processes determine both. However, we cannot falsify the hypothesis that more exotic bird species are found on larger islands simply because more exotic bird species were introduced there. In other words, the SPAR for exotic birds may require no biological explanation whatsoever (see also Leprieur, et al. 2008; Chiron, et al. 2008).

Several studies have shown that islands and countries that have lost more native bird species to extinction have also tended to gain more exotic bird species through introduction (Case 1996; Sax, et al. 2002; Sax, et al. 2005a). Sax, et al. (2005a) used this information to assess how these changes fitted with the predictions of dynamic models of species richness. Island biogeography theory posits that the species richness of an island will be a dynamic balance between rates of colonization and extinction on the island (MacArthur and Wilson 1963; 1967). It predicts that richness will increase in response to increasing colonization rates, and that cutting off the supply of colonists will lead to decreases in richness. A species capacity model suggests that there is a maximum species richness of

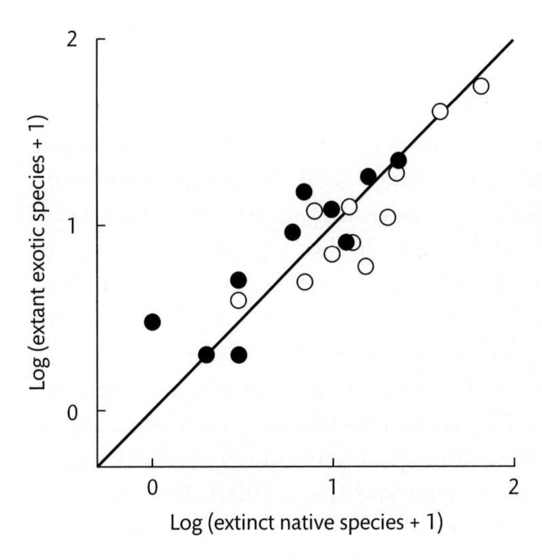

Fig. 7.2 The relationship between the number of extant exotic and number of extinct native bird species, for a sample of islands around the world. Open circles indicate islands that were inhabited before European colonization, while filled circles are islands that were not. Reprinted with permission from Sax, et al. (2002). © 2002 University of Chicago.

any given taxon that an area can support, determined by features of the environment (Sax, et al. 2005a). This model predicts that introductions will bring species richness up to this capacity, and that removing the supply of colonists will not reduce richness below this level. Stochastic niche theory views patterns of colonization and coexistence to be primarily a function of niche partitioning (Tilman 2004). It predicts that species can in principle be added indefinitely to a location, albeit at ever-decreasing rates as competition amongst previous colonists makes it harder for later species to establish. Removing the supply of colonists will not reduce richness below the resulting level. Sax, et al. (2005a) argue that exotic birds suggest a species capacity model. In support, they note that the number of established exotic bird species closely matches the number of extinctions across their twenty-three islands, with points clustering around the 1 : 1 line on a plot of numbers of extinctions versus numbers of successful introductions (Figure 7.2). Nevertheless, they present no evidence for the response of richness to cessation of exotic bird introductions.

However, if the exotic bird species richness of islands is indeed simply a consequence of patterns of introduction, this also has implications for the interpretation of the pattern found by Sax, et al. (2005a), and for other richness relationships observed in exotic species. For example, relationships between the numbers of bird species extinctions and established exotic species may simply be a consequence of larger areas having had more bird species to lose to extinction, thus having lost more in total (Blackburn and Gaston 2005; Karels, et al. 2008), and coincidentally having had more exotic species introduced. Similar arguments could explain strong positive correlations between numbers of native and exotic bird species in States of the USA (see section 5.4.1; Stohlgren, et al. 2006).

McKinney (2006b) assembled data on the species richness of exotic plants, birds, mammals, and herpetiles in a range of areas. These data show that the richness of exotic groups is inter-correlated, and that richness increases with area with the same rank order of model intercepts as compared to natives (and apparently with similar slopes, albeit that this was not tested). The data also show that area is confounded with native plant species richness and human population density (see also Sax, et al. 2002). McKinney (2006b) argued that these latter variables are both better predictors of exotic species richness than is area. Similarly, Blackburn, et al. (2008) showed that human population size, but not island area, entered the minimum adequate model for the number of exotic bird species on islands: human population size scaled in direct proportion to area in their data (estimate $\pm$ s.e. $= 1.02 \pm 0.16$, $N = 35$). We suspect that areas with larger human populations may on average experience more species-introduction events because they receive greater volumes of trade (Taylor and Irwin 2004; Westphal, et al. 2008) and have higher populations of captive animals. Both of these characteristics will elevate the opportunities for deliberate and accidental introductions. The same areas may then also incur more extinctions either because of the effects of larger human populations in areas with more native species to lose to extinction (Karels, et al. 2008) or because even random extinction will tend to remove more species from richer islands (Blackburn, et al. 2008).

Case (1996) argued that the relationship between the number of native bird extinctions and exotic bird richness is not a consequence of cause and effect but arises because environmental disturbance by human colonists affects both variables. The conditions under which such interactions could cause the kind of relationship shown in Figure 7.2 deserve further study, and would allow more rigorous assessment of the plausibility of this mechanism. In particular, it would be interesting to know why number of introduced bird species scales with area as $z = 0.20$, similar to the z-value expected for the native species richness of islands.

Island biogeography theory classically suggests that the richness of islands should be a function of their isolation as well as their area, because of isolation's effect on the probability of colonization from the mainland species pool (MacArthur and Wilson 1963; 1967; see also Kalmar and Currie 2006). It seems unlikely that introduction effects would produce exotic species-isolation relationships, because colonization by exotics is much less constrained by distance. In fact, Blackburn, et al. (2008) found that more isolated islands have had more exotic bird species introduced on to them than islands closer to mainland areas, and these same islands have more exotic bird species established (Figure 7.3). These relationships persist when controlling for island area. Thus, the exotic bird colonization-isolation relationship is exactly opposite to that pertaining naturally, because of the peculiarities of human transportation mechanisms (e.g., acclimatization societies). Nevertheless, this relationship concurs with the explanation for the natural pattern in terms of colonization

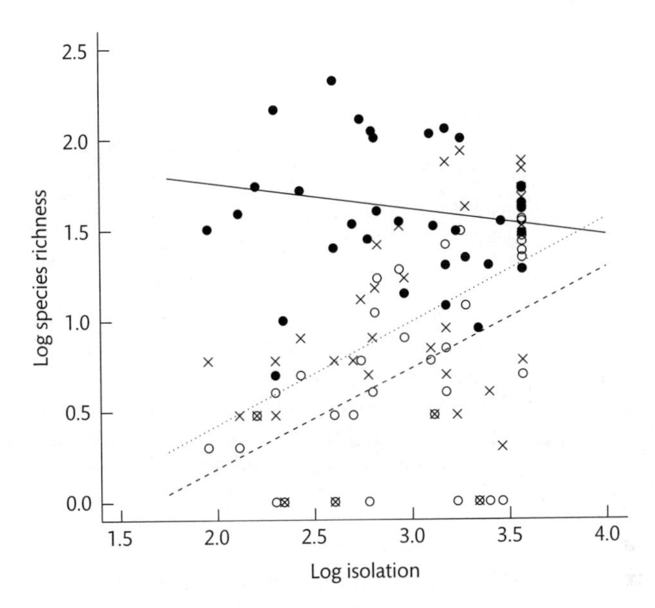

Fig. 7.3 The relationship between $\log_{10}$ (x + 1) bird species richness and $\log_{10}$ isolation (km from the nearest continental region) for native species (solid line, filled circles), established exotic species (dashed line, open circles), and introduced exotic species (dotted line, crosses) on thirty-five oceanic islands around the world. Reprinted with permission of John Wiley & Sons Ltd from Blackburn et al. (2008).

rates: it is islands with higher human-mediated colonization rates (more exotic bird species introduced) that end up with more established exotic bird species (see also Lockwood 2005). In contrast, Chown, et al. (1998) found that distance to the nearest continent was negatively related to the number of exotic land bird species on islands in the Southern Ocean, as expected based on island biogeography theory. However, they suggested that this classic relationship arose because many of these islands were secondarily colonized by species originally introduced to New Zealand, which then dispersed to the other islands under their own steam.

Along with area, latitude ranks as one of the most frequently cited correlates of species richness in assemblages of native species. The exact reason for this association remains to be resolved, but seems likely to be due to the effects of temperature or availability of energy on population dynamics (Hawkins, et al. 2003; Whittaker, et al. 2003; Davies, et al. 2007; see also Kalmar and Currie (2006) in the context of island biogeography). There is evidence that exotic bird species richness also varies with latitude. Sax (2001) showed that exotic bird species richness increases from the poles to the tropics in the northern (Figure 7.4) and southern hemispheres, but attains relatively low levels throughout the tropics. He argued that the environmental factors that determine native species

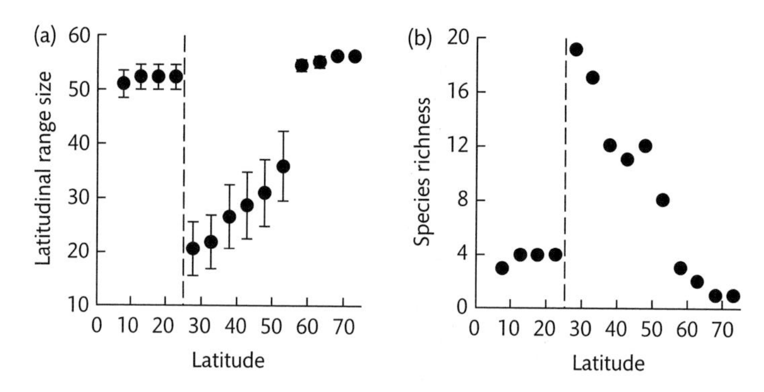

Fig. 7.4 The relationship between (a) mean latitudinal range size and latitude, and (b) species richness and latitude, for exotic bird species established in North America. Error bars indicate ± 1 standard error, and the dashed line indicates the Tropic of Cancer. Reprinted with permission of John Wiley & Sons Ltd from Sax (2001).

richness outside the tropics also determine the richness of exotic bird species, which must respond to these factors in the same way. The low exotic bird richness of the tropics, in contrast, may result from high biotic resistance there preventing the establishment of most bird species (Sax 2001; see also Sax and Gaines 2006).

Here again, however, it is possible that human placement effects have not been fully eliminated. Sax (2001) argues that the latitudinal pattern of introduction events for birds introduced in North America does not match the latitudinal pattern of established bird species richness. However, those latitudes where islands hold higher numbers of established exotic birds are also those latitudes that have seen most exotic bird species introduced (Figure 7.5). The frequency distribution of latitudes of introduction for all avian introduction events shows a similar pattern (Blackburn and Duncan 2001b). That said, at a more restricted scale, Evans, et al. (2005b) have shown that the richness of exotic bird species in Britain relates to energy availability, as indeed does the richness of native bird species. This pattern does not seem to be a consequence of an influence of human population density on exotic bird richness, even though people tend disproportionately to occupy high-energy areas of Britain (see also Balmford et al. 2001; Araújo 2003; Gaston and Evans 2004; Evans and Gaston 2005). We suggest that more detailed analyses of data for introduction and establishment across regions would be useful in elucidating whether, or the extent to which, latitudinal patterns in exotic species richness are more than a consequence of human whim.

7.3 Changes in Diversity Patterns across Space after Invasion

The number of species occupying a region (gamma diversity) is a function of the mean number of species occupying local sites within the region (alpha diversity)

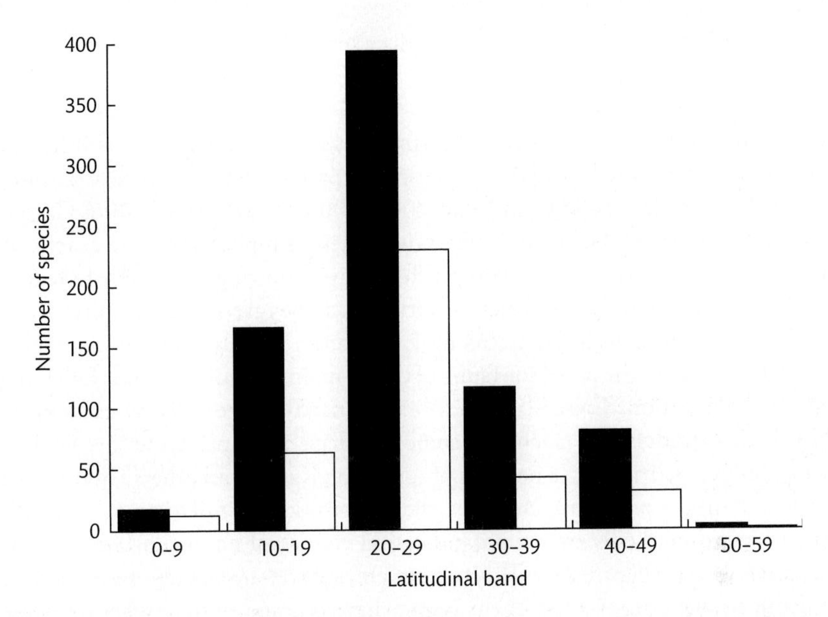

Fig. 7.5 The frequency distribution of the number of exotic bird species introduced (filled bars) and established (open bars) on oceanic islands in different latitudinal bands. From data analysed by Cassey, et al. (2005b) and Blackburn, et al. (2004).

and the extent of turnover in the identity of species between sites (beta diversity). Compared to alpha and gamma diversity, the study of beta diversity in native species assemblages has been somewhat neglected, even though it is a key component of global biodiversity (but see Gaston, et al. 2007). In contrast, turnover has received relatively much more attention in the exotic species literature in the context of biotic homogenization (or differentiation): the process by which the genetic, taxonomic, or functional similarities of regional biotas increase (or decrease in the case of differentiation) over time (McKinney and Lockwood 1999; Lockwood and McKinney 2001; Olden and Rooney 2006).

Spatial turnover is typically calculated as a function of three components *a*, *b*, and *c*, where *a* (continuity) is the total number of species shared by two areas; *b* (gain) is the number of species present in the other area but not in the focal one; and *c* (loss) is the number of species present in the focal area but absent from the other one (Koleff, et al. 2003; Gaston, et al. 2007). Continuity, gain, and loss in assemblages of native species may be expected to depend on the relative position of the sites, their biogeographic histories, the spatial extent of the sites, and on the dispersal ability of the species concerned (Nekola and White 1999; Steinitz, et al. 2006). The establishment of exotic species clearly has the potential to affect any or all of these three components, and so change beta diversity. However, understanding how exotic species will affect spatial turnover in species identity (beta diversity) is complicated by the fact that multiple introductions can affect *a*, *b*, and *c* simultaneously. Moreover, *a*, *b*, and *c* can also be affected by extinction, while the quantitative effects of extinctions and exotic species establishment on homogenization will depend on the spatial pattern in beta diversity pertaining in the first place (Olden and Poff 2003; 2004; Cassey, et al. 2006; Cassey, et al. 2007; Spear and Chown 2008). Given these complications, the exact way in which assemblages homogenize or differentiate following the establishment of exotic species is difficult to predict a priori, but can be investigated empirically.

This last point is well illustrated by Lockwood's (2006) study of taxonomic homogenization in the passerine fauna of the six main Hawaiian Islands following human colonization. These islands have experienced a range of native extinctions and exotic introductions, such that some now have more passerine species than before human arrival (e.g., a net gain of species on Kauai) and others fewer (e.g., a net loss of species on Maui). However, the compositional similarity of the islands has increased markedly over time, indicating consistent homogenization of the passerine faunas (Figure 7.6). This pattern emerges because many native species endemic to one or only a few of the islands have become extinct, whereas exotic bird colonization has largely proceeded in a 'hub and spoke' manner, whereby species have been introduced and established on one island (often Oahu) and subsequently spread to others. Overall, the total number of passerine species inhabiting the archipelago has fallen by eighteen species, and the distinctiveness of the passerine fauna of each island has also declined (Lockwood 2006).

The situation on Hawaii cannot be viewed as typical, however, in terms of how island bird assemblage similarity changes—indeed, it seems that there is no

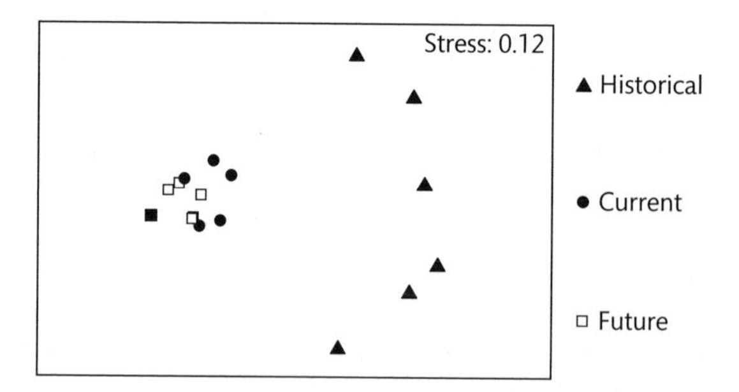

Fig. 7.6 Non-metric multidimensional scaling plot of the passerine composition of the six main Hawaiian Islands at three time periods. The historical values measure assemblage similarity before human colonization, while the future values predict what assemblage similarity will be like if all populations of passerine species currently threatened with global extinction actually disappear. The greater the distance between points, the lower the compositional similarity of the faunas. From Lockwood (2006).

such thing as typical (Cassey, et al. 2007). While bird assemblages in Hawaii, New Zealand, and the Mascarenes have homogenized, those in other Pacific and Atlantic island groups have tended to differentiate (Figure 7.7). Archipelagos also differ in whether their bird assemblages become more or less similar to those on other islands in the same ocean. Invasions and extinctions make the Hawaiian Island bird assemblages less similar to assemblages on other Pacific islands, but make the Mascarene bird assemblages more similar to those on other islands in the Indian Ocean. Indeed, all combinations of within and between archipelago changes in similarity are expressed in island bird assemblages (Figure 7.7)! To date, these patterns seem to be driven more by exotic introductions than by native extinctions (Cassey, et al. unpublished analysis).

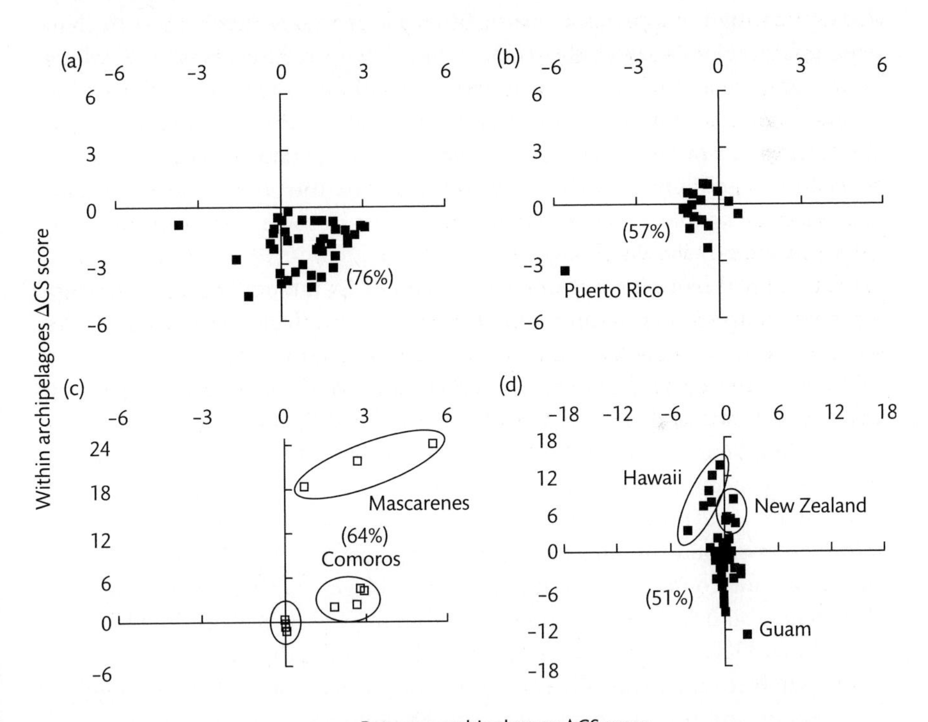

Fig. 7.7 Bivariate scatterplots of the relationship between changes in similarity (ΔCS) among islands within an archipelago and ΔCS of islands between different archipelagos within the Atlantic (a), Caribbean (b), Indian (c), and Pacific (d) oceans. Each of these plots can be divided into four quadrants: islands above the horizontal axis are those that homogenize relative to others in the same archipelago; islands right of the vertical axis are those that homogenize relative to others in different archipelagos within the same ocean basin. Percentages relate to the quadrants with the greatest number of member islands. Reprinted with permission of John Wiley & Sons Ltd from Cassey, et al. (2007).

In theory, the higher the initial similarity of two assemblages, the less likely it is that exotic species will homogenize them (Olden and Poff 2004; Spear and Chown 2008). In the limit, the similarity of identical communities cannot be increased. However, island bird assemblages again show disparate patterns in how similarity changes in relation to initial similarity (Cassey, et al. 2007; see also Spear and Chown 2008). For example, islands in the Indian Ocean show the expected pattern of high initial similarity leading to less homogenization when islands within archipelagos are compared, but the relationship reverses at the larger spatial scale of islands between archipelagos (Cassey, et al. 2007). If there is a tendency for islands or archipelagos that are in close proximity to receive similar suites of exotic species (see also McKinney 2005), then potentially homogenization could increase with increasing initial similarity. A relatively high proportion of exotic bird species tend to derive from the biogeographic region in which the recipient location is sited (section 2.2.2), which could underlie any tendency for proximal locations to receive similar sets of species. This relationship suggests that the kind of biogeographical factors that determine the similarity of native faunas, such as separation (isolation) or distance to a source pool (Nekola and White 1999; Steinitz, et al. 2006), may also influence the similarity of exotic faunas, at least in some cases (Cassey, et al. 2007; Spear and Chown 2008).

The study of biotic homogenization is still in its infancy. In birds we see a range of changes in community similarity in response to human impacts on the gain, loss, and continuity components of turnover. Yet, results from birds may differ from those of other taxa (Lockwood and McKinney 2001; Marchetti, et al. 2001; Olden and Poff 2004; McKinney 2004; 2005; McKinney and Lockwood 2005; Smart, et al. 2006; Spear and Chown 2008), which will differ in important aspects of their life histories, such as vagility, that will influence initial similarity and how it changes. Analyses in all taxa have so far been restricted largely to taxonomic homogenization, leaving questions about how changes in the taxonomic composition of different assemblages translate into changes in their genetic and functional composition (but see Smart, et al. 2006; Devictor, et al. 2007; Devictor, et al. 2008). The tendency for extinctions and invasions on the whole to derive from different bird taxa (Lockwood, et al. 2000) suggests that genetic and functional changes will be no less significant. Nevertheless, as yet we can make no general statements about the factors that drive some island avifaunas to homogenize while others differentiate, or that cause these responses to differ at different spatial scales. Important clues seem likely to come from studies that adopt a modelling approach (e.g., Olden and Poff 2003, 2004; Cassey, et al. 2006), or comparative analyses of the relationship between changes in turnover and the biogeographical and cultural histories of different locations. Part of the problem may be that as yet we have a relatively poor understanding of the mechanisms underlying turnover in natural assemblages (Gaston, et al. 2007).

7.4 Large-scale Patterns in the Distribution and Abundance of Exotic Birds

Data on the presence and absence of species at different locations provide information about patterns in their richness and turnover, but also allow consideration of patterns in the extents of their distributions (Gaston and Blackburn 2000). It has long been appreciated that species vary considerably in distributional extent (Hesse, et al. 1937), and that this variation shows spatial patterning (Lutz 1921; Rapoport 1982). However, Stevens (1989) was the first to suggest that a tendency for range size to increase with latitude, known as Rapoport's rule, may be causally linked to the inverse gradient in species richness. He proposed that the narrower range of environmental conditions that tropical species are required to tolerate restricts their distributions. These conditions can also promote higher levels of interspecific coexistence due to immigration of individuals into communities outside their restricted microhabitats. Although it is now clear that Rapoport's rule is a phenomenon largely restricted to Nearctic and Palaearctic biotas, it is nevertheless true that species richness is inversely related to geographic range size in the global bird assemblage (Orme, et al. 2006). Indeed, since the species richness of any given area is the sum of the number of species whose geographic distributions overlap that area, it follows that the factors that determine geographic range sizes are also likely to influence species richness.

Sax (2001) presents evidence that the latitudinal range sizes of bird species introduced to North America increase with latitude north of the tropical zone, but are consistently large through the tropics (Figure 7.4). This pattern broadly conforms to that of variation in range sizes observed in the native avifaunas of these biogeographic regions (Blackburn and Gaston 1996b; Orme, et al. 2006). Sax (2001) argued that exotic range sizes can be explained in terms of the response of species to contemporary ecological conditions, and used variation in the high and low latitude range limits of the native and exotic distributions to assess likely causes. He found that while the highest latitudes in the native and exotic ranges tend to be correlated across species, species often exceed their natural high-latitude range limits in their exotic range (Figure 7.8a). In contrast, species only rarely exist at lower latitudes in their exotic range than they do in their native range (Figure 7.8b). Sax further argued that because exotic bird species established on islands tend to have range limits that extend further towards the equator than do exotic bird species established at continental locations, these low latitude limits are set by biotic resistance.

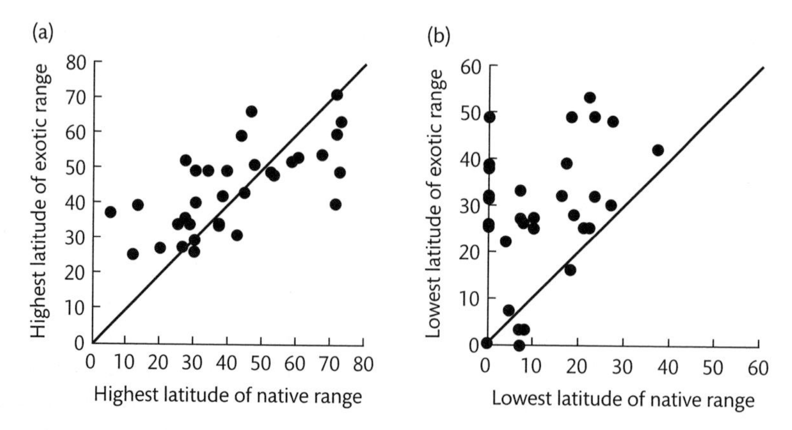

Fig. 7.8 The relationship between (a) the highest latitude of the native range and the highest latitude of the exotic range, and (b) the lowest latitude of the native range and the lowest latitude of the exotic range, for individual species of exotic bird on continents in the northern hemisphere. The solid diagonal is the 1 : 1 line on each plot. Reprinted with permission of John Wiley & Sons Ltd from (Sax 2001).

Once again, however, human placement effects may play a role. The range of values spanned by the highest latitudes in the exotic ranges of birds is less than that for the ranges of native birds (Figure 7.8a), as is the standard deviation of these latitudes (exotic species = 12.3, native species = 18.8). Species with high northern native range limits apparently tend to have exotic northern range limits at lower latitudes, and vice versa for species with lower northern native range limits (Figure 7.8a). Southern range limits tend to be more northerly in the exotic than the native ranges (Figure 7.8b). As a consequence, both high- and low-latitude exotic bird range limits are clustered at mid-latitudes. Since most exotic bird introductions to continental locations in the northern hemisphere tend to have been clustered at the same mid-latitudes (30° and 40°, Blackburn and Duncan 2001b), the pattern found by Sax (2001) is consistent with the idea that exotic bird populations have not subsequently expanded their ranges to occupy the full extent that they can potentially occupy, as indexed by their native range limits. This result suggests in turn that exotic bird range sizes are influenced at least in part by where these species were introduced. Nevertheless, it is possible that the pattern we note could occur if exotic ranges are responding to contemporary ecological conditions. In that regard, it would be useful to compare latitude of the site of introduction, the latitudinal midpoint of eventual exotic distribution, and the latitudinal midpoint of the native range, to determine the extent to which patterns in range limits are a consequence of patterns of introduction.

Patterns in introduction, however, do not seem likely to explain the tendency for geographic range size to increase with latitude in exotic birds (Figure 7.4). This explanation would require exotic birds established in high latitudes to have been introduced either more times or more widely, for which there is no evidence. One possibility is that latitudinal range size variation in exotic birds is a consequence of one cause of establishment success—environmental matching. As we noted in section 5.2, bird species are more likely to establish exotic populations if introduced into environments that match those they experience in their native ranges. If the non-native range limits of exotic birds are positively correlated with the array of environmental conditions they can tolerate (Stevens 1989), or if the spatial extent of climatic zones is correlated across regions, then we should expect to see exotic birds established in areas they can easily tolerate recapitulating native distribution patterns. In this respect, there is evidence from New Zealand and Australia that bird species that are widespread in their native ranges end up with larger ranges following establishment (section 6.2.2). In these cases, species have attained larger exotic ranges if there are larger areas of preferred (New Zealand) or climatically suitable (Australia) habitat.

Native species that are widespread across a region tend also to attain relatively high abundances, whether measured as population size at the scale of the region or as population density at sites within it (Brown 1984). The end result is a positive relationship between abundance and distribution that, like the species-area relationship and latitudinal richness gradient, is another of the more general patterns in ecology (Gaston 2003; Blackburn, et al. 2006). There are currently few studies of this relationship in exotic bird species, but those that exist find positive relationships between population size and range size that seem to match reasonably closely the same patterns in the associated native avifauna. For example, Holt and Gaston (2003) found that neither the slope nor the intercept of abundance-range size relationships differed between native and exotic British birds (nor indeed between native and exotic British mammals). Mean abundance and occupancy also showed no differences between natives and exotics. Similarly, Labra, et al. (2005) found no differences in the slopes of abundance-range size relationships for a sample of North American exotic bird species and samples of similar (in terms of phylogeny, ecology, and life history) randomly chosen native species. Blackburn, et al. (2001) found that the density-range size relationship for passerine species introduced to New Zealand was similar to the relationship for the same species in their native range of the United Kingdom, but differed markedly from the relationship for native New Zealand species for which data were available.

That native and exotic species seem to follow the same scaling of population size on range size suggests that the same processes influence both categories of species. Moreover, although it has been suggested that propagule pressure may influence which species become abundant and widespread, it is hard to see how human-mediated processes could be responsible for these commonalities in scaling. The similarity in scaling also suggests that as exotic species spread, they increase in both population and range sizes in a way that keeps them close to the interspecific scaling relationship for native species. If they did not, then exotics would be identifiable as outliers. Labra, et al. (2005) were able to assess the form of intraspecific abundance-distribution relationships in their sample of North American exotics, where each data point is the population size and range size for the species in one of the years between 1966 and 2002. They found that these intraspecific relationships were uniformly positive, but with slopes that were steeper than for the interspecific relationship across exotic species (mean ± s.e. = 0.836 ± 0.125 for intraspecific relationships for nine species, but in the range 0.408–0.520 for interspecific relationships calculated every ten years from 1970 to 2000). This result suggests that as exotic bird populations increased in number, their range sizes increased faster than would be expected from the interspecific relationship, at least in the period 1966–2002.

7.5 Impacts of Exotic Birds on Native Species and Communities

Large-scale patterns in the richness, distribution, and abundance of exotic species generally recapitulate patterns observed in native species. We have argued above that in some cases, such as the species-area relationship, this is an interesting coincidence between the results of human placement effects on exotic species and the factors that drive native species richness. Nevertheless, in other cases, such as the commonalities in the scaling of abundance and distribution, it seems more likely that the same processes that determine patterns in native species also determine them in exotics. This begs the wider question of what determines the structure of natural ecological assemblages—a subject of considerable and unresolved debate.

One long-standing question in this debate is the extent to which biotic interactions influence ecological assemblage structure (Clements 1916; Gleason 1926). Broadly, one can distinguish between niche-assembly and dispersal-assembly views of such structure (Hubbell 2001). The former argues that interspecific interactions and niche differentiation determine the set of species that can coexist, whereas the latter argues that local assemblages are more stochastic collections of species determined by immigration from the wider pool of species inhabiting the region within which the assemblage sits. While truth probably lies somewhere between these extremes, the extent to which species interactions are likely to determine community composition continues to exercise ecologists (cf. stochastic niche theory and species capacity versus island biogeography models in section 7.2).

Evidence for interactions between exotic and native species has the potential to provide important information about their effects on assemblage structure.

The studies we have reviewed to date seem to imply that the effects of such interactions are weak. As we saw in the previous two chapters, considerable controversy has surrounded the question of whether or not interactions between exotics, or between natives and exotics, drive the establishment or rates of spread of exotic bird species. However, we have not directly examined evidence for the possibility that exotic birds have impacts on the native species or communities in which they have become established. Generally, it would seem that these effects are not perceived as large (although see Cheke and Hume 2008). This is evinced by the scarcity of exotic birds within lists such as the 100 Worst Global Invaders <www.gisp.org>, and by the lack of published compilations of ecological impacts of exotic birds (although the latter could also be used to infer that these impacts are badly understudied; Pyšek, et al. 2008). Nevertheless, there are reasons to believe that the impacts of exotic birds on native assemblages may sometimes be substantial.

Parker, et al. (1999) formalized the total ecological impacts of an exotic species, I, as the product of the range size of the species R, the average abundance of this species per unit area across that range A, and the per-individual effect on components of the native ecosystem (e.g., individuals, species, communities) E, such that $I = RAE$. Thus, an increase in any one of R, A, and E will increase the overall impact of the exotic species, and these terms combine to determine total impact in a linear fashion (although Parker and colleagues acknowledge non-linear dynamics as a possibility). While most exotic bird species remain relatively restricted in their distribution (Figure 6.1), and therefore probably relatively rare (see above), there are a number of examples where R and A are relatively large (section 6.2.2). Thus, we might expect that exotic birds would have impacts on the structure and function of avian assemblages beyond those accruing from their simple presence in tallies of species numbers. However, the extent of this impact will also depend on the magnitude of E. In fact, most studies of the impact of exotic species on native assemblages concern E, or some unknown combination of A and E (Parker, et al. 1999).

In this section, we review the evidence for the likely magnitude of the per-individual effects of exotic species on components of the native assemblage, organizing the effects according to standard species interaction categories of competition, predation, mutualism, and parasitism (Begon, et al. 1996). Based on existing evidence concerning which exotic species have the strongest ecological effects (e.g., Davis 2003; Blackburn and Gaston 2005; Salo, et al. 2007), we should expect exotic birds generally to show relatively strong per individual unit effects when they act as novel predators (and perhaps mutualists) on oceanic islands, or other isolated ecosystems. However, we note that nearly all studies of E are comparative, contrasting aspects of the native assemblage (dependent variables) before and after the establishment of the exotic bird species, or between habitats invaded and uninvaded by the exotic bird species. This trend is in keeping with the broader literature on exotic species' impacts (Parker, et al. 1999), but tends to produce weak inferences, as it is difficult to rule out confounding factors influencing the dependent variable(s).

7.5.1 Competition with Native Birds

Most exotic bird species establish in highly modified habitats, such as farmland and urban areas, which native birds do not use as readily (section 5.4.1.1). This observation has often led authors to conclude that competition between native and exotic bird species should be of little consequence, and indeed studies that have documented fitness effects of competition between native and exotic bird species are rare. Even in cases where exotic and native birds appear to share very similar ecological niches, these niches may overlap only superficially when critically assessed.

Kawakami and Higuchi (2003) examined interactions between native Bonin Island *Apalopteron familiare* and exotic Japanese *Zosterops japonicus* white-eyes in the Bonin Islands. The species overlap in distribution in secondary forests, have very similar ecologies, and there is evidence that the native species adjusts its foraging behaviour in the presence of the exotic. However, Kawakami and Higuchi could detect no negative effect of the exotic on the native, and notably densities of the native species have not declined. They suggested that the native white-eye would only decline to the benefit of the exotic in the face of habitat conversion, as the exotic species prefers modified (open) habitats while the native prefers primary forest.

Similarly, based on congruence in body size, habitat, and foraging strategy, Moeed (1975) suggested that the native New Zealand pipit *Anthus novaeseelandiae* and the exotic European skylark *Alauda arvensis* may directly compete for food. However, when their diets were quantitatively compared, significant differences were found in both the relative volumetric importance and the occurrence of all food groups most frequently consumed by each species (Garrick 1981). Finally, in Haleakala National Park on the island of Maui, Hawaii, a small population (<200 birds) of the endangered nene *Branta sandvicensis* is localized in several small habitat patches that it shares with the exotic ring-necked pheasant *Phasianus colchicus*. The pheasant and nene appear similar in their food habits, but the abundance of favoured food items throughout their distribution suggest that food quantity is not limiting nene abundance, and there is no evidence that competition for food items exists (Cole, et al. 1995).

Competition between bird species for food resources may be less intense than is competition for nest sites, particularly amongst cavity nesting species. Nest holes are a scarce and limiting resource, especially for species that cannot excavate their own holes (Newton 1998). Harper, et al. (2005) studied nest box occupancy in twenty remnant patches of indigenous vegetation in Melbourne, Australia. No Australian bird species excavates its own cavity, and competition for them extends across animal classes; many of the boxes were occupied by native marsupials. Of those used by birds, 131 out of the 139 nests found belonged to the exotic common myna. This result is certainly consistent with the hypothesis that native species are losing out to the exotic in competition for nest sites, but Harper and colleagues could not rule out the alternative explanation that the nest boxes are unsuitable for native species, which may prefer cavities with different properties (e.g., that are deeper, and not smooth sided).

The European starling *Sturnus vulgaris* is also a cavity-nesting species that uses tree holes or natural crevices for nest sites, often aggressively evicting primary cavity excavators or other species of secondary cavity nesters (Wiebe 2003). In the USA, nest-site competition with starling populations has been widely implicated in population declines of native cavity-nesting birds (see references in Wiebe 2003), although the proportion of nesting attempts lost to starlings is less than the proportion that is lost to predators (e.g., Fisher and Wiebe 2006). Nevertheless, in some cavity-nesting communities, the majority of nests belong to exotic starlings (Wiebe 2003), and native species that are behaviourally less aggressive can lose as many as two-thirds of their nesting attempts to starlings through being physically evicted (Ingold 1998). Somewhat surprisingly therefore, Koenig (2003) concluded that European starlings have yet unambiguously to cause population declines in any species of North American cavity-nesting bird, with the possible exception of sapsuckers. He did, however, provide the caveat that significant declines may yet be detectable if population densities of starlings continue to increase. This situation may also be occurring in Australia where starlings compete with (and evict) native parrot species and small arboreal marsupials for nesting sites (Blakers, et al. 1984), although quantitative data on population effects of this observed behaviour are lacking here too.

If native and exotic bird species rarely appear to compete directly for resources it may be because they more frequently do so indirectly via apparent competition (Holt 1977; 1984; Holt and Lawton 1994). Apparent competition defines the situation where two species do not compete via a shared resource, but via a shared natural enemy. There is now abundant evidence, mainly from oceanic islands, that many native bird species declined dramatically, and frequently to extinction following the introduction of exotic predatory mammal and reptile species (Fritts and Rodda 1998; Roff and Roff 2003; Blackburn, et al. 2004). Exotic bird species have typically evolved in the presence of such predators, and so may be able to maintain stable or increasing population sizes where native bird species cannot. Moreover, the exotic birds may provide an alternative food source for the exotic predators, maintaining both the predator populations and predation pressure on the native birds. These indirect effects of exotic birds on native species via their effect on predator populations may facilitate the decline of the native species and explain why direct competition between natives and exotics is rarely observed. To our knowledge no one has explored this possibility.

7.5.2 Exotic Bird Predation on Native Vertebrates

A small number of exotic bird species are predators, frequently introduced with the motive of biocontrol. The introduction of the little owl *Athene noctua* to New Zealand was specifically to combat the negative impacts of exotic frugivorous finch species on horticultural stocks (Thomson 1922). Similarly, the swamp harrier *Circus approximans* was evidently introduced to Tahiti in the 1880s to help control rats *Rattus sp.* (Long 1981), while the Chimango caracara *Milvago chimango* was apparently introduced to Easter Island to act as a scavenger (Lever 2005). Predation by exotic birds has been implicated in the declines of several native bird species. Thus, the Chimango caracara is reported as having a marked effect on colonies of nesting birds on Easter Island by preying on their young (Harrison 1971), while the swamp harrier is implicated in the probable extinction of the Polynesian imperial pigeon *Ducula aurorea* on Tahiti (Thibault 1988; Thibault and Cibois 2006).

The exotic common myna has also frequently been suspected of causing both clutch failure and population reduction of native island birds (Holyoak and Thibault 1984; Thibault 1988). This species is a widespread exotic, reaches relatively high abundances where it establishes, and on oceanic islands can act as a novel predator. Following eradication of cats on Ascension Island, common mynas were the dominant nest predators of breeding sooty terns *Onychoprion fuscata* (25% of failed clutches; cf. 20% attributed to introduced ship rats *Rattus rattus*: Hughes, et al. 2008). Similarly, in 1998 a recovery programme was initiated to conserve the endangered endemic Tahiti flycatcher *Pomarea nigra*. At this time it was supposed that the main threat to the flycatcher was nest predation by ship rats. However, Blanvillain, et al. (2003) found that more encounters between the flycatchers and common myna (and red-vented bulbul *Pycnonotus cafer*) were observed in flycatcher territories that experienced nest failure or fledgling death compared with those that experienced reproductive success. The higher number of encounters and interactions between common myna and Tahiti flycatcher around nests containing eggs and chicks strongly suggested the occurrence of nest predation by common myna.

7.5.3 Mutualisms

The most likely types of mutualistic interactions involving exotic bird species are those with food plants, where the latter benefit from pollination or seed dispersal functions from birds, and the bird itself gains a meal of nectar or fruit. Birds play a large role in the dispersal of plant seeds, and a lesser one through their still significant role in pollination (Landsborough Thomson 1964). Birds in general have such broad foraging behaviours that interactions with individual native or exotic plant species will be weak. This relationship will probably be exaggerated in exotic birds because dietary generalism is one of the life-history traits that has been shown to increase the probability that an exotic bird population will establish (section 4.3.3.1). It therefore seems unlikely that tight mutualistic associations between individual plant and bird species will develop in the exotic ranges of many bird species.

Looser exotic bird–plant mutualisms may nevertheless have significant impacts upon the recipient environment. If exotic birds prey upon fruits and then disperse seeds that differ in composition or relative proportion to those eaten by native species, they may alter the floristic composition of environments, especially among regenerating or successional habitats. The possibility that exotic bird species preferentially interact with exotic plant species to create 'invader complexes' (D'Antonio and Dudley 1993), where the spread of each exotic species is facilitated by the other(s), is a major concern. Although in general there is no shortage of pollinators or vertebrate dispersers for fleshy fruited exotics, the urban environments where alien birds are often most abundant also include the highest variety of ornamental plant and exotic orchard species in comparison to other landscapes (Day 1995; Pyšek, et al. 1998). Exotic plants may thus comprise the majority of food items in the diet of exotic frugivorous species (Mandon-Dalger, et al. 2004; Williams and Karl 1996), promoting co-invasion by plants and birds.

There are several examples of exotic birds acting as seed dispersers for exotic plants (see Richardson, et al. 2000). For example, the red-whiskered bulbul *Pycnonotus jocosus*, native to continental India, disperses the seeds of several invasive exotic plants on islands in the Indian Ocean (Macdonald, et al. 1991). Mandon-Dalger, et al. (2004) found that exotic plants compromised more than 80% of food items in the diet of this species on La Réunion, and that in the majority of cases the birds actually facilitated germination of the invasive plants. The bulbul is also established in Florida, where it consumes and disperses the seeds of more than twenty-four exotic plant species (Simberloff and Von Holle 1999).

While exotic plant species may preferentially benefit, native plant species may suffer from the replacement of native bird species by exotics. Several studies have found that exotic birds do not take over the pollinating or dispersal roles previously filled by native species. Anderson, et al. (2006) studied seed set in the New Zealand bird-pollinated forest shrub *Rhabdothamnus solandri* (Gesneriaceae) at two mainland sites where only a single species of native pollinator is present but exotic species are abundant, and on an offshore sanctuary where all extant native bird pollinators are abundant but exotic species are rare. They performed three pollination treatments: un-manipulated flowers accessible to birds, bagged flowers from which pollinators were excluded, and hand-pollinated flowers. Fruit set in unmanipulated flowers (67%) on the island sanctuary was almost as high as for hand-pollinated flowers (70%), whereas unmanipulated fruit set (average 16%)

at both mainland sites was substantially lower than for hand-pollinated flowers (83%). Direct examination of flowers showed that 83% of flowers had been visited by bird pollinators on the island, compared to 20% on the mainland (Anderson, et al. 2006). Seed production at sites with exotic birds but few natives was only 10% of that at sites at which pre-human native bird densities exist. Native seed dispersal was also shown to be significantly lower at the mainland sites than on the island sanctuaries (McNutt 1998; Anderson, et al. 2006).

7.5.4 The Role of Exotic Birds in Disease Transmission

When diseases arrive and establish in a novel location, their effects can be as large as those of novel exotic predators, sometimes causing rapid declines in native species' populations. Exotic birds facilitate the expansion and persistence of bird (and indeed human) diseases by acting as reservoirs for the disease organisms, and also as transmission vectors that bring the disease to a location where it too is exotic. The largest relative impacts come from exotic bird species that perpetuate a disease that is novel to the native bird community. The emergence of several major wild bird disease episodes in recent decades has been attributed to the rapid global movement of exotic avian hosts and pathogens (McLean 2003), coupled with major changes in the size, quality, and continuity of natural habitats.

The classic example of exotic birds intensifying the magnitude of disease effects comes from the Hawaiian Islands. Avian malaria *Plasmodium relictum,* a parasitic disease transmitted by mosquito vectors, has been implicated in the widespread decline and possible extinction of many species within the endemic Hawaiian honeycreeper radiation (van Riper III, et al. 1986). Migrant birds infected with malaria have probably long visited the Hawaiian Islands, but the absence of a suitable native mosquito vector meant that the disease was never transmitted to resident species. Unfortunately, a suitable vector (*Culex quinquefasciatus*) was introduced to the islands in 1826 (Warner 1968), since when native Hawaiian birds have been exposed to reservoirs of parasites harboured by the many resident exotic bird species. Van Riper III, et al. (1986) demonstrated that a higher incidence of malaria occurred in native birds compared with exotics, while experimental inoculation with infected blood revealed that native birds were highly susceptible to the exotic malarial parasite. Indeed, while mortality from the parasite in exotic birds is negligible, mortality in many endemic birds can range from 50% to 90% (Jarvi, et al. 2001). Hawaiian native bird species rapidly declined to extinction within the altitudinal range inhabited by the exotic mosquito, but maintained populations at altitudes too cold for the mosquito to persist (Warner 1968). The extreme susceptibility of Hawaiian endemic birds to malaria has been explained by their long evolutionary history (*c.*3–4 million years) in the absence of the parasite.

Avian malaria may not be a problem for all oceanic Pacific islands, however, which may have to do with the composition of the local exotic avifauna and its competency as a reservoir of malaria. On American Samoa, native mosquitoes, stable bird populations, and chronic malarial infections all point to malaria being indigenous and having a long history of co-evolution with the native bird species (Jarvi, et al. 2003). Indeed, malaria was not detected in the two exotic bird species tested, although sample sizes were small in at least one (common myna, $N = 2$; red-vented bulbul, $N = 25$). Steadman, et al. (1990) found no blood parasites in fifty-five samples from nine native bird species, or in twenty-four samples from one exotic bird species (common myna) in the Cook Islands.

A second disease of birds, the viral infection avian pox, has also been implicated in native bird declines in Hawaii (Warner 1968; van Riper III, et al. 2002). The transmission of avian pox is enhanced by higher vector and host densities, and can be transmitted by direct contact, via biting insects, and even through aerosol transmission, particularly in confined situations (e.g., aviaries). Atkinson, et al. (2005) surveyed native and exotic forest birds at three different elevations on leeward Mauna Loa volcano for the prevalence of both avian malaria and pox. They reported significant declines in prevalence of both diseases with increasing elevation. Native species at all elevations had the highest prevalence of pox-like lesions. By contrast, pox-like lesions were not detected in any individuals of four non-native species (red-billed leiothrix *Leiothrix lutea,* Japanese white-eye *Zosterops japonicus,* northern cardinal *Cardinalis cardinalis*, and house finch *Carpodacus mexicanus*) suggesting that these exotic species may serve as reservoirs for pox.

The influence of exotic birds in spreading disease is not limited to island ecosystems. For example, the infectious disease mycoplasmal conjunctivitis was first reported in wild exotic house finches from eastern North America in 1994 (Fischer, et al. 1997). The ocular lesions are the result of infection with *Mycoplasma gallisepticum*, a non-zoonotic pathogen of poultry that had not been previously associated with disease in wild songbirds. The detection of mycoplasmal conjunctivitis in house finches was correlated with significant population declines and the estimated loss of tens of millions of individuals (Nolan, et al. 1998). By late 1995, mycoplasmal conjunctivitis had spread to a native species, the American goldfinch *Carduelis tristis*. Unlike their native western counterparts, house finches within the eastern exotic range may have evolved partial migration up to several hundred miles (Belthoff and Gauthreaux 1991), and therefore possess the ability to disseminate the infectious agent over a large geographic area.

Psittacine beak and feather disease (PBFD) is the most significant infectious disease among psittacine birds (Raidal, et al. 1993), and in many cases is fatal (Pass and Perry 1984). PBFD was first described in various species of Australian cockatoos and is thought to have originated in Australia, but through transportation and trade is now found worldwide (Heath, et al. 2004). Almost two-thirds of parrot species have been transported outside their native distribution

(Cassey, et al. 2004b) and Thomsen and Mulliken (1992) estimated that nearly four million individual parrots were being harvested from the wild each year in order to supply the burgeoning pet bird industry. The risk to endemic populations of parrots is unclear, but potentially important to wildlife conservators.

Deaths from pathogens spread by exotic birds may not be restricted to native birds. In a large number of regions, exotic rock doves *Columba livia* and house sparrows *Passer domesticus* contaminate grains stockpiled for human or livestock consumption through their faeces. It is widely reported that these two species have assisted in spreading diseases transmitted to humans such as *Salmonella*, tuberculosis, and *Escherichia coli*. A study of rock doves from urban and dairy farm sites in the USA revealed that 8% of birds carried some type of virulence marker gene associated with hemorrhagic disease in humans and 9% of birds from dairy environments carried pathogenic *Salmonella* (Pederson, et al. 2006). During 1984, twenty-three outbreaks of Newcastle disease were confirmed in chickens in the United Kingdom. Nineteen of these outbreaks occurred either directly or indirectly as a result of spread from diseased exotic rock doves infesting food stores at Liverpool docks (Alexander, et al. 1985). Exotic bird species have also been implicated as carriers of West Nile virus in North America (Lanciotti, et al. 1999; Jozan, et al. 2003; Kilpatrick, et al. 2006).

7.6 Conclusions

There are many important questions in ecology to which the answer remains the subject of debate. Why do large areas house more species? Why is there large-scale spatial variation in species richness? Are local or regional communities saturated with species? How important are interspecific interactions in structuring animal assemblages? What are the relative roles of contemporary and historical factors in ecology? Why are some species common and others rare? Why are widespread species also more abundant? The evidence we have reviewed in this chapter suggests to us that exotic species shed light on these key questions, albeit through the way in which their distributions often represent the impact of human, rather than natural, processes.

The true extent of human impacts has, we believe, been underestimated in treatments of exotic-species distribution and richness relationships to date. Species-area relationships, species-isolation relationships, and latitudinal variation in richness for exotics, as well as correlations between numbers of extinct and introduced species, may all be explained by where we expect exotic species to be as a result of the action of humans. Latitudinal patterns in range limits, and possibly in the size of exotic geographic ranges, may also have the same explanation. Homogenization by definition has a large human component, albeit that large-scale patterns in the distribution of native populations will have a modifying effect.

These results imply that the influence of exotic birds in changing biodiversity patterns lies primarily in changes to entry rules (*sensu* Kunin 1997). Specifically, changes in biodiversity patterns seem to vary in location and magnitude largely in response to human-induced changes in the location and rate of immigration (Blackburn, et al. 2008; Chiron, et al. 2008). This is perhaps most graphically demonstrated by the species-isolation relationship for island birds: native bird species richness is a negative function of distance to the mainland, but the function for exotic bird species richness is positive because of where exotic species were introduced (Blackburn, et al. 2008). A significant role for dispersal is an increasingly common feature of models of community assembly (e.g., MacArthur and Wilson 1963; 1967; Hubbell 2001; Tilman 2004). We suspect that exotic species will have a major role to play in elucidating the precise relationship between patterns of dispersal and community structure—for example, through studies like that by Sax, et al. (2005a) matching the predictions of different hypotheses to the observed structure of communities in the presence of exotics.

Exotic species not only change entry rules, however. It is likely that biotic homogenization at least has also been influenced by changes to exit rules. The role of exotics in influencing transformations seems to us to be an area that would benefit from further exploration. For example, the tendency for exotic species to lie on the same abundance-range size relationships as native species suggests to us that exotics and natives follow the same transformation rules about how their abundances and range sizes change. Whether the response of exotics to those transforming processes has knock-on effects to other species in the assemblage (native or exotic) is an open question. Certainly, there is at present rather little evidence that exotic bird species strongly influence native bird species through competition or predation (or indeed vice versa: section 5.4). The effects of exotic birds may be more significant through their role as reservoirs for diseases, or in terms of their consequences for plant invasions and altering whole ecological communities through their complex interactions with habitat modification (Didham, et al. 2007). Nevertheless, the distribution of total ecological impacts, *I* (Parker, et al. 1999), across exotic bird populations seems likely to be highly right-skewed.

The ecology of exotic bird species in their recipient environments also provides information about aspects of the invasion process itself. First, it suggests that human-mediated processes primarily drive the richness of exotic bird species. Thus, in this case the obvious answer is likely to be the correct one—exotic bird species *are* distributed by and large where humans wanted them to be (Lockwood 2005).

Second, there is evidence that the processes that influence establishment also influence the ecological patterns that exotic species express. For example, the fact that North American exotic bird ranges conform to Rapoport's rule seems to us to reflect the influence of environmental matching, with species introduced to zones that they can tolerate consequently recapitulating native distribution patterns. Conversely, the relatively weak influence of competitive interactions between exotic and native species in determining establishment (section 5.4.1) is reflected in the apparent lack of competitive effects between such species following establishment. As we have argued in previous chapters, the sequential nature of the invasion process means that filters acting at earlier stages influence patterns expressed at later stages.

Third, if all exotic species do is recapitulate patterns in the native range, that could be taken as evidence that studying large-scale patterns of exotic bird species is largely a waste of time. However, this recapitulation may itself be informative. If widespread, abundant species stay widespread and abundant in their exotic ranges, and likewise for narrowly distributed and rare species, one possible reason is that common factors limit population expansion in both native and exotic areas. This in turn suggests to us that these common factors are more likely to be abiotic than biotic, because exotic species are more likely to have common responses to climate in native and exotic regions than they are to competitors and predators (which is not to say that they will not have *any* responses to competitors or predators: section 5.4). This is also concordant with the dispersal-limitation view of community structure, which plays down the limiting effects of biotic interactions on coexistence relative to the limitations imposed by being able (or not) to reach a suitable location.

Overall, the ecology of exotic species in the novel environment leads us to speculate that there may exist a reciprocal association between invasion filters and native ecology. Most species are rare (Gaston 1994). Yet, it is widespread, abundant species that are more likely to be transported and introduced. They are also more likely to be introduced in larger numbers, and hence to have higher propagule pressure and concomitantly higher establishment success. Environmental matching also influences establishment success. The response to climate, either through physiological tolerances or the extent of suitable zones, may then determine the extent to which established exotic populations grow and spread. Thus, the factors that influence transport, introduction, and establishment are the factors that influence the success of the species in its native range, and these factors then cause the native patterns to be recapitulated in the exotic range. This may go some way to explaining why there are 'few winners replacing many losers' in the current mass extinction (McKinney and Lockwood 1999).

8

The Genetics of Exotic Bird Introductions

> Thus, we are left with the conclusion that in the last 100–120 years, bottlenecks and random drift have promoted rapid genetic shifts in isolated populations of common mynas equal to those among different subspecies of birds.
>
> A. J. Baker and A. Moeed (1987)

8.1 Introduction

The invasion process begins with the transport of individuals out of their native range and their deposition somewhere new. In Chapter 2, we reviewed the many reasons why birds are transported and introduced, and showed that these processes are non-random with respect to a variety of features of species and location. Notably, widespread, abundant species are more likely to be transported and introduced than are narrowly distributed, rare species. However, transport mechanisms necessarily sub-sample individuals from the broader native population, and this sub-sample is typically very small: indeed, fewer than 50 individuals are usually introduced in any single introduction event (Table 3.3). In Chapter 3, we reviewed the influence of this number on establishment probability, and found amongst other things that there is a general increase in establishment success with propagule size. This is because of the adverse effects of demographic stochasticity, environmental stochasticity, and Allee effects on small populations: the larger the initial introduced population, the more likely it is to avoid these deleterious effects, and so also to avoid extinction.

An additional problem faced by small populations is the potential for the loss of genetic diversity. Genetic diversity is defined as the variety of alleles and genotypes within a population. It is generally considered to be advantageous for a population to have high genetic diversity because this is the raw material for evolutionary change, and hence allows a population to respond to changes in selection regimes. It is thus also better for a species to have a larger population size, because large populations tend to have greater genetic diversity than small populations (Frankham 1996). Since introductions by default involve only a few individuals founding a new population, it seems likely that the transport and introduction of exotic birds (and all other non-native organisms) will leave the incipient introduced population genetically impoverished (Roman and Darling 2007; Vellend, et al. 2007). Moreover, the genetic diversity of small exotic populations may continue to decline after initial introduction through random processes. These populations may thus be more vulnerable to factors such as disease or environmental change, with knock-on consequences for the subsequent processes of establishment or spread. Nevertheless, a reduction in genetic diversity is not the only possible outcome of exotic introductions. A surprising number of case studies of invasive species have shown only modest reductions in genetic diversity, and in some cases diversity has increased in the exotic as compared to native range through inter- and intra-specific hybridization (Roman and Darling 2007; Wares, et al. 2005).

In this chapter, we discuss theoretical predictions relating to how transport patterns and subsequent population growth of exotic birds influence genetic variability within their non-native populations, and the empirical evidence that tests these predictions. We also explore spatial patterns in genetic variation across exotic bird populations. Finally, we address whether changes in genetic variability influence establishment success or the degree of range expansion among exotic birds, including a consideration of the role of hybridization between species or between distinct genetic lineages within the introduced range.

8.2 Population Genetics in the Context of Exotic Bird Introductions

There is a wealth of theory addressing mechanisms of increase or decrease in genetic variability within populations. Most was developed in an effort to understand non-adaptive evolution, but it has subsequently been applied to the conservation of rare and endangered species (Frankham, et al. 2004). It is also relevant in the context of invasions. Genetic variation in an incipient exotic population may be reduced through founder effects, population bottlenecks, and genetic drift (Cox 2004) and can increase through hybridization, mutation, and migration (gene flow). In this section, we review each of these processes, and consider how they relate to the circumstances of exotic bird introductions.

Genetic diversity can be measured in a range of different ways (Frankham, et al. 2004). Most of the empirical studies we review below consider three metrics: proportion of polymorphic loci, average heterozygosity, and allelic diversity. The proportion of polymorphic loci is measured as the number of loci with two or more alleles divided by the total number of loci sampled, average heterozygosity is calculated as the sum of the proportions of heterozygotes at all loci divided by the total number of loci sampled, while allelic diversity is defined as the average number of alleles per locus (all definitions follow Connor and Hartl 2004). Changes in any of these metrics will result if allele frequencies (the relatively frequency of a particular allele in a population) differ between the native source population and the exotic populations it spawned, or between two or more exotic populations of the same species.

The extent of genetic diversity in a population is most often measured using molecular markers. The technology behind these markers has improved substantially over the past few decades, and thus so too has the resolution with which we can measure genetic diversity. The vast majority of case studies of exotic birds we review below were conducted in the mid-1980s and early 1990s, when the prevailing tool for measuring diversity was protein electrophoresis (Avise 2004). Although an extremely important tool that advanced population genetics, electrophoresis evaluated protein variation and not segments of the DNA itself, as do the currently more popular methods such as microsatellite and mtDNA analyses. Therefore, the results from these early studies likely were limited in the amount of genetic variation they could uncover. It is important to keep this in mind when considering the broader conclusions we draw. For example, one consistent theme is that genetic diversity of exotic bird populations is not always lost through founder effects, genetic drift, and bottlenecks. However, this result could be overturned were the same suite of species re-examined using microsatellites instead of protein electrophoresis (Avise 2004; Wares, et al. 2005), and indeed we show one example where this has happened (Hawley, et al. 2006).

8.2.1 Reductions in Genetic Variation

The most obvious way that the transport and introduction of exotic birds will ensure reduced genetic variation in an incipient introduced population is via founder effects (Connor and Hartl 2004). By chance alone, the set of colonist individuals may show allele frequencies that differ markedly from the population from which they originated. The degree to which allele frequencies are likely to differ depends primarily on the number of founding individuals. Thus, the greater the propagule size, the less likely it is that allele frequencies within the founding population will differ from those of their source region, and vice versa. Allele frequencies should differ most clearly in the prevalence of alleles that are rare in the native range. These rare alleles are unlikely to be 'captured' in the founding propagule and thus may be missing in the exotic range. Missing rare alleles will tend to lower allelic diversity. Whether they have a strong effect on the proportion of polymorphic loci or average heterozygosity depends on the number of rare alleles lost in the founding event (Frankham, et al. 2004).

The proportion of initial heterozygosity remaining after a founding event, or after a single generation population bottleneck, can be calculated as $H_1/H_0 = 1-(1/(2N))$, where N is population size, H_1 is the heterozygosity immediately after founding event (or single-generation bottleneck), and H_0 is that before. Thus, if ten (unrelated) individuals found an exotic bird population, these founders will capture 95% of the original heterozygosity from the source (native) population (Frankham, et al. 2004). From this equation, it should also be apparent that genetic diversity will decline through time if the population remains at low numbers for multiple generations, but will rebound quickly if the population increases after founding. The rate at which diversity declines via multi-generational bottlenecks is much faster in persistently very small populations (<25 individuals) but is essentially nil in populations that average 100 or more individuals through time (Frankham, et al. 2004). Since the modal propagule size for exotic birds is fifty (Chapter 3), we should a priori expect the loss of heterozygosity due to founder effects typically to be quite small in these populations, especially if they enjoyed rapid population growth after founding.

Small founding populations will also lose alleles after initial introduction through genetic drift. Drift is caused by random fluctuations in allele frequency through time as a result of sampling error. The alleles that are passed on to the next generation after founding reflect the random sorting of genes to make gametes and then the random mating of individuals. Genetic drift results in the loss of some alleles and fixation of others due to these random processes, thus reducing overall variation within the population (Connor and Hartl 2004). The rate at which fixation occurs is strongly dependent on population size such that variation is lost much faster in populations that persist at small numbers for more than one generation.

Even if the founding population is reasonably large (>100 individuals), genetic variation can be lost through low effective population sizes (Frankham, et al. 2004). The effective population is the idealized population size that would experience the same magnitude of drift as the actual population (Connor and Hartl 2004). Effective population size is reduced if there is an unequal sex ratio, if there is sexual or natural selection such that all individuals in the population do not have an equal chance of mating in each generation, or if the population attains a very small size for one or more generations (i.e., a bottleneck). It is quite conceivable that all of these factors influence genetic variability within incipient populations of exotic birds. The rate at which genetic diversity is lost, however, can be attenuated even within small populations if there is non-random mating such that heterozygous individuals are preferred (Frankham, et al. 2004).

A variety of ecological mechanisms can serve to keep an incipient population small for long periods (Chapters 5 and 6). Even exotic bird populations that eventually reach very large numbers can show periods of reduced population size after initial introduction (e.g., house finch *Carpodacus mexicanus* in eastern North America; Veit and Lewis 1996) or experience boom and bust cycles where the population goes through one or more population bottlenecks (Simberloff and

Gibbons 2004). Any delay or reduction in population growth after founding will reduce the effective population size, with a greater loss of diversity the longer the delay or the narrower the bottleneck.

Given the preference people have shown for introducing bright or showy birds, it seems likely that the initial propagule could also be biased toward the more showy gender, which in most cases is the male. A strongly skewed sex ratio will reduce genetic variation, especially at small population sizes. Even if equal numbers of each sex were introduced initially, highly dichromatic species can exhibit strong sexual selection. Thus, for several generations after introduction, one to a few males may be responsible for most young produced. This will reduce the effective population size and further drive down genetic diversity. There is some evidence that male house finches were very much more common in eastern US markets than females owing to their brighter red colouration (females are a dull red-brown; Hawley, et al. (2005) and references therein). Sexually dimorphic bird species may also be less likely to establish exotic populations (see section 4.3.2).

There is very little theoretical or empirical information on the likely effective population sizes of birds, much less exotic birds. Thus, we have no means of judging the relative importance of these various mechanisms in reducing effective population size, or even what the average ratio of effective to census population size may be. This topic is thus a productive area for future research, especially as it likely modifies how we interpret the role of propagule pressure on establishment success (Chapter 3). All measures in birds to date have used census population size as the measure of propagule size even though we should perhaps expect effective population size to be quite a bit lower than this number, and that some species' life-history traits may make the difference between the two measures larger (e.g., sexual colour dimorphism).

8.2.2 Increases in Genetic Variation

There are three mechanisms by which an exotic bird population may show increases in genetic diversity through time: migration, mutation, and hybridization. Mutation is the ultimate source of all genetic variation, but mutation rates are, on average, very low (Connor and Hartl 2004). Most calculations put the probability of mutation at any given locus within one generation in the range of one in ten thousand to one in a million. Nevertheless there are regions of the genome in most animals that mutate faster than others (e.g., regions that code for disease resistance), and there are likely to be species differences in overall mutation rate (Graur and Li 2000). Most mutations are deleterious and are typically lost from a population through selection. There is theoretical evidence that suggests that mutations may have a higher probability of being retained within rapidly expanding populations, which is obviously a characteristic of invasive species (Wares, et al. 2005). Little is known about the mutation rates experienced by exotic bird species, but it is safe to assume that mutations are indeed occurring and that these will change gene frequencies, albeit on a much longer time scale than drift, migration, or hybridization.

A more likely candidate mechanism for increases in genetic diversity among exotic bird populations is migration. At first glance we might assume that migration between exotic populations or between the native and non-native ranges of one species are low. After all, from Chapter 2 we get the clear impression that the transport and introduction of exotic birds is idiosyncratic through space and time. For most exotic bird species, we can probably assume that migration between populations (be they native or non-native) is likely to be very rare, as the introduction event was a one-time occurrence while most exotic populations of the same species are located far from one another (e.g., on different continents or islands).

In some cases, however, and particularly for species commonly transported as pets or for hunting purposes, this assumption may be false. Here, it is quite common for regular shipments of individuals to arrive at any one location over a period of subsequent years. Individuals transported by hunting groups are immediately released into the wild and may interbreed with individuals that were released earlier, or with the offspring of these early releases. In some cases, captive-reared birds (e.g., ring-necked pheasant *Phasianus colchicus* in Britain; Draycott, et al. 2008) continue to be intentionally released every summer. Any individuals that escape from confinement as pets have the opportunity to breed in the wild as well. The initial colonists may have been few in number, such that the initial founding population is likely to have experienced some loss of genetic variability relative to the source population. In this context, later releases or escapees can be considered 'migrants' as they bring with them a set of alleles at each locus that may, or may not, already be found in the incipient exotic population (Novak and Mack 2005). It would not take many of these later releases to counteract any erosion of genetic variability due to drift in the original founding population, as one migrant per generation is enough to counteract this loss even at very small population sizes (Connor and Hartl 2004). In larger populations (e.g., effective population size of fifty individuals or more), variation is not lost rapidly due to drift and thus a much lower rate of migration is needed to counteract it.

Given the market dynamics underpinning the pet bird industry (Robinson 2001), we may expect species that are easily kept and bred in captivity to experience 'migration' frequently enough to counteract any effects of drift. Indeed, we may predict when 'migrants' are more likely to arrive if provided with enough information on changes in market prices for a species. Nevertheless, subsequent arrivals of escaped pet birds may not appreciably raise genetic diversity if captive populations themselves show low genetic variability. We have little information on the degree to which bird populations kept in captivity for the pet industry are genetically depauperate, although Forstmeier (2007) shows that zebra finches *Taeniopygia guttata* that are kept as captive populations for behavioural research show very little loss of alleles as compared to wild populations. If the same is broadly true for captive pet populations of birds, then their continued introduction into established exotic populations can provide novel genetic material.

Finally, genetic variation may increase in an exotic population relative to its native source via intra- or interspecific hybridization. Introductions of exotic birds can bring together individuals from genetically distinct native populations or from related taxa. In the former, the degree of differentiation between native populations is low enough that individuals from these different groups can, and will, mate with each other in the exotic range. Indeed, the offspring from these matings may enjoy fitness advantages through heterosis. In the latter instance, individuals of one species may mate with congeners that are native to where they were introduced, or with previously established exotic congeners. Typically, although not always (see below), the young produced from these matings will suffer from outbreeding depression. Nevertheless, intra- or interspecific hybridization will serve to introduce new genes to an exotic population. In some cases, intraspecific hybridization produces higher within-population genetic diversity than can be found anywhere across the native range of an invasive species (Lockwood, et al. 2007). Novak and Mack (2005) liken this process to the 'migrant pool' scenario of Slatkin (1977), whereby colonists are pulled from across a species' range and then coexist as a single incipient population. This sampling process tends to introduce a broad array of alleles into the incipient population, and the diversity of alleles that results may supersede that of any single native population, especially if the native population shows moderate to high levels of spatial structure. The alternative is the 'propagule pool' where colonists are drawn from one source only, and that source is genetically well-mixed so that each individual has a fairly constant probability of carrying common alleles (Slatkin 1977).

8.2.3 Empirical Examples of Change in Genetic Variation in Exotic Bird Populations

In the mid-1980s exotic birds were seen as novel systems for answering basic questions in population genetics, and thus several authors have considered how often, and to what extent, exotic bird populations lose genetic diversity after founding (Merilä, et al. 1996, and sources therein). Typical of these articles are studies on three exotic bird species (chaffinch *Fringilla coelebs*, European starling *Sturnus vulgaris*, and common myna *Acridotheres tristis*), which were introduced to New Zealand by acclimatization societies, and are now widely distributed there having increased in population size relatively quickly after founding (Ross 1983; Baker and Moeed 1987; Baker 1992). These authors report that four hundred chaffinches from Britain were introduced between 1862 and 1877. The introduction event that gave rise to most of the common mynas in New Zealand was made in 1875 and 1876 and included about 100 individuals, most likely from exotic populations in Australia. The European starling was repeatedly introduced during the 1860s and 1870s, with most introductions including in excess of 100 individuals, and all likely transported from Great Britain.

Electrophoretic studies on allozyme variation between the source and New Zealand exotic populations of these three species are remarkably consistent in their findings and are broadly consistent with theoretical predictions reviewed above. All three exotic populations in New Zealand tended to have fewer rare alleles as compared to their source regions (Ross 1983; Baker and Moeed 1987; Baker 1992). The common myna had fewer polymorphic loci than their native Indian populations (Baker and Moeed 1987), but there was no similar statistical difference in this metric for either starlings (Ross 1983) or chaffinches (Baker 1992) in New Zealand versus Britain. None of the exotic populations had significantly lower average heterozygosity than observed in the native populations. Despite the lack of statistical evidence for reductions in the proportion of polymorphic loci and average heterozygosity, each exotic population did have lower scores for these metrics than in their native range.

Such results are to be expected if the native populations are essentially panmictic, and thus show little geographic structuring in genetic variability (i.e., conform to the propagule pool model above). There does indeed appear to be very little genetic geographic structure within native populations of chaffinches, mynas, and starlings (Ross 1983; Baker and Moeed 1987; Baker 1992). In such cases, the process of collecting individuals within any locale in the native range will randomly sub-sample nearly all available genetic information. The founding individuals are nevertheless more likely to possess alleles that are common in the native source population, and less likely to possess alleles that are rare, as was shown to be the case in the three species tested. Thus, while there is evidence for a founder effect in their New Zealand populations, enough individuals of each species were introduced to ensure that most common alleles were represented, and that these alleles were present at their 'native' frequencies. Consequently, there were rarely statistically significant drops in average heterozygosity or the proportion of polymorphic loci within the exotic populations, even though they lost rare alleles. The rapid rate of population increase each of these populations experienced after introduction was also argued to contribute to the observed lack of a reduction in genetic variability (Ross 1983; Baker and Moeed 1987; Baker 1992). If any of these populations had lingered at low numbers after initial introduction, reductions in genetic variation may have occurred despite their propagules being large enough to ensure no large initial changes in allele frequencies. Other studies on house sparrows *Passer domesticus* in North America, greenfinches *Carduelis chloris* in New Zealand, and Eurasian tree sparrows *Passer montanus* in North America all show broadly similar patterns (Parkin and Cole 1985; St Louis and Barlow 1988; Merilä, et al. 1996).

The existence of several studies evaluating largely the same question also allowed Merilä, et al. (1996) to conduct a comparative analysis of the findings. The amount of genetic variability lost in a newly established population will vary approximately linearly with the number of individuals in the founding group, the number of founding events, and the rate at which the incipient population increases through time. Thus, low numbers for each variable should be correlated with reductions in any measure of genetic variability. Merilä, et al. (1996) compiled information from seven studies, including some of those listed above, to test for relationships consistent with these theoretical expectations (Table 8.1). They used percentage of polymorphic loci, average heterzygosity, and number of alleles per locus as their measures of genetic variability and calculated absolute differences in these scores between the native and exotic populations of each species. Note that most species were introduced successfully to several non-native locales so there are more than seven independent data points with which to establish a relationship.

Merilä, et al. (1996) found that low propagule size, slow post-establishment growth rate, and fewer independent introduction events led to reductions in genetic variability in exotic bird populations. Nevertheless, when they graphed these relationships (Figure 8.1), it was clear that these measures of genetic variability often did not differ between the exotic and native ranges. Only rarely was there a drop in any metric of genetic variability associated with the introduction of exotic birds. Indeed, on some occasions variability was actually larger in the exotic range as we might expect following the migrant pool scenario (see above): we will return to this observation in more detail below. The results presented in Merilä, et al. (1996) suggest that most measures of genetic variation will not be substantially reduced unless the founding population is below 500 individuals, remains at this or a lower number for more than ten generations, or is introduced fewer than five times.

There are some interesting questions left unanswered by Merilä, et al. (1996). For example, Figure 8.1 does not address how these three mechanisms of genetic change interact with one another. Comparing the slopes for each mechanism could help determine the relative effect on genetic variation of increasing numbers of introductions versus increasing propagule size versus increasing the rate of post-establishment population growth. Such comparisons have value if there is evidence that a lack of genetic variability can inhibit establishment in the first place or stall growth rates of invasive bird populations (see below).

Although there are sufficient studies on genetic changes within exotic bird populations to allow comparative analyses, we still have a relatively limited understanding of how conditions at introduction and establishment affect genetic variation. There are more than 700 known established exotic bird populations, representing more than 200 species, but only twenty to thirty populations from seven species have been surveyed for changes in genetic variability. Merilä, et al. (1996) point out that the sample of studies available to them consists of a

Table 8.1 Comparison of genetic variability in native and introduced bird populations. n_p, number of populations; n_1, average number of individuals scored per locus; *A*, average number of alleles per locus; *P*, percentage polymorphic loci; H_E, average expected heterozygosity; intro, year of introduction; Abun, year when the species was recorded as abundant; n, minimum number of introduced individuals; #times, minimum number of introductions. AU = Australia, E-USA = eastern USA, FI = Fiji, HA = Hawaii, NZ = New Zealand, SA = South Africa. From Merilä, et al. (1996).

				Genetic parameters			Introduction history			
Species	**Area†**	n_p	n_1	*A*	*p*	H_E	**intro.**	**Abun.**	*n*	**#times**
European starling *Sturnus vulgaris*	Native	6	24.0	1.42	28.5	0.033	–	–	-	-
	NZ	6	24.0	1.35	33.8	0.043	1867	1878	653	13
Common myna *Acridotheres tristis*	Native	7	28.4	1.43	31.5	0.060	–	–	–	–
	NZ	5	57.4	1.24	19.5	0.042	1862	1912	100	2
	AU	2	43.9	1.30	23.1	0.060	1862	1883	252	5
	HA	1	37.6	1.20	18.0	0.060	1865	1879	–	–
	FI	2	25.9	1.30	20.6	0.050		–	–	–
	SA	2	38.3	1.15	12.9	0.030	1888	1930	6	2
Chaffinch *Fringilla coelebs*	Native	10	28.7	1.34	24.5	0.048	–	–	–	–
	NZ	8	26.1	1.39	28.3	0.066	1867	1870	377	13
House finch *Carpodacus mexicanus*	Native	6	40.6	1.48	27.4	0.045	–	–	–	–

	E-USA	6	38.5	1.37	23.7	0.042	1940	–	–	–
	HA	4	20.0	1.33	21.0	0.055	186?	–	–	–
Greenfinch *Carduelis chloris*	Native	14	33.6	1.19	14.1	0.025	–	–	–	–
	NZ	7	35.1	1.22	16.8	0.025	1862	1868	64	5
Eurasian tree sparrow *Passer montanus*	Native	2	55	1.45	33.3	0.098	–	–	–	–
	USA	3	83	1.33	23.9	0.078	1870	–	20	1
	AU	1	15	1.3	23.1	0.074	1863	–	–	–
House sparrow *Passer domesticus*‡	Native	14	60.7	1.63	36.9	0.097	–	–	–	–
	NZ	4	51.4	1.42	36.2	0.080	1859	1869	418	7
	AU	3	36.7	1.50	33.3	0.103	1863	1906	494	8

‡ Only loci scored in all populations included

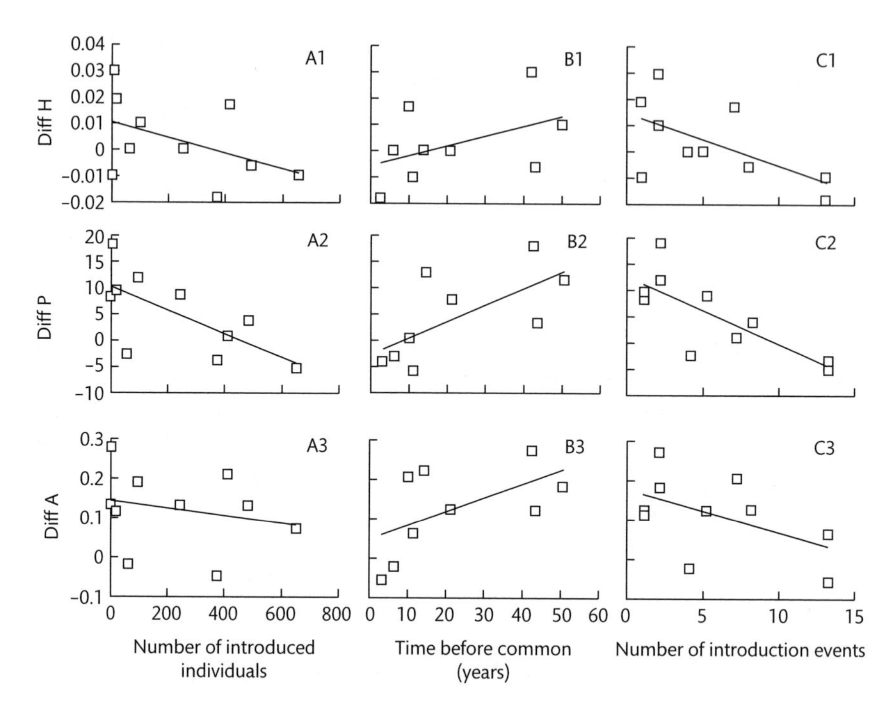

Fig. 8.1 Differences in the levels of genetic variability (Diff A = difference in average number of alleles per locus, Diff P = difference in percentage of polymorphic loci, Diff H = difference in average heterozygosity) between native and exotic populations of birds as a function of number of introduced individuals (A1–A3), time elapsed between the introduction of a species and when it becomes common (B1–B3), and the number of introduction events (C1–C3). From Merilä, et al. (1996).

highly non-random subset of all exotic birds. Most studies of changes in genetic variability are on exotic bird populations that were introduced in large numbers relative to most such introductions (see Chapter 3). The introductions also concern populations that are now very large. As shown in Chapter 6, the set of exotic bird species that are widespread is a non-random set of those that established. There is a variety of exotic bird populations that have not appreciably spread beyond their initial introduction points. The lack of information about genetic changes in initially and/or persistently small exotic populations represent large missing pieces in our knowledge of how genetic variability is altered via the introduction process: they may show considerable reductions in all measures of genetic variability.

The ability to compare rates of genetic erosion between exotic bird populations that are now widespread to those that are not may be especially relevant when considering endangered species management and reintroductions (Briskie 2006).

It is the group of exotic populations that did not expand for some period after initial establishment that probably most accurately reflects the situation of imperiled bird species. It is thus this set of exotic populations that will repay close study in terms of documenting the effects of low genetic diversity on fitness and population growth rates (see below).

There is one final caveat needed for this review of changes in genetic variation within exotic bird populations. All of the above analyses relied on allozyme electrophoresis, which evaluates variability in protein structure. Yet, there is good recent evidence that measurements of variability in protein structure are less variable than direct measurements of segments of DNA (Avise 2004). Modern molecular techniques that isolate microsatellites are particularly favoured for studies of population variation as these segments of DNA are highly variable and individual genotypes can be directly inferred (Frankham, et al. 2004). The use of modern genetic tools may reveal more fine-grained changes in genetic diversity, and perhaps provide evidence for reductions where none was found before.

A clear example of the influence of changing techniques on our understanding of genetic loss via colonization comes from studies of house finch populations established in the eastern USA from releases by pet traders in the 1940s (Elliott and Arbib 1953). The release event that is thought to have led to the eastern population may have had an effective population size of as few as twenty-four individuals. Given this introduction history, and the results presented by Merilä, et al. (1996), we would expect the eastern house finch population to show evidence for loss of genetic variability. Yet, two studies evaluating amplified fragment length polymorphism (AFLP) found equivalent levels of diversity within the native source population in the western USA and the introduced population in the eastern USA (Wang, et al. 2003; Hawley, et al. 2006). However, using ten highly variable microsatellites, Hawley, et al. (2006) showed evidence for significantly lower allelic diversity and average heterozygosity within the exotic than the source population. This example suggests that our understanding of how initial propagule size (i.e., number of founders), number of introduction events, and rate of population growth after introduction influence genetic variability within exotic bird populations will need updating as modern molecular genetic techniques are increasingly applied.

8.3 Studies of Genetic Variation between Exotic Bird Populations

The above section considered changes in genetic variability within single exotic bird populations following translocation. However, there are a number of species that have been introduced to several different locations, while many exotic populations have substantially expanded in geographic extent over time. These latter populations may in the extreme span entire continents, and may be viewed as existing as several connected sub-populations. This aspect of exotic bird invasions inevitably leads to consideration of the genetic structuring of populations across space, or more formally, the degree of between-population genetic variation. If most exotic populations do not experience significant drops in within-population genetic variation (above), does this mean that they also do not exhibit large differences in between-population genetic variation?

The answer to this question will depend in part on the degree to which exotic populations of the same species exchange individuals through migration. As long as gene flow (i.e., the movement of individuals from one population to another) is high, exotic populations will not show substantial between-population variation. The reverse is also true: we expect between-population variation to be high if there is no gene flow between exotic populations. In this latter case, between-population differences in allele frequencies also depend in part on founder population size: they may be higher if only a few individuals in each location independently founded the non-interacting exotic populations. Founder effects dictate that each exotic population should receive a random sub-sample of alleles that are relatively rare in the native source range, and that these differences will be exaggerated through time by drift (Connor and Hartl 2004).

Based on these expectations, we can define the extremes on each end of the drift versus migration spectrum for exotic bird populations. At one end, we would expect that a large well-mixed population founded by a single introduction event would show very low between-population variation—perhaps lower than in the native range. Cabe (1998) provides an example of this scenario among European starlings introduced into North America. About 100 European starlings were purposefully brought to New York City from Britain in 1890 and 1891 and released in Central Park. These founding individuals established a breeding population that has so increased in size and geographic extent that by the early 1980s Feare (1984) estimated that 200 million European starlings inhabited the continent. Cabe (1998) collected starlings from across North America and, using enzyme electrophoresis, compared within- and between-population genetic variability to those of the likely source population in Britain. He measured between-population variability using F_{st}, which measures the proportional reduction in metapopulation heterozygosity due to differentiation between subpopulations (Connor and Hartl 2004). F_{st} is usually calculated by averaging heterozygosities across all subpopulations, and one of the benefits of the metric is that it can be directly compared across geographic regions. For birds, highly structured populations, such as those spread across islands, can show F_{st} values of 0.1 or higher whereas unstructured populations typically show F_{st} values <0.01 (Barrowclough 1983).

Cabe's (1998) results for the within-population variability of European starlings in North America very closely matched those described in the previous section, in that rare alleles were lost but there was no significant reduction in average heterzygosity or the percentage of polymorphic loci. He also found remarkably little between-exotic-population variability across North America ($F_{st} = 0.0226$). In other words, European starlings collected in California had very similar allele frequencies to those collected in Vermont, despite the two locations being separated by almost 5,000 km. Within Britain, European starlings show an equally modest amount of between-population genetic structure ($F_{st} = 0.0101$). For comparison, Ross' (1983) work on European starlings in New Zealand showed more between-population structuring than in either North America or Britain ($F_{st} = 0.0316$). Thus, starlings in North America exhibit low genetic variation both within and between populations. We also have the interesting situation where the same species established as an exotic in two different places (New Zealand and North America) from the same source stock has evolved different levels of geographic genetic structure over approximately the same time period (about 100 years). Counter-intuitively, the location that shows more structure (New Zealand) is by far the smaller of the two non-native populations (Cabe 1998). This may be because European starlings were liberated in at least five different acclimatization districts in New Zealand (Thomson 1922; Long 1981), versus the single release site in North America. None the less, the amount of between-population genetic structure of European starling populations anywhere (exotic or native range) is comparatively small.

At the other end of the drift versus migration spectrum, we would expect that several geographically very distinct exotic populations (e.g., on oceanic islands) founded by independent sets of founding individuals should have very high between-population variation—perhaps substantially higher than in the native range. Baker and Moeed (1987) provide a classic example of this scenario for exotic populations of the common myna on islands. We described above how Baker and Moeed found little evidence for reductions of within-population genetic variation among mynas in New Zealand, but these authors also sampled individuals from Hawaii, Australia, South Africa, and Fiji. The populations in Hawaii, Fiji, and Australia were all founded from individuals taken from India. The New Zealand population probably was founded from the Australian population. Baker and Moeed (1987) found that within-population genetic variation within all exotic populations except South Africa showed relatively little reduction. The South African population was founded by only a few individual mynas taken from Mauritius, which population was itself founded by only a few individuals introduced from India. Thus, the South African myna population experienced two strong bottlenecks and, as expected from this, shows a significant reduction in all measures of within-population genetic variation.

If we consider all the exotic common myna populations as part of a single metapopulation, then we can calculate F_{st} values just as did Ross (1983) and Cabe (1998). Baker and Moeed (1987) estimated between-exotic-population F_{st} for common mynas to be 0.123, which is an order of magnitude above values obtained for European starlings in North America and New Zealand. It is also well above the calculated F_{st} values for common mynas in their Indian native range (0.032; Baker and Moeed 1987). Such levels of geographic genetic structure are on the order of that seen among recognized subspecies of birds. This observation led Baker and Moeed (1987) to suggest that bottlenecks and random drift had resulted in rapid genetic structuring across the common myna's exotic range within about 100 years of introduction.

Nearly every author that has reported on changes in allele frequencies within exotic populations has also reported the degree of geographic structuring between populations in the exotic range. Although no one has adequately reviewed studies of between-population differentiation as Merilä, et al. (1996) did for within-population genetic changes, we can see some broad-brush patterns when comparing published results (Table 8.2). Notably, the degree of geographic structuring in allele frequencies has little to do with the ultimate size or extent of the exotic bird population. Very large populations can be genetically homogenous if gene flow is high and the number of independent introduction events is low (or if only one event resulted in successful establishment). The key to understanding when we are likely to see strong between-population differences is, first, to identify clearly independent introductions to locations that are very unlikely to exchange individuals through dispersal and, second, to recognize how individuals of a single species spread over a large area are responding to available habitat. If habitat is patchy and the space between patches is unsuitable, some degree of between-population differentiation is likely. This seems to be the case for house sparrows introduced to Australia. Even though house sparrows are now widespread across this continent, they show comparatively high between-population differentiation, which Parkin and Cole (1985) attribute to the fact that they find the natural habitats that lie between human habitations in Australia somewhat unsuitable.

It seems likely that there will be many situations where exotic bird populations show some level of geographic differentiation, but there is a dearth of studies that explicitly consider how the degree of dispersal between habitat patches (or lack of dispersal) structures exotic populations. This echoes the need to study dispersal distances within expanding (or static) populations of exotic birds in order better to predict when, and how quickly, these species will spread across the non-native environment (Chapter 6). The need for more studies on this topic, and especially comparative studies, becomes clear when we consider what we can learn from exotic bird populations that may be of use to managing threatened species in fragmented habitats. It does not take much imagination to recognize that the two key issues for understanding the degree of differentiation among exotic populations

Table 8.2 Across-locality genetic similarity scores (F_{st}-values) for exotic bird populations and their native counterparts. s.d. = standard deviation.

Species	Status	Locations	F_{st}	Number of loci	Date of introduction	Number of individuals introduced	Number of release events
European starling[1]	Exotic	USA	0.0067	23	1890–91	100	>2
Sturnus vulgaris		New Zealand	0.0316	24	1862	100–1000	>5
	Native	Europe	0.001	–	–	–	–
Common myna[2]	Exotic	Australia	0.123	39	1862–1872	200+	>4
Acridotheres tristis		New Zealand			1870s	100+	>2
		Fiji			1890 to 1900	unknown	unknown
		Hawaii			1865	unknown	unknown
		South Africa			1888–90	<50	>2
	Native	India	0.032	39	–	–	–
Eurasian tree sparrow[3]	Exotic	Australia and USA	0.054	39	Australia–1863	Australia–30 to 40	Australia–1
Passer montanus							
Native species[4]	Native	Survey of 25 native species	Avg = 0.049 s.d. = 0.05	1 to 14	–	–	–

[1] Cabe 1998; Ross 1983.

[2] Baker and Moeed 1987.

[3] St Louis and Barlow 1988.

[4] Barrowclough 1983.

mentioned above—the degree of population isolation, and the distribution of available habitat—relate also to the situation for most threatened and endangered species (Hanski and Gaggiotti 2004). There is such a variety of exotic bird populations that most examples of population structuring encountered among native species will surely be represented. We just await the analyses.

8.4 The Role of Genetic Variation in Establishment Success and Range Expansion

8.4.1 Reductions in Genetic Variation Lower Establishment Success and Range Expansion

Do any of the changes in genetic diversity of the sort documented in exotic bird populations influence probability of establishment or how fast the population expands its exotic range? Clearly the moderate loss of some genetic diversity has not stopped invasions of starlings, house finches, common mynas, or any of the other species considered above. But as we have already pointed out, we should perhaps expect greater losses of genetic variation within populations founded by only a few individuals or that have not rapidly increased in population number since initial introduction. Certainly, species introduced in small numbers are most likely to have reduced genetic variation, and they also have documented lower rates of establishment (Chapter 3). Unfortunately, it is impossible to know whether any of the hundreds of exotic bird populations that failed to establish self-sustaining populations after introduction did so because of a lack of genetic variation.

There is a robust literature on problems associated with low genetic variability, nearly all of which comes from the study of imperiled species (Frankham, et al. 2004). There is good evidence that bird populations that exist at very low numbers have relatively low genetic diversity. Indeed, most investigations of the genetic composition of these populations indicate that individuals show some level of inbreeding and thus share a relatively large fraction of alleles through common descent (e.g., Bouzat 1998). High levels of inbreeding can manifest as low reproductive output or reduced adult and juvenile survival (Keller and Waller 2002). For birds, the effect is typically low sperm viability, low hatching success, or reduced clutch size (Briskie 2006). Any population experiencing these effects is more likely to become extinct than those that are not, simply through a reduction in annual fecundity. We currently lack studies that directly explore how the loss of genetic variability affects establishment success or fitness in exotic species, although Briskie (2006) makes a strong argument that such studies are worthwhile for the insights they will provide for threatened bird species management.

However, the suite of exotic birds that are most valuable in this regard is precisely the set of species for which we have no genetic information. Highly imperilled bird species have typically existed at small population sizes for several generations, and may only survive today via successful reintroduction attempts that necessarily included one or more founding events. Nearly all published genetic assays of exotic bird populations involve species that are now common, were introduced in relatively large numbers (over fifty), or underwent rapid population increases fairly soon after introduction. As shown above, and according to expectation, these species have lost only marginal amounts of genetic diversity. Although we cannot rule out some level of inbreeding, it is probably far below the level recorded for imperiled species. For detailed demographic studies of exotic birds to inform avian conservation efforts, especially related to the influence of low genetic diversity, the field must shift towards a focus on species that are newly introduced (and thus may eventually fail to establish) or that are known to have persisted at small population sizes for a number of generations.

There is, however, one study that suggests that low genetic variability may be associated with fitness effects in exotic birds. For a range of exotic species established in New Zealand, Briskie and Mackintosh (2004) compared the rates of hatching failure in eggs laid in the exotic populations to the equivalent rates in the native populations. They found that the magnitude of the difference in these hatching failure rates was negatively related to propagule pressure. Relative to the native range, exotic species had increasingly higher hatching failure rates as the number of individuals introduced declined (Figure 8.2). This provides strong evidence that there is a fitness consequence to exotic species of reduced population size, presumably caused by the loss of genetic variation as the population passed through a bottleneck at introduction. The extent to which these fitness consequences may have influenced the failure of species that did not establish of course cannot be gauged, but Briskie and Mackintosh's results do suggest that reduced genetic variation can be a significant jeopardy to the establishment of an exotic species.

Although not addressing low genetic variability per se, Carrete and Tella (2008) suggest that genetic effects may influence the establishment success of wild caught versus captive bred escapees. Establishment success in exotic bird species in Spain was higher for species more commonly available on the pet bird market (high propagule pressure), and higher for wild-caught than captive-bred species. Carrete and Tella suggest that the latter effect may be because of inbreeding depression, loss of genetic diversity, and artificial selection by aviculturalists. Formal tests of genetic variation in captive bred versus wild caught individuals would be one way to assess whether this variation is related to establishment success.

There is even less information on the role of genetic variation in determining the size of the geographic range or rate of range expansion among well-established exotic populations. No studies yet explicitly link these processes in exotic birds. Nevertheless, there is one example that shows how such effects might work in

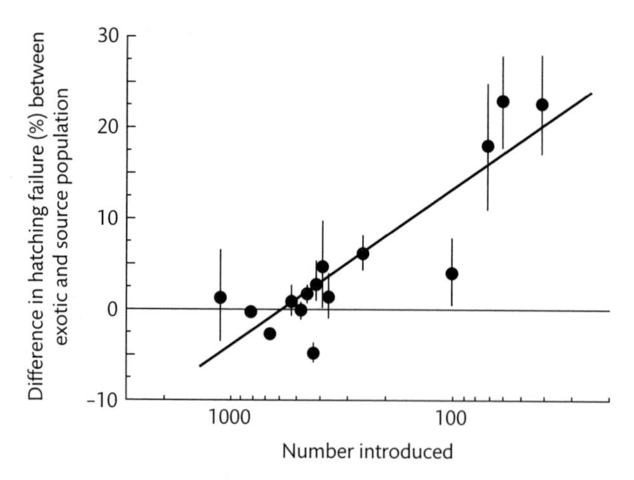

Fig. 8.2 The effect of propagule pressure on hatching failure in exotic bird species established in New Zealand. Species introduced in smaller numbers show a larger increase in hatching failure rates in the exotic relative to the native population. Reprinted with permission from Briskie and Mackintosh (2004).

principle. It concerns reductions in genetic variation experienced by exotic house finches in the USA, how this plays a role in their resistance to the novel strain of the common poultry pathogen *Mycoplasma gallisepticum* (MG) that appeared in house finch populations in the USA in 1994 (section 7.5.4), and, by extension, how this may have mediated a decline in population size.

MG causes severe conjunctivitis in house finches, and infected birds within their exotic range have been shown to have lower over-winter survival rates (Fischer, et al. 1997). The infection reached epidemic proportions within 3 to 4 years and large decreases in exotic house finch numbers followed (Hill 2002). The MG pathogen spread quickly across eastern populations and finally made its way into the native western populations in 2002. This situation allowed Dhondt, et al. (2006) to compare the rate of expansion and degree of prevalence of the MG pathogen within the introduced and native populations of the house finch. They found that MG spread was faster and prevalence rates were higher within the introduced range as compared to the native range.

Hawley, et al. (2006) drew a connection between these differences in the effects of MG and genetic variation. They captured house finches within their exotic range (New York) and conducted controlled experiments to see if individuals with lower levels of multi-locus heterozygosity were more susceptible to the MG pathogen. They found a clear pattern whereby the more genetically diverse individuals in the exotic population were less susceptible to MG infection (Figure 8.3). Given that the history of the house finch introduction into the eastern United States has

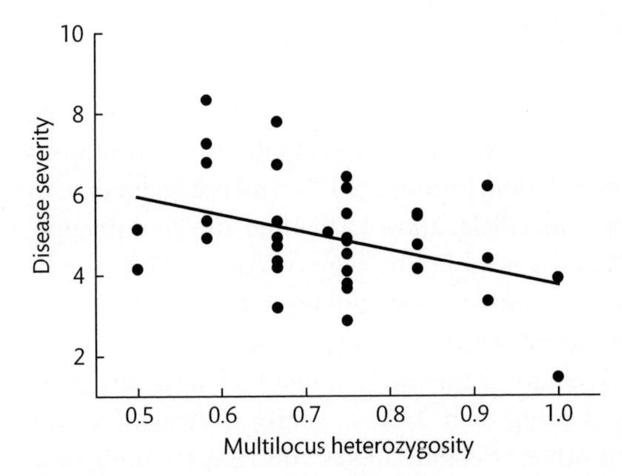

Fig. 8.3 The relationship between disease severity and multilocus heterozygosity in house finches *Carpodacus mexicanus* from the eastern USA infected with *Mycoplasma gallisepticum*. Disease severity was scored as the sum of the severity of inflammatory symptoms for seventy days post-inoculation. Reprinted with permission of John Wiley & Sons Ltd from Hawley, et al. (2006).

lead to genetic erosion there (see above), there is the distinct possibility that the MG pathogen spread more rapidly, and had more of an impact on population numbers, in the exotic range because of an overall reduction in genetic diversity (Dhondt 2006; Hawley, et al. 2006). Briskie (2006) presented further evidence that reductions in genetic variability may have measurable effects on disease resistance in birds, and the spread of disease may have profound impacts on bird populations that exist in low numbers. Thus, there seems the possibility that some exotic bird populations may be kept at low numbers and in small ranges, or were possibly driven extinct, because of the negative effects of disease on population growth rates, and that this dynamic is mediated by a drop in genetic variation associated with the founding event.

8.4.2 Reductions in Genetic Variation and Increases in Establishment Success and Range Expansion

Can the reverse happen, where a decrease in genetic variability allows a greater probability of establishment or invasion success? Population bottlenecks and founding events can lead to linkage disequilibrium and a change in pleiotropic relationships between genes (Connor and Hartl 2004). These changes may allow the emergence of a novel trait that has a fitness advantage within the incipient population. Such 'genetic revolutions' have been argued as necessary precursors to speciation (Mayr 1963) and it is possible that such changes could lead to increases in establishment success and geographic spread among exotic species (Cox 2004; Lindholm 2005). Although to date this has never been directly investigated within exotic bird populations, the changes in the genetic structure of populations that undergo bottlenecks and founding events may set the stage for successful establishment or rapid range expansion.

A well-known example of how reductions in genetic variation positively influenced range expansion in an exotic species comes from the invasion of Argentine ants *Linepithema humile* in North America (Tsutsui, et al. 2000). The Argentine ant achieved very high population densities and wide geographical distributions in part because the founding population experienced an intense reduction in allelic diversity. In this species, specific alleles allow individual ants to differentiate between their 'home' colony and a neighbouring 'away' colony. When individuals from two colonies come into contact, they will often fight to the death. The founding of the North American Argentine ant population resulted in a massive loss of alleles, such that individuals in the exotic range no longer recognize individuals from other colonies as being such. In fact, nearly all Argentine ants in North America consider each other as colony mates and thus form gigantic super-colonies, which allow them to dominate native ants and expand their range widely (Suarez, et al. 2001).

There is no equivalent in the bird world for eusocial ants. Nevertheless, many bird species are highly social and the genetic basis for the various mating systems, song dialects, and courtship behaviours they exhibit are not well understood. How the loss of rare alleles, or genetic revolutions, affect the breeding behaviour of exotic bird populations is thus a complete unknown.

8.4.3 Increases in Genetic Variation and Increases in Establishment Success or Range Expansion

One further aspect of the genetics of exotic populations that has received almost no attention within birds is the possibility that increases in genetic variability within the introduced range result from admixtures or hybridization. Species known to hybridize with native or already established exotic species may experience greater success in establishment or spread because of any resulting increase in genetic diversity (e.g., acquisition of a rare allele that confers tolerance to some abiotic or biotic factor). There is evidence for such effects among exotic plant species, with the best-known case being hybridization between species of *Spartina* in Britain and San Francisco Bay (Lockwood, et al. 2007). However, while the diversity generating potential of hybridization is commonly recognized amongst botanists, the opposite view prevails amongst zoologists (Seehausen 2004). The genetics of plants and animals is quite different (mainly the ability to reproduce asexually: Ellstrand 2000; Seehausen 2002) suggesting that the mechanisms for increased success are not directly transferable when discussing the role of hybridization in the establishment and spread of exotic birds. Nevertheless, hybridization is common in the bird world and can happen even between species in different genera (McCarthy 2006). Hybridization seems especially common within the Anatidae (e.g., the mallard *Anas platyrhynchos* or ruddy duck *Oxyura jamaicensis*) and Psittacidae, which are clades that include unusually high numbers of exotic species (Chapter 2).

In general, hybridization appears to be more frequent in populations that are at the edge of their range, where population densities are low, mates are rare, or selection against hybrid offspring by competition with parents is weak (Seehausen 2004). Even though hybrid offspring often have lower fitness than parental forms in either of the 'parental niches', some hybrids do better in a 'third niche' (see Seehausen 2004 and references therein). For example, Grant and Grant (2008) provide evidence that hybridization introduces the genetic variability needed to adapt to dramatic annual changes in resource availability among Galapagos finches. These studies raise the distinct possibility that exotic birds might realize establishment success or fast spread rates within their non-native range through hybridization and the resultant creation of novel genotypes, some of which may be more suited to the non-native environment.

Drake (2006) presents a compelling scenario by which such increases in genetic variability due to hybridization (specifically, heterosis) elevate establishment success. He posits a 'catapult effect' whereby the increase in genetic variation in the generation (F2) that follows hybridization between two distinct lineages (F1 or parental lines) will confer a fitness advantage to individuals of the F2 generation. If this increase in fitness coincides with the very early stages of establishment, when population sizes are typically low, then the larger number of offspring produced by the F2 generation will increase the likelihood of population persistence (Figure 8.4). The mechanism for this increase involves the simple observation that populations with more individuals typically persist longer than those with relatively few individuals since they are buffered against demographic and environmental stochasticity. Drake (2006) found evidence in favour of the catapult effect in ring-necked pheasants released across sites in the USA. Pheasant populations founded by hybrid individuals were 2.2 times more likely to establish than populations founded by individuals that were of the parental subspecies. This effect was not confounded with propagule pressure, and indeed Drake (2006) found little evidence that propagule pressure by itself influenced establishment across ring-necked pheasant releases (but see section 3.2.1 for a possible reason for this lack of association).

The increased fitness associated with hybridization tends not to persist past the F2 generation in animals, due to segregation in sexual reproduction (Lee 2002). Thus, the advantages of hybridization are likely to be short lived, and may only influence establishment success via a quick boost in population size that reduces population extinction probabilities (hence 'catapult'). To our knowledge, no one has investigated the possibility that the advantages to hybridization persist beyond the F2 generation in exotic bird populations.

Finally, the possibility exists that exotic bird populations have been founded by individuals derived from distinct genetic lineages or sub-populations in the native

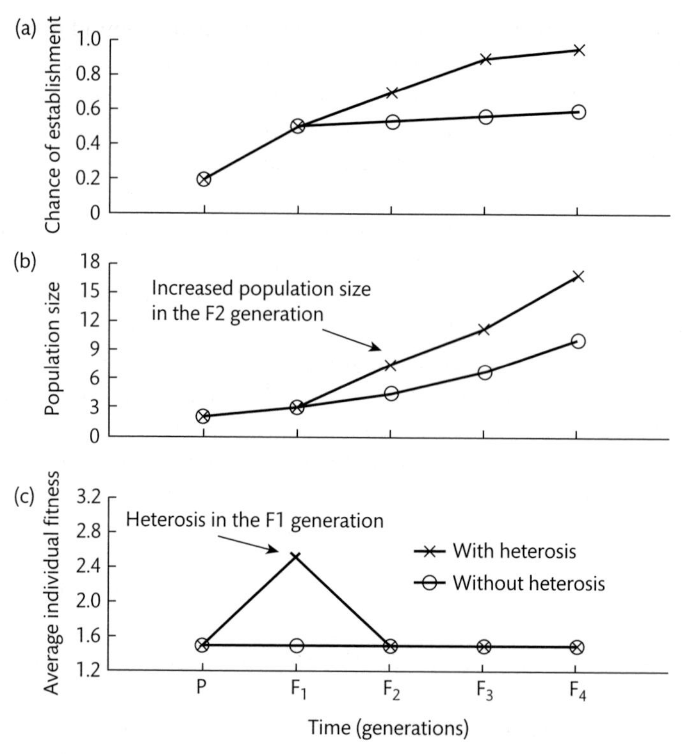

Fig. 8.4 A graphical model of the catapult effect. The trajectory of a population exhibiting heterosis (crosses) is compared with that of a population not exhibiting heterosis (circles). In the F1 generation, the population exhibiting heterosis shows an increase in average individual fitness (c). In populations of small size, the increase in fitness in the F1 generation of the heterotic population results in an increase in the population size in the F2 generation, relative to the non-heterotic population (b). If the chance of establishment is defined as the complement of the chance of extinction conditioned on the size of the F2 generation, then the chance of establishment for the heterotic population is also greater (a). Sexual reproduction returns the average individual fitness in the heterotic population to the level of the parent generation by the F2 generation (c), but as long as the relative difference in population size obtained in the F2 generation is maintained, the average conditional chance of extinction will generally be greater in the non-heterotic population. Reprinted with permission of the Royal Society from Drake (2006).

range. These individuals would not normally have coexisted, but could have been brought together to make a single founding population in the non-native range through the introduction process. There are now two good examples within invasion biology of the creation of such admixtures within exotic populations, including Cuban anoles in Florida (Kolbe, et al. 2004) and cheatgrass in North America (Novak and Mack 2005).

We have very little understanding of what the generation of relatively high intra-population genetic variation will do to the establishment probability or spread rates of birds. Nevertheless, there are reasons to believe that the generation of admixtures is not common within exotic bird populations. Birds seem to be picked for introduction based on their availability in the native range, and there is a propensity for people to capture birds near exporting cities (i.e., birds are not typically captured across their native range; Chapter 2). In addition, the few exotic bird species studied to date (e.g., European starlings, house sparrows, and common mynas) have very little genetic geographic structure in the native range (section 8.2.3). Of course, these species comprise the most non-random sample one could draw from the 200 or so bird species with known exotic populations. We can certainly identify a few likely candidates for the production of admixtures. For example, the Java sparrow *Padda oryzivora* is found as a native across several Indonesian islands and thus probably shows potential for some geographic structure, while Java sparrows have been introduced widely via the pet trade. Thus, we consider the prevalence of admixtures within exotic bird populations to be an open question.

8.5 Introgressive Hybridization of Exotic Birds with Natives

Hybridization between exotic and native individuals may not only increase the probability that the exotic population will establish, but may also drive the original pure-bred native species to extinction (Rhymer and Simberloff 1996). Such a scenario will occur when either the individuals of the F1 generation preferentially backcross with individuals of only the exotic parental species, or if the parental exotic species is very common whereas the native parental species is rare. In either case, the genes from the native species are introgressed into the exotic population through time and may result in the loss of all genetically pure native individuals. The risk of such 'genetic extinction' depends on the strength (or weakness) of reproductive barriers between taxa, the vigour and fertility of hybrids, the relative and absolute sizes of parental populations, and their relative competitive ability (Levin 2002).

Avian hybridization is increasingly recognized as commonplace (McCarthy 2006) and potentially occurs as a by-product of global change. However, the extent to which exotic species are likely to hybridize with natives depends on the evolutionary distance between the two. In many cases this distance may be great enough that hybridization is unlikely. For example, thirty-five exotic landbird species from sixteen families breeding on the mainland of New Zealand are exotic, but only five of these families (Anatidae, Columbidae, Psittacidae, Strigidae, Alcedonidae) have any native representatives, and there is only one New Zealand genus (*Anas*) that includes both native and exotic species. The lack of evolutionary similarity between native and exotic bird species in this case may in part be explained by the huge geographic distance between New Zealand and the source populations for many of the exotics (mostly Britain; see Chapter 2).

Nevertheless, the single genus that does contain both New Zealand native and exotic species provides probably the best-documented example of the threat to native species from introgressive hybridization, in the form of the mallard. Mallards have been widely and deliberately introduced for hunting, and the establishment of exotic breeding populations has contributed to the decline of the closely related native species *Anas superciliosa* in New Zealand (Rhymer, et al. 1994). The mallard has had a similarly negative effect on *A. superciliosa* in Australia, *A. rubripes* and *A. fulvigula* in North America, *A. wyvilliana* in Hawaii, *A. undulata* in South Africa, and *A. melleri* in Madagascar, all due to introgressive hybridization (Rhymer and Simberloff 1996). In most of these cases the native is on the verge of disappearing as a distinct species.

Examples of introgressive hybridization are not limited to the genus *Anas*. In Europe, hybridization between the endangered European white-headed duck *Oxyura leucocephala* and the exotic American ruddy duck is a significant threat to the former (Hughes 1996). A programme of ruddy duck extermination has now been initiated in the United Kingdom (Smith, et al. 2005), which holds the main European population of this species (Hagemeijer and Blair 1997). Interbreeding of populations differentiated at the level of subspecies has been enough to render the Seychelles turtle dove *Streptopelia picturata rostrata* functionally extinct in its insular distribution (Cade 1983). This loss has occurred over the past 100 years or so, following introduction by humans of the Madagascar turtle dove *S. p. picturata* to one or more of the Seychelles. Most of these islands now have populations of *S. p. picturata,* some have hybrids, and on only two islands can individuals showing any characteristics of *S. p. rostrata* be found. Although alleles of the native Seychelles turtle doves may still be present on most if not all islands, the insular subspecies, once an independent evolutionary entity, no longer exists.

In at least two other cases hybridization with exotic bird species that have been domesticated by humans have either obscured the status of the truly wild population or extirpated it all together. Although conservation agencies do not formally consider either the red junglefowl *Gallus gallus* or the rock dove *Columba livia* endangered in any part of their range (Birdlife International 2000) there are probably few areas where populations can live outside contact with domesticated breeding populations. In the red junglefowl, only one (increasingly rare) captive line still exhibits a complete annual eclipse moult and other morphological and behavioural traits considered to be indicative of pure-wild birds (Brisbin 1995). More recently, the phenotypic characters generally used as indicators of purity were all found to appear to some extent in domestically contaminated progeny (Brisbin and Peterson 2007).

8.6 Conclusions

Our understanding of the genetic composition of exotic bird populations is frustratingly patchy. There is relatively good information on how often within-population genetic diversity declines amongst exotic birds. From these studies, we can confirm the prevailing wisdom that low founding population sizes, slow rates of population growth after establishment, and few independent introduction events will lead to genetic erosion within the non-native population. We can also confirm that exotic birds, on the whole, tend to show no or moderate decreases in genetic diversity, despite a priori expectations to the contrary, which is consistent with recent comparative reviews of the genetics of biological invaders (e.g., Vellend, et al. 2007).

Most of the information on these topics was generated in the 1980s, largely in response to then simmering controversies in evolutionary biology. It would be well worth revisiting these questions using modern techniques, and expanding the repertoire of exotic bird species examined such that it includes populations that have established but remained at low population sizes, or have undergone boom-and-bust cycles. This effort should reward us with a more complete understanding of when losses of genetic diversity are likely and to what extent they may occur. We also suggest a thoughtful examination of how between-population genetic diversity is generated and maintained across exotic bird populations. To produce more than rules of thumb in this regard, a careful consideration of the spatial scale over which an exotic population has extended, and the biological mechanisms driving gene flow between locations, are required.

In contrast, we know very little about what the loss or gain of genetic diversity means for establishment probability or rates of geographic spread within exotic bird populations. In instances where genetic diversity is severely reduced, we may find evidence for a loss of fitness. Low fitness could either lead to a subsequent high likelihood of a population crash, or to small and variable increases in population growth rates. Equally plausible, however, is the generation of hybrid individuals and admixture populations. There is limited evidence indicating that the increase in genetic diversity that results will increase establishment probability. Given the number of currently established exotic bird populations, our detailed knowledge of the history of founding for many of these populations, and the sophisticated state of genetic analyses, answers to these questions are well within reach. What is clear, however, is that at least some exotic bird groups readily hybridize with native species, and this process, given the right conditions, can lead to the loss of native species one gene at a time.

9
The Evolution of Exotic Birds

> The evolutionist and ecologist both could find rich problems for study here, since the firmly established exotics constitute populations isolated from the parental ones under peculiar environmental circumstances.
>
> J. W. Hardy (1973)

9.1 Introduction

Most discussions and research on biological invasions consider exotic species as static entities that have a standard set of life history, morphological, ecological, physiological, and other traits. Nevertheless, we know that many of the traits possessed by species are in general not constant, but rather evolve in response to a variety of selective forces. We have not tended to observe significant evolution in most species, because the selective forces acting on them do not change substantially at the scale of human lifetimes (although rapid climate change may alter this). However, the abiotic and biotic interactions an exotic species experiences in its non-native range may differ substantially from those in its native range, so that the introduction process essentially acts as a rapid alteration to its selection regime. We thus might expect to see relatively fast evolutionary changes in exotic species, and indeed exotic birds (and other non-native species) have been used as 'probes' into the mechanisms underlying natural and sexual selection for over 100 years. This more dynamic perspective on exotic species has recently also begun to enter into the wider debate about their impacts (Cox 2004; Lambrinos 2004; Vellend, et al. 2007).

Despite the lack of observable evolutionary changes in most native species, it is nevertheless now well documented that such changes can occur over less than a few hundred years (Stockwell, et al. 2003). Examples include the evolution of morphology in response to changes in food availability after a climatic disturbance (e.g., Galapagos finches: Grant 1986), the evolution of life-history traits that decrease a species' susceptibility to predation pressure (e.g., Trinidadian guppies *Poecilia reticulata*: Walsh and Reznick 2008), and the evolution of lower maturation age in response to size-specific harvesting regimes (e.g., grayling *Thymallus thymallus*: Haugen and Vollestad 2001). In general, such responses should be common when an increase in the intensity of directional selection affects traits with high heritability (Stockwell, et al. 2003), but even under typical selection strengths and moderate heritabilities, evolutionary changes of one standard deviation can occur as quickly as within twenty-five generations (Stockwell, et al. 2003). However, we would expect even more rapid evolutionary changes to occur under some conditions, including when species are introduced into a novel environment. In these instances, selection strength may be much higher than is typical for populations existing near their optimal conditions. In addition, and unlike many instances of contemporary evolution in native populations, the selection regime is far more likely to be temporally constant and thus produce noticeable adaptations to local conditions (Carroll, et al. 2007).

For contemporary evolution to occur within bird populations, there must be enough genetic variation for selection to act upon (Connor and Hartl 2004). This is not a concern for most native species, except perhaps for those with fragmented or over-harvested populations (Stockwell, et al. 2003). The issue is key, however, when considering the evolution of non-native species (Gilchrist and Lee 2006). As detailed in section 8.2.1, some exotic bird populations are likely to show reduced genetic variation in their non-native range due to founder effects and drift. The loss of genetic variability may serve to constrain evolution in response to novel selection forces. On the other hand, there is some evidence that founder effects and hybridization can increase genetic diversity or allow the formation of new associations between genes (Connor and Hartl 2004). Some of the genetic novelty that is created may prove beneficial given the conditions experienced in the non-native range. There are sufficient examples of instances where genetic variation has not been appreciably eroded in the establishment process (section 8.2.2), and indeed has sometimes been enhanced (section 8.2.3), that we have little reason to suspect that a lack of variation will in general limit exotic bird populations from evolving in response to novel selection forces.

Another key aspect to the evolutionary potential of exotic bird populations is the nearly complete reduction in gene flow between the native and non-native ranges. Our definition of 'exotic' requires that individuals of one species be physically transported (via human mechanisms) from their native range and introduced somewhere clearly geographically separated from this native range. Thus, the possibility of natural dispersal between the native and non-native range is typically almost nil. We saw evidence for this in section 8.3 with the often quite large F_{st}-values observed between several isolated populations of the same exotic species (e.g., common myna *Acridotheres tristis*). There remains the possibility that individuals from the native range continue to be released into an exotic population through time, usually through the pet trade or the actions of hunting organizations. However, with some obvious exceptions (e.g., ring-necked pheasants

Phasianus colchicus in the United Kingdom), the inflow of novel genes into the exotic population via these mechanisms is likely often to be of insufficient magnitude to counteract the changes in gene frequency brought about through natural or sexual selection. Thus, exotic species should have free rein to evolve, unlike native populations in marginal areas where adaptive changes may be prevented by gene flow from core populations experiencing different selection pressures (e.g., Keitt, et al. 2001).

Exotic birds are subject to a host of evolutionary pressures that may either maintain the status quo or act to change mean trait values leading to rapid evolution toward a new adaptive peak. These agents may include climatic events, the physical or chemical composition of the environment, associations with new species, and the loss of association with species from the native range. This chapter describes examples of evolution in the face of these types of selection agents. Most research to date has considered the effects of climatic conditions on the morphology and life history of exotic bird populations, and we begin with a review of this work. There must also be a role for interspecific interactions in exotic bird evolution, however, and although there has been considerably less published on this topic, we explore some possibilities in the second half of the chapter. It is worth noting at the outset that to prove that the differences we review below are due to natural selection, it is necessary to show that (1) the differences in mean trait values across locations are heritable and, even if this is the case, that (2) the differences did not arise due to drift or as a response to divergent environmental conditions (Merilä and Sheldon 2001). We note here that many of the traits that have been shown to differ between the populations of exotic birds below show relatively high heritabilities (e.g., morphology; Merilä and Sheldon 2001). However, to our knowledge no one has performed the necessary common garden or transplant experiments that would prove that these differences indeed represent genetic evolution.

9.2 Evolution in Response to Climatic Shifts

In Chapter 5, we concluded that the match between an exotic species and the novel environment was a better predictor of establishment success for birds than were features of the environment or of the species in isolation. Thus, establishment is more likely when the exotic and native environments differ but little. It follows that only limited evolution in exotic bird species in response to environmental shifts might be expected.

Work by Bob Holt and colleagues has explicitly considered the evolutionary potential of exotic species in response to the match between the ancestral and exotic locations. In particular, Holt, et al. (2005) contrast the response of populations of an exotic species introduced to areas that differ with respect to that species' ecological niche response surface (Figure 9.1). They note that if conditions in the non-native range are very near the ideal conditions experienced in the species' native range (point A in figure 9.1), the exotic population can grow rapidly without the need for evolutionary adaptations, and hence no such adaptations would necessarily be expected. On the other hand, if conditions in the non-native environment lie well outside the species' fundamental niche (point D) the exotic population is likely to decline to extinction too quickly for selection to act. Thus, it is when climatic conditions in the exotic environment are not ideal, and are far enough (but not too far!) from ideal to present a challenge to reproduction and survival, that selection may be strong or directional enough to precipitate evolutionary change. If an exotic population is introduced to a location with conditions outside the normal range within which it can achieve positive population growth (e.g., point C), the population will ultimately die out if it does not adapt. It is in such situations that rapid evolution is likely to be observed, especially given that any populations that do not so adapt will disappear. Adaptive changes will also enhance the persistence of species introduced into an area where population growth would otherwise only be slow (point B), and which would otherwise only gradually escape from the dangers inherent in small population size (Chapter 3). Simulation modelling confirms the importance of adaptation to the persistence of

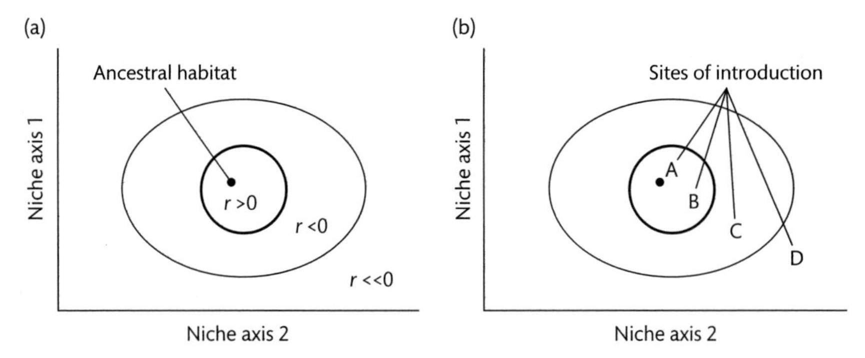

Fig. 9.1 A schematic representation of an ecological niche response surface.
(a) The inner circle represents the fundamental niche, and thus an environment in which the population has a positive growth rate. The outer circle represents conditions whereby on the inside, the population slowly declines, but on the outside it rapidly declines.
(b) The interplay between this niche response surface and the fate of an introduced exotic population. If niche conditions at the location of introduction are matched perfectly (Point A), or sufficiently well (Point B) with those of the ancestral habitat, the population is likely to establish. If the conditions are far outside the fundamental niche requirements of the species (Point D), the population is likely to fail to establish. It is when conditions are outside the fundamental niche, but still within the realm of only slow population decline (Point C) that evolutionary responses are most likely. From Holt, et al. (2005).

populations for which conditions at the exotic location are close to or beyond the edge of a species' fundamental niche.

The work by Holt, et al. (2005) shows that there is no incompatibility between the conclusions of climate-matching studies and the expectation that exotic species will evolve in their non-native range. Indeed, climatic conditions within the non-native range have regularly been shown to select for particular traits that allow exotic bird populations to increase survival and reproductive rates.

9.2.1 Evolution in Exotic House Finches

One classic example of rapid evolution in exotic bird populations comes from work on the house finch *Carpodacus mexicanus*. House finches have been successfully introduced to eastern North America and the Hawaiian islands. They have expanded their native range northward in the western USA without direct human intervention, but, more remarkably, have also developed migratory movements within the exotic range in the eastern USA. Historical records and AFLP methods show that the source population of eastern house finches is located in southern California (Wang, et al. 2003). While a few individuals in the California population migrate (operationally defined as moving more than 80 km between sites of capture and recapture in ringing studies: Able and Belthoff 1998), most are sedentary. Nevertheless, the individuals from the California population that established the eastern population showed a tendency to migrate very soon after establishment. Able and Belthoff (1998) used ringing recoveries to track the direction and distance of all movements over 80 km. They found that the earliest ringing recoveries for the eastern house finch populations showed a strong directionality in movement oriented towards the south-west after breeding, and towards the north-east after over-wintering. Since the 1950s, the percentage of eastern house finches that migrate has hovered between 30% and 50%, whereas over the same time span the number of house finches in the source population that show the same magnitude of movement is between 2% and 3%. Thus, within twenty years of initial establishment, the eastern population had evolved towards partial migration.

Able and Belthoff (1998) showed that young birds generally migrate farther than adults, although some adults will move very long distances, and that females tend to over-winter further south than do males. However, it is not clear whether differences between individuals in propensity to migrate are genetically or environmentally induced, or a combination of these as suggested by the threshold migration model of Berthold (2003). In addition, it is not clear if the appearance of migratory behaviour in the eastern populations represents genetically based evolution in reaction to the novel climatic conditions faced in the east or whether pre-existing migratory behaviours were induced by the eastern environment. Nevertheless, migration does require behaviours that facilitate successful movements, such as building up fat reserves prior to migration, while correct orientation had to develop rapidly in the eastern population. Most recent explorations of the genetic controls over migration patterns have shown a clear genetic link, often including genetically correlated syndromes of traits (see review by Berthold 2003).

Some evidence for a genetic basis to these changes in migratory pattern is also provided by variation in the propensity to migrate and in migratory distance. Both have increased through time within the founding north-eastern population (Able and Belthoff 1998), and this propensity has moved with the individuals that colonized other parts of the exotic range in eastern North America (Figure 9.2). Indeed, the distance that individuals migrate has tended to increase as the population has expanded out of the north-east (Figure 9.2), and has done so at a faster rate than the observed increase in migration distance through time within the founding population. This should be expected to some extent if migrating individuals were those that founded these new populations and if migration behaviour was genetically determined. If the environment induced migration in some individuals, we would expect more southerly exotic populations not to migrate as far as those living farther north; the data depicted in Figure 9.2 shows this not to be the case.

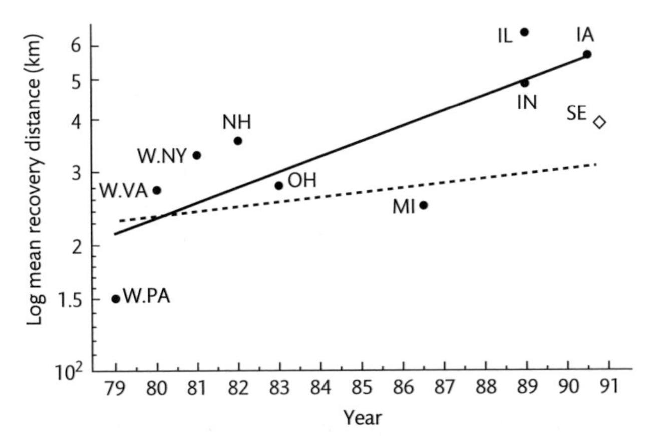

Fig. 9.2 Relationship between the mean distance of earliest year of ringing recoveries of exotic house finches *Carpodacus mexicanus* across States of the USA (solid line) and year of recovery. As the house finch population expanded its range, migratory distance increased, such that areas colonized later were characterized by longer initial migration distances. For comparison, the dashed line represents this relationship calculated for only the 'core' eastern populations, which correspond to areas that were colonized early in the range expansion of house finches. W. PA = Western Pennsylvania, W. VA = Western Virginia, W. NY = Western New York, NH = New Hampshire, OH = Ohio, MI = Michigan, IN = Indiana, IL = Illinois, IA = Iowa, SE = South-eastern states. Reprinted with permission of the Royal Society from Able and Belthoff (1998).

House finches also vary in morphological traits across their native and non-native ranges. House finches in Hawaii are generally smaller than other populations, whereas house finches in New York are generally larger (Hill 1993). Body size is thought often to vary in response to variation in climate (see below). However, the situation in house finches seems likely to be more complicated, as different morphological traits change in different exotic populations (Badyaev and Hill 2000). Thus, despite an expectation that various traits showing covariation within populations, such as wing length and body mass, should evolve in tandem, each morphological trait seems to respond to local selection pressures in its own unique way (i.e., the within-population covariance patterns do not hold across populations). In some locations, house finch tails lengthened, in some locations wings became wider, while in others bill length decreased. The changes observed are modest, as we might expect given the short time available for divergence between

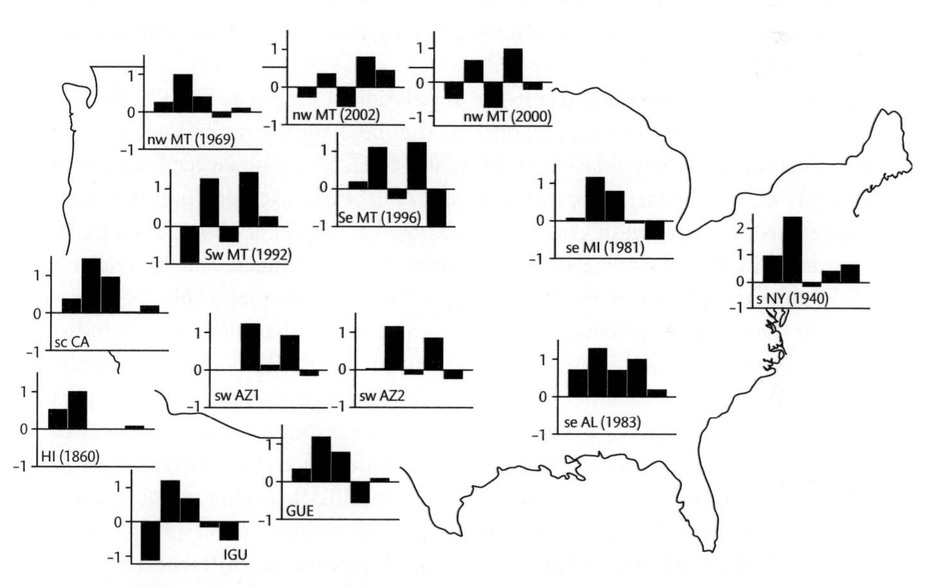

Fig. 9.3 Illustration of the idiosyncratic phenotypic shifts in sexual dimorphism across study populations of the house finch *Carpodacus mexicanus*. Each panel represents the difference between male and female trait size (in standard deviations of female trait). The bars from left to right are: bill length, wing length, tail length, tarsus length, and body mass. MT = Montana, CA = California, HI = Hawaii, AZ = Arizona, AL = Alabama, MI = Michigan, NY = New York, IGU = Isla Guadalupe, Mexico, GUE = Guerro, Mexico. Other letters represent cardinal directions (e.g., se = south-eastern). House finches are native to western North America, which here includes the California, Arizona, and Mexican study populations. House finches have expanded their range naturally into Montana, such that nw MT, sw MT, and se MT are newly established. All other sites represent exotic populations that originated from NY, including the ne and nc MT populations. From Badyaev (2009).

populations. Nevertheless, the changes seem likely to have a genetic basis as the traits involved have moderate to high heritabilities (Merilä and Sheldon 2001).

Figure 9.3 reveals complex changes in sexual size dimorphism between house finch populations. In their native range, house finches show distinct size dimorphism (as illustrated by the morphology scores for California in Figure 9.3), but patterns in dimorphism change across the non-native populations. Badyaev, et al. (2000) showed that house finches experience selection for increased size dimorphism in some populations but not in others, and that selection pressure seems to be particularly strong on the males across all populations. Male morphological traits were shown to be less constrained, as males tended to experience relatively larger inter-population differences in the values of given traits than did females. This effect was driven by greater maternal effects in sons as opposed to daughters, which translated into higher survival of sons, and thus the perpetuation of pre-selection variation in morphology in males but not females. The greater morphological variation in males is then subjected to local environmental selection forces, and thus manifests as locally structured sexual dimorphisms (Badyaev 2005).

The development of migration in the population of the house finch in the eastern USA seems likely to be a response to the more continental climatic conditions encountered there relative to the southern Californian source location. In contrast, while the morphological responses of exotic house finch populations to novel environmental conditions are interesting, their apparent idiosyncrasy makes it harder to identify a clear adaptive hypothesis for the changes. The driver(s) may not be abiotic. Badyaev, et al. (2000) suggest that selection on morphology in the south of the exotic range tends to act upon components that influence fecundity, but on components that influence survival in the north. They further hypothesize that this pattern may be due to the relatively high prevalence of ectoparasites in the southern parts, which should have their strongest effect on successful reproduction. Badyaev and colleagues have also identified that newly established house finch populations in Montana have sex-specific clustering of oocytes, sex-biased egg-laying order, and increased phenotypic variation in offspring morphology (Badyaev, et al. 2006). These environmentally induced differences between native and recently established populations are slowly replaced by genetically based locally adapted morphologies over time (Badyaev and Oh 2008). This suggests a developmental mechanism for the creation of morphological differences between exotic and native bird populations, if not an understanding of the underlying driver of these differences.

9.2.2 Evolution in Exotic House Sparrows

In 1898, Hermon C. Bumpus, a professor at Brown University, was the recipient of a set of house sparrows *Passer domesticus* that were barely alive after a relatively ferocious winter storm had swept through Providence, Rhode Island, the night before (Bumpus 1899; Calhoun 1947; Johnston, et al. 1972). About half the birds Bumpus received died within a day of their arrival in his lab. To Bumpus this situation presented an unusual opportunity to learn about natural selection, since he had unprecedented access to a set of individuals that survived *and* the set that died in response to what he supposed was a strong selective agent (i.e., the winter storm). He quickly made specimens of all of the birds received, and meticulously recorded several morphological traits from each specimen. He then visually compared the frequency distributions of these traits in males and females that survived the storm to those that did not. Because of the elegant nature of his work, and because Bumpus did not statistically evaluate any of his comparisons between survivors and non-survivors, his results have been reanalysed several times, with ever-more sophisticated techniques (e.g., Harris 1911; Calhoun 1947; Grant 1972; Johnston, et al. 1972; Pugesek and Tomer 1996).

The conclusions from all these evaluations, however, are broadly congruent, and suggest three overall trends evinced by Bumpus's house sparrows: (1) larger males survived the stress of the winter storm relatively better than did their small counterparts; (2) more females of intermediate size survived when compared to smaller and larger females; and (3) small sub-adult males died more often than adult males of the same size class. There is also a clear indication that the stress of the winter storm tended to reinforce existing differences in body size between male and female house sparrows. Johnston, et al. (1972) showed that females that had male-like morphological proportions were more likely to have perished in the storm, and males with female-like proportions were also more likely to die. This latter result suggests that sexual size dimorphism in house sparrows results not only from fitness advantages related to reproduction, but also to differential survival in the face of severe physiological stress.

The differential survival of different-sized house sparrow individuals in response to environmental severity suggests that this species would have been expected to evolve distinct morphological clines in response to climatic gradients as it spread across North America. This question has been addressed in a series of papers produced by Richard F. Johnston and his colleagues—notably Robert Selander—between the 1960s and early 1980s (e.g., Johnston and Selander 1964; 1971; 1973; Selander and Johnston 1967; Johnston and Fleischer 1981). Johnston and colleagues collected house sparrows across a north–south temperature cline that spanned the southern provinces of Canada to Oaxaca, Mexico, and along a rainfall cline that spanned the humid eastern states to the deserts of California and Arizona (Figures 9.4a–b). They also collected house sparrows living in insular situations on Bermuda and Hawaii. They carefully measured a variety of morphometric traits of the individuals they collected, as well as scoring their plumage coloration. Their collective results represent probably the most exhaustively documented examples

(a)

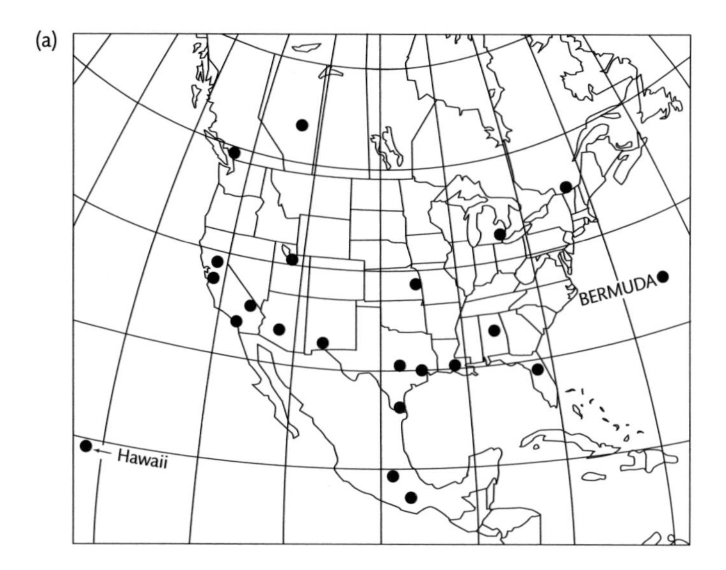

(b)

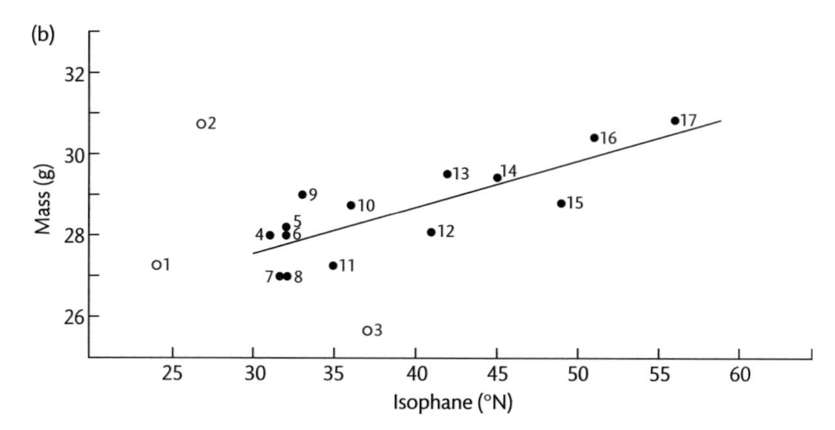

Fig. 9.4 (a) Locations across North America at which Johnston and colleagues collected house sparrows *Passer domesticus* for their study on phenotypic divergence. (b) Relationship between mean body masses of adult male house sparrows and climatic isophanes. As average temperature decreases (higher isophanes), body size increases. Numbers correspond to the following sampling locations; 1. Oaxaca, Mexico; 2. Progreso, Texas; 3. Mexico City, Mexico; 4. Houston, Texas; 5. Los Angeles, California; 6. Austin, Texas; 7. Death Valley, California; 8. Phoenix, Arizona; 9. Baton Rouge, Lousiana; 10. Sacramento, California; 11. Oakland, California; 12. Las Cruces, New Mexico; 13. Lawrence, Kansas; 14. Vancouver, British Columbia; 15. Salt Lake City, Utah; 16. Montreal, Quebec; 17. Edmonton, Alberta. Specimens were also collected from Bermuda and Hawaii. The regression line is based on points 4 to 17. From Johnston and Selander (1964). Reprinted with permission from the American Association for the Advancement of Science (AAAS).

of rapid evolution of exotic birds to date (although the work on the house finch runs a close second), and reveal two striking patterns.

First, Johnston and Selander (1964; 1971; 1973) demonstrated that house sparrows show spatial variation in colour patterns. Individuals collected from northern and Pacific coastal areas had comparatively dark plumage, with the most extreme plumage coming from Vancouver, British Columbia. In contrast, those collected from southern California east to Texas were relatively pale in appearance, with the extreme on this end of the spectrum coming from Death Valley, California (Figure 9.5). Individuals collected in the middle of the continent (e.g., Salt Lake City, Utah) showed intermediate plumage colouration. The variation in plumage colour recognized within North America follows a well-documented ecogeographic 'rule' named after C. H. Gloger. 'Gloger's Rule' specifies that endothermic vertebrates tend to be paler in hotter and less-humid environments in order to ease thermo-regulatory demands, whereas individuals tend to be darker in cold and wet environments in order to increase the rate of heat acquisition. This rule had been documented in North American birds and mammals prior to Johnston and Selander's work, and the demonstration that a non-native species had evolved parallel clines in less than a century was, and indeed still is, considered a remarkable example of contemporary evolution. The colour differences observed

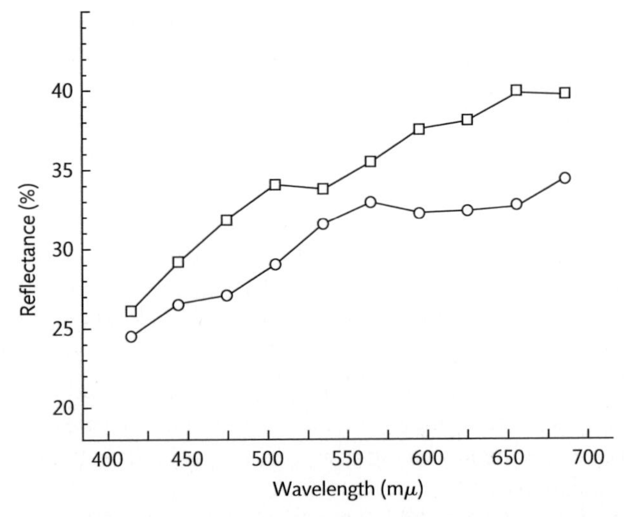

Fig. 9.5 Spectral reflectance of breast feather colour for female house sparrows *Passer domesticus*. Open squares represent specimens collected in Death Valley, California, and open circles from Vancouver, British Columbia (n = 5 from each locality). Hotter climates are associated with feathers with a higher intensity of reflectance. From Johnston and Selander (1964). Reprinted with permission from AAAS.

across localities were not the product of small clinal differences across space, but rather were so stark that Johnston and Selander (1964) could categorize specimens to locality based solely on plumage colouration.

Second, Johnston and Selander (1964; 1971; 1973) demonstrated that house sparrows show spatial variation in body size. On average, house sparrows in the northern sections of North America are larger-bodied than those found in the south-west of the continent, with individuals residing between these geographic extremes showing intermediate body size. In Figure 9.4, this relationship is expressed as body mass as a function of isophane, which is a measure of gross climatic features (based on latitude, longitude, and altitude). Low isophane scores reflect mild or austral climates, while high isophane scores reflect boreal climates with severe winter cold (Johnston and Selander 1964). It is clear that North American house sparrows have evolved body sizes that are distinct to particular climatic conditions, being small in milder, less-seasonal climates, and larger in colder, more seasonal climates.

This pattern is consistent with another ecogeographic rule, named after Carl Bergmann. 'Bergmann's Rule' proposes that large body sizes are favoured in climates that experience regular cold stress, since heat loss will be reduced within individuals with lower surface area-to-volume ratios, although the generality of this rule, the applicability of the mechanism, and indeed the taxonomic level at which it applies (within vs across species) are all the subject of contention (e.g., Bergmann 1847; James 1970; Blackburn, et al. 1999; Ashton 2002; Meiri and Dayan 2003; Cruz, et al. 2005; Meiri and Thomas 2007; Meiri, et al. 2007). Regardless of this debate, however, the presence of such a pattern in North American house sparrows strongly suggests that selection has shaped morphological traits within a species, and that the observed gradient has developed in less than a century. Moreover, the differences in morphological characters observed between study populations in North America were well outside the bounds observed in the species' native range (Johnston 1973) and rival the differences observed between named North American subspecies (Selander and Johnston 1964). The larger body sizes apparent in northern climates are, in part, a product of cold weather events reinforcing sexual size dimorphism (Johnston and Fleischer 1981), as suggested by Bumpus' data (Johnston, et al. 1972). These cold-weather events tend to select for larger body size in males only. Thus, climate in this case is acting on only one of the two sexes and producing an average upward shift in body size across all northern populations.

A similar cline in size and inter-locality differentiation has been shown by Baker (1980) for the house sparrows of New Zealand. The amount of differentiation here is less than that seen in North America despite the fact that the two exotic populations have been established for about the same amount of time. There are clearly differences between populations that live on islands, even large sets like New Zealand, and those that live on continents, but, nevertheless, it is interesting to note that in each case there is fairly clear evidence for contemporary evolution of morphological traits in response to prevailing local conditions.

9.2.3 Evolution in other Exotic Bird Species

Clines in body size have also been documented for common myna populations in New Zealand, and exotic rock dove *Columba livia* populations in the United States. Domestic pigeons were derived from rock doves by artificial selection perhaps 5,000 years ago (Sossinka 1982). Exotic populations originated from escaped captive birds and in the USA are genetically closer to domestic birds than to European exotics or wild rock doves (Johnston 1992). Nevertheless, North American exotic rock doves are closer in skeletal size and shape to European exotics and wild birds than to domestics. Evidently, natural selection has been reconstituting wild size and shape phenotypes in exotic populations (Johnston 1990; 1992). In addition, over the past 400 years, North American exotic rock doves have reproduced a clinal size variation qualitatively consistent with that found in European populations (Johnston 1994). Because variation in the size of exotic rock doves has a significant relationship to genetic variation, Johnston (1994) concluded that differential survival and productivity of size classes at different localities may somehow have been causally involved.

Baker and Moeed (1979) showed significant differences between localities across New Zealand in seventeen out of twenty-eight morphological traits measured in male common mynas, and for thirteen of the traits in females. As was the case for house finches (see above), the set of traits that varied was largely different for males and females, indicating sexual differences in geographic variation and possible differences between the sexes in response to a similar set of selective forces. Although inter-locality differences were relatively small, there was a trend for increased body size in the northern reaches of New Zealand, and this was especially pronounced in males. In section 8.2.1, we reviewed evidence that the New Zealand common myna population had passed through a genetic bottleneck. It would seem that the concomitant reduction in genetic diversity did not remove the potential for morphological differentiation. These mynas provide another example of contemporary evolutionary change by an exotic species in a novel environment, as they had only been established in New Zealand for around 100 years at the time of Baker and Moeed's work. In fact, some populations had been established for considerably less time than this, as the founding population of mynas occurred on the south end of the North Island, with their northward spread only reaching the northern extent of the island around 1965 (Baker and Moeed 1979). The mynas differ from house sparrows in North America in that body sizes are greater in the relatively warmer and wetter climates found in the north of New Zealand. This cline is exactly the opposite expected from Bergmann's Rule, showing that exceptions to the rule are present in exotic species as well as in natives (Ashton 2002; Meiri and Dayan 2003).

There is also some evidence that novel climates may influence changes in exotic bird life histories. One of the most striking spatial patterns in ornithology is that avian clutch sizes tend to decrease towards the equator, both within and across species (Lack 1947; 1948; Skutch 1949; Cardillo 2002). However, this gradient is asymmetrical, such that clutch sizes are larger at northern than equivalent southern latitudes (Moreau 1944; Yom-Tov, et al. 1994; Martin 1996; van Zyl 1999). Niethammer (1970) was the first to present evidence that exotic birds translocated from northern to southern latitudes showed differences in mean clutch sizes. The latitudinal gradient in clutch size predicts that clutches should decrease in size following translocation, and this is indeed what Niethammer observed. Using data for nine passerine species translocated from Europe to New Zealand, Niethammer (1970) showed that all nine species had smaller mean clutch size in New Zealand, albeit only marginally in the case of the dunnock *Prunella modularis*, and that the difference was significant for six of them. Niethammer (1970) invoked the high population densities of these passerines in New Zealand, and hence decreased food availability for each breeding pair, as the likely cause.

Evans, et al. (2005a) used much larger samples of passerine clutch sizes provided by nest record schemes in New Zealand and the United Kingdom to test hypotheses for the likely mechanism driving the clutch size gradient. These data showed reduced clutch sizes in all eleven species analysed, of which nine decreases were statistically significant (including the dunnock), albeit that the mean decrease across species was only 0.3 of an egg per clutch. However, more importantly, the decreased clutch sizes tended to be accompanied by increases in the length of the breeding season, and less marked temporal variation in clutch size throughout the breeding season in New Zealand relative to the United Kingdom. These changes are consistent with a hypothesis for the latitudinal gradient in clutch size proposed by Ashmole (1963). This hypothesis posits that at higher latitudes, where environments are more seasonal, colder winters result in reduced carrying capacity and increased mortality, and so lower the number of individuals that survive to enter the breeding season. There are thus fewer individuals present the following spring to take advantage of the flush of resources at the start of the breeding season, and hence higher per capita resource availability for those individuals that do survive. These individuals can in consequence raise larger clutches. The longer breeding seasons and lower temporal variation in clutch sizes in New Zealand are indications that the exotic passerine species are experiencing lower resource fluctuations, and hence less seasonality, than in their native ranges. These observations are at least consistent with Ashmole's hypothesis (Evans, et al. 2005a).

The number of eggs laid is only one way in which a bird can modify its investment in a clutch. Cassey, et al. (2005d) present evidence that the egg sizes of exotic passerines have also changed since introduction, decreasing in volume by as much as 16% in the case of the chaffinch *Fringilla coelebs*. Egg size is known to correlate positively with clutch size in many passerines (Greig-Smith, et al. 1988; Flux 2006), so that we would expect egg size decreases given that clutch size has declined since introduction. Contrary to expectation, however, the magnitude of the decline in egg size is not correlated with the proportional decrease in clutch size across these exotic passerines (Cassey, et al. 2005d). Interestingly, combining data on egg size and clutch size shows that the change in clutch volume is negatively related to the change in temporal variation in clutch sizes in New Zealand. This observation is not consistent with Ashmole's hypothesis, which predicts that those species experiencing the greatest change in seasonality should rather show the greatest decreases in investment in the clutch. This observation suggests that changes in egg size and clutch size may trade off, such that selection for a reduction in investment in the clutch has been achieved by a decrease in one or the other, but not in both.

Overall, evidence is accumulating for phenotypic change in exotic birds following their introduction to novel environments, and that that change is in some cases correlated to climatic variation. Whether there is a genotypic basis to all of these changes remains to be proven, but the studies by Bumpus (1899) and others are at least indicative of the action of selection by the climate on natural variation within the exotic population. Yet, we think that this is an area of study where the benefits of exotics are not being exploited to anything like their potential. In particular, exotic species in the process of spread offer an unusual and exciting opportunity to explore population changes in morphology and life history at the edge of the advancing wave front. Questions about the characteristics of colonizing individuals, changes in those characteristics in local populations in the process of securing the bridgehead (Chapter 4), and the development of clines, can all be studied as dynamic processes, as they happen.

9.3 Evolution in Response to Novel Interspecific Associations

The novel biotic interactions exotic bird populations experience should also lead to evolution, and there is no reason why the rate of change should not be equally as rapid as that seen in response to abiotic challenges. Birds are known for their co-evolutionary associations with other species (e.g., plants, insects, and other birds). Thus, when birds find themselves embedded in a new assemblage these interactions either have to be generated *de novo*, or the exotic population must survive without them. There is little evidence that exotic bird populations fail to establish because of an inability to establish interspecific interactions (Chapter 5). However, there is a distinct need for research on the converse issue of whether established populations evolve in response to either the lack of associations or the establishment of novel associations. Below, we review the expected responses of exotic birds to changes in interspecific interactions (summarized in Table 9.1), providing examples of existing research where appropriate or suggesting situations that are ripe for detailed study. In each case, we are simply expanding on the topics initially presented in Chapters 5 and 7, except that now we are considering the longer-term response of exotic birds to these same influences.

9.3.1 Competition

As reviewed in Chapters 5 and 7, the long and controversial debate over the role of interspecific competition in ecology and evolution has partly been played out in studies of the establishment of exotic bird populations. However, the argument about whether or not interspecific competition influences establishment has largely overshadowed (and perhaps stunted) research into the possibility that such interactions could lead to evolutionary shifts in species once established. Studies have addressed whether competition can structure entire exotic bird communities via competitive exclusion, but more specific research on the existence or form of competition between two or more exotic species has been sparse (but see Moulton, et al. 1990; MacLeod, et al. 2005a). Furthermore, to our knowledge, no one has considered whether the presence or absence of competition in the exotic range of a bird species has led to character displacement or character release.

Part of the reason why, beyond the studies reviewed in Chapters 5 and 7, competition has been neglected, is that proving competitive effects in field situations is notoriously difficult. This difficulty grows when one is dealing with birds, as they are vagile and wide-ranging. These traits make it difficult experimentally to reduce the

Table 9.1 A list of interspecific interactions that exotic birds may experience and how they may respond evolutionarily to an increase or decrease in the level or intensity of such interactions. Possible examples are given, but few if any of these evolutionary responses have been considered or researched.

Interaction	Increase	Decrease	Avian examples
Competition	Character Displacement	Competitive release	*Lonchura* species in Hawaii; Pacific white-eyes
Predation	Evolve anti-predator defences	Loss of defence mechanisms	Mongoose predation and exotic birds on islands; exotic birds of New Zealand
Parasitism	Increased resistance	Reallocation of resources	Orange-cheeked waxbills and pin-tailed whydahs on Puerto Rico; village weavers on Mauritius and Hispaniola
Mutualism	Co-evolution with new associates	Alternative resource acquisition mechanism	Frugivourous birds such as red-vented bulbuls and Japanese white-eyes

population size of a putative competitor and observe the response of the 'released' species. These concerns are mirrored to some extent when exploring the possibility that an exotic bird species has evolved a new morphology or life history in response to an increase or decrease in competition, with the further complication that one must now additionally consider the length of time it would take for such a change to evolve. The best evidence in this regard must come from 'home-and-away' comparisons of exotic birds that are clearly either coming into contact with potentially strong competitors or are escaping such competitors in the exotic range.

Home-and-away comparisons are relatively easily performed, although we could find no published examples that relate specifically to a change in competitive situation. Nevertheless there are myriad situations in which such an investigation could prove fruitful. For example, species in the genus *Lonchura* are all similarly sized and tend to share a preference for open, grassy habitats (Restall 1997). This suggests that when they co-occur they may compete intensely for food and nesting resources. They are also very popular in the cage bird trade because of their small body size and predilection for eating grass and reed seeds. Consequently, they have established non-native populations all over the world, with several species coexisting on islands such as Hawaii and Puerto Rico (Table 9.2).

Table 9.2 *Lonchura* species established in Puerto Rico or the Hawaiian Islands, their source populations, and current distributions. PR = Puerto Rico, Ka = Kauai, Ma = Maui, Mo = Molokai, Oa = Oahu, Ha = Hawaii Island. From information in Raffaele (1989) and Pratt, et al. (1987).

Name	Exotic Distribution	Source Population	Date of Introduction	Exotic Range Status
Warbling silverbill *L. malabarica*	PR	India, Arabian Peninsula, c. Africa	1960s	Common in metro areas
	Ha, Ma, Mo, Ka	Africa	1970s Ha, spread from there	Abundant locally Ha; rare Ma, Mo, Ka, expanding
Bronze mannikin *L. cucullata*	PR	Africa	1800s	Common
Nutmeg mannikin *L. punctulata*	PR	India, se. Asia	1960s	Fairly common, may be declining
	Mo, Ha, Ma, Ka, Oa	India, se Asia	1865	Abundant and widespread on all islands
Chestnut mannikin *L. malacca*	PR	India, se. Asia	Recently introduced	Common
	Oa, Ka	India, se. Asia	Prior to 1959	Abundant Oa, common locally Ka

In many cases, the coexistence of *Lonchura* on these islands represents the first time two or more of the species have existed in sympatry. In other instances, the degree of range overlap is much higher in the non-native than in their native range (especially on islands). Moreover, there is some evidence consistent with the idea that exotic *Lonchura* species compete. Moulton, et al. (1990) showed that the abundance of nutmeg manikins *L. punculata* in the grassy margins of Hawaiian canefields was much higher in the absence than in the presence of chestnut manikins *L. malacca*, although they were unable to rule out habitat differences as a cause of this difference. Yet, no one to our knowledge has systematically looked for character or reproductive divergence amongst a suite of *Lonchura* that are coexisting or hyper-competing for the first time in their non-native ranges. Because the introduction of exotic birds is taxonomically non-random (Chapter 2), there are several other instances where we might expect that competition has intensified in the non-native range, and thus where we might expect some degree of divergence to evolve if the species are to continue to coexist. Good candidate groups include the pigeons, mynas, and bulbuls, as species from each group have established non-native populations in sympatry. There is some indication that mynas and bulbuls compete in their non-native ranges (see sections 6.2.3 and 6.4).

In other cases, we may expect that exotic birds are escaping the effects of interspecific competition. This situation may be especially prevalent for species that were introduced during the heyday of acclimatization societies but are not regularly sold as pets. Such species typically were introduced from mainlands to islands (Blackburn and Duncan 2001b), and thus will tend to exist in situations where there are fewer putative competitor species. For example, most of the exotic bird species now established in New Zealand are native to Europe. Despite the fact that there is some evidence that the suite of exotic birds in New Zealand compete with one another (Duncan and Forsyth 2006), they almost certainly are competing with fewer species than their conspecifics in Europe (MacLeod, et al. 2005a), and there is no evidence that the exotic birds of New Zealand are competing strongly with the remaining native birds (Diamond and Veitch 1981). One of the reasons given for why clutch sizes have evolved to be smaller amongst the exotic birds of New Zealand is that they exist in much higher densities there then they do in their native range (section 9.2.3). For example, MacLeod, et al. (2005a) found that yellowhammers *Emberiza citrinella* attained territory densities of 0.40 ha^{-1} in New Zealand farmland, versus 0.12 ha^{-1} in high-quality farmland habitats in the United Kingdom. Such high densities may directly result from the lack of interspecific competition experienced in New Zealand (MacLeod, et al. 2005a). At the very least, there is no evidence of higher resource availability or lower nest predation in yellowhammers in New Zealand versus the United Kingdom (MacLeod, et al. 2005b).

Although not relating to introduced populations as we define them here (Chapter 1), the study of Pacific island silvereyes *Zosterops lateralis* provides a relevant example of how competitive release in a newly colonized location may induce a rapid evolutionary response. The silvereye is the most successful colonizer of Pacific islands compared to all other passerine groups, and has continued to colonize new islands in the wake of human transformations of habitat (Clegg, et al. 2002). The unusually detailed history of the latest colonization episodes (all occurring within the past 170 years) provide an exceptional opportunity to track how conditions on each island have influenced the speed and manner of morphological differentiation across populations. Using a variety of methods, Clegg, et al. (2002) showed that the silvereye has evolved larger body sizes on islands as compared to mainland populations, and that this change in morphology is most likely the result of contemporary evolutionary diversification and not drift. These authors also presented evidence that observed shifts in body size could be the result of an increase in generalist foraging behaviours within populations of silvereyes that experience a reduction in interspecific competition in insular situations (Scott, et al. 2003). These results were somewhat equivocal, however, and thus there is still plenty of room to explore alternative explanations for observed changes such as reduced predation pressure (see next section), increase in intraspecific competition (Robinson-Wolrath and Owens 2003), or reduced opportunities for dispersal. There are numerous examples of exotic bird populations in the same mould, which could easily be studied using the same approach as taken with silvereyes in the Pacific.

9.3.2 Predation

Predation in birds is a strong evolutionary structuring force: it is the primary manner in which individuals perish, and is the leading cause of reproductive failure. It should thus be no surprise that a variety of life-history traits in birds have evolved to minimize the ill-effects of predation (Bennett and Owens 2002; Martin 1995). Indeed, in Chapter 5 we presented evidence that predation pressure can be strong enough to lead to the failure of exotic bird populations to establish. Based on this evidence, we should expect to see numerous research papers on the role of predation, or lack thereof, in the contemporary evolution of exotic bird populations. However, we are aware of none that directly implicates predation pressure as a reason for morphological or life-history evolution amongst exotic birds, and only one that mentions predation as a possible indirect explanation for observed differentiation patterns (Scott, et al. 2003).

One explanation for this lack of research is that it reflects a fundamental reality about the probability of exotic birds escaping predation pressure, or encountering a novel predator in the exotic range. In other words, perhaps predation pressure is not that different in the non-native range of exotic bird species, and thus there is simply nothing very interesting to study. However, we suggest that predation pressures are in fact very likely to have shifted in two ways within the non-native ranges of exotic bird species, and thus that the lack of research represents a hole in the ornithological literature.

First, given the number of birds that have been introduced to islands from a continental mainland area, there are likely to be several examples of species that have experienced reduced predation pressure in their exotic range as compared to their native source areas. Despite the relatively large number of extinctions of large predators on mainlands, these areas still tend to hold a full complement of potential bird predators including several species of birds of prey, large reptiles, and variously sized mammals. In contrast, oceanic islands in particular are naturally depauperate in predators of birds, most notably lacking predatory terrestrial mammals (Whittaker and Fernandez-Palacios 2007). While many of these islands have subsequently also become home to exotic mammal species, even in these cases, the contemporary list of potential predators on islands is often restricted in comparison to continental areas, mainly comprising a few (albeit destructive) exotic mammal species. Furthermore, most islands have only a subset of this group (Blackburn, et al. 2004). Thus, we suggest that predation pressure may be lower on many islands both through a reduction in the numbers of predator species and, since those that are present are mainly terrestrial mammals, through a reduction in predator functional diversity. Whether the birds now inhabiting islands currently experience reduced predation pressure as compared to their mainland source populations is an open question, although as we noted in the previous section, nest predation is not lower in exotic New Zealand versus native United Kingdom populations of yellowhammers (MacLeod, et al. 2005a). The extent to which this may be a special case is of course impossible to judge at present.

Second, there are several examples of exotic birds living in sympatry with predators in whose presence they did not evolve. For example, one of the primary predators of birds (native and exotic) in Hawaii (and other islands) is the small Indian mongoose *Herpestes javanicus*. Some of the exotic bird species that established on Hawaii probably encountered mongoose predation in their native range, if not by this species then by one of its congeners. However, roughly half the exotic birds that now exist on Hawaii evolved where there are no mongooses. Tamar Dayan and colleagues have shown rapid morphological and behavioural shifts in non-native island populations of the small Indian mongoose as compared to its native source population (Dayan and Simberloff 1994; Simberloff, et al. 2000). Thus, even in those cases where exotic birds did evolve with mongooses, whatever worked before in terms of escaping their predation may not work now. Harkening back to the work of Holt, et al. (2005), the species that currently coexist with mongooses or any other novel predator must have had sufficiently well-developed defence mechanisms to allow them to remain extant in the face of novel predation pressure, but the shift in predation pressure may also have presented enough of a change to cause rapid evolutionary responses.

9.3.3 Parasitism

Interest in the role of parasites as a selective force on bird populations has grown dramatically in the last twenty-five years with the realization that, while most parasites do not kill their hosts, they may none the less have important effects on host fitness by reducing the probability that individual birds will reproduce, or by reducing the reproductive output of infected populations. We saw in Chapter 5 that the opportunity provided by the introduction process potentially to escape from parasites in the native range is viewed as one likely determinant of introduction success. However, we also saw there that while non-native species may lose parasites from their native range via the introduction process (Torchin, et al. 2003), they also tend to gain novel parasites within their non-native range relatively quickly (Colautti, et al. 2005). Therefore, net parasite loads may increase or decrease following introduction. Moreover, because hosts may be parasitized by a wide variety of pathogens, the effect of any given parasite is likely to be small (Møller, et al. 1999; Garamszegi, et al. 2003). The loss of one or a few parasites hence may not have noticeable impacts on the characters of the host. Instead, evolution after introduction into a non-native habitat is likely only if the total immune response is substantially lowered, thus indicating the loss of enough parasites, or if the loss of a particularly virulent parasite reduces the need to mount large immune responses. Nevertheless, we would expect the change in selection pressure from changes in parasite identity, richness, or prevalence to produce evolutionary responses in exotic populations. Yet, to date, no information exists on how exotic bird populations may have evolved in response to the loss or gain of key parasites.

One of the most-studied host–parasite systems is the effect of parasites on the sexual characters of birds. Hamilton and Zuk (1982) suggested that if females chose mates on the basis of secondary sexual characteristics, and the expression of these is limited by parasites, then males with more exaggerated secondary characteristics should be more attractive to females as they are signalling their resistance to parasitic infection. If there is indeed a role for immunity to parasites in driving sexual selection, then we would expect that exotic birds that lose some subset of their 'native' parasites will respond via the evolution of sexually selected traits, including plumage colour, feather ornamentation, or song. Our own observations lead us to suspect that aspects of coloration differ between native United Kingdom and exotic New Zealand populations of several farmland bird species, although it is premature to speculate about the causes of any such differences before their existence has been established. Thus, to date, no one has directly tested this hypothesis.

There is at least evidence for changes in a sexually selected trait in the Eurasian tree sparrows *Passer montanus* introduced from Germany to the USA. Lang and Barlow (1997) showed that the number of distinct elements in tree sparrow song (i.e., memes) showed a marked reduction within the exotic population. North

American and German tree sparrows also share relatively few song syllables. Lang and Barlow (1997) attributed this difference to a founder effect and drift, and indeed the North American population has much reduced genetic variation relative to its German source population. Just as with rare alleles (see section 8.2.1), rare song elements are typically lost when only a few individuals from the source population found a new population, while any rare song syllables that were contained in the founding population are likely to be lost through random demographic processes (i.e., drift). Also consistent with the genetic analogy is evidence that the North American population has acquired new song syllables that were distinct from the German source. This is likely to be due to 'mutation', or the development of unique song elements by chance. Cultural evolution happens at a much faster rate than genetic evolution, and following from this there is evidence for a relatively high rate of unique song elements within the North American population (Lang and Barlow 1997). There is no suggestion that the changes in song were a consequence of changes in selection pressure due to parasites, and while this possibility was not investigated by Lang and Barlow (1997), any study of such a link would have to rule out more prosaic explanations based on founder effects and drift.

In fact, it is possible that the reduction of genetic variability associated with founder events may be enough to decrease the ability of an exotic bird population to respond to even a reduced parasite load (Apanius, et al. 2000). In this case, we should expect the ill effects of parasites (either from the bird's native or non-native range) to be much worse than those experienced in the native range. This may put a premium on females choosing to mate with rare uninfected males, as such males are likely to be able to provision her and their nestlings at substantially higher rates than can infected males. If this dynamic persisted for even a few generations, we would expect to see very rapid evolution in sexual characters that are honest signals of male parental investment (e.g., plumage colour; Siefferman and Hill 2003). On the other hand, most studies show that genetic diversity is not always lower after introduction in exotic birds (section 8.2.2), and in cases where it has been reduced by a founder effect, the recovery of genetic diversity through time may happen quickly enough to limit the importance of this evolutionary mechanism. Again, however, no one has explored these questions.

One example of a host–parasite relationship that has been investigated using exotic bird species is that of obligate brood parasitism. Brood parasites lay their eggs in the nests of other species, which bear the burden of incubating, brooding, and feeding the parasite's young. Thus, host species suffer drastic reductions in fitness in the face of brood parasitism, which can be large enough to threaten a host population with extinction when it first encounters the parasite (Pease and Gryzbowski 1995). This situation sets up an evolutionary race between brood parasite and host, and such races have resulted in some spectacular cases of co-evolution (e.g., Old World cuckoos and their hosts; Payne 1977). Further, because males in brood parasitic species have no parental investment beyond the contribution of sperm, there is typically also strong sexual selection within these species. The end result can be extravagant male plumage attractive not only to females of the species, but also to cage bird enthusiasts worldwide. Given their visual appeal to humans, it should be of no surprise that several species of obligate brood parasites have been introduced to novel geographic locations. Once these species are released in this new location, the suite of host birds is sometimes completely new. Thus, for these species to establish and thrive in their non-native range they must begin to parasitize the nests of the bird species around them. If new hosts are found, however, the hosts themselves are subject to very strong selectional gradients in reproductive fitness, and would be expected to evolve the ability to discriminate parasitic eggs in their nests and to eject these eggs or abandon parasitized nests.

One strategy hypothesized to be selectively advantageous under brood parasitism is for hosts to evolve high uniformity of eggs within a clutch (Davies and Brooke 1989) while at the same time maximizing across-population variability in egg appearance (Lahti 2006). These variations in egg appearance increase a host's ability to recognize the egg of a brood parasite within its own clutch and thus to reject it (e.g., Roberts and Sorci 1999). They also decrease the ability of the brood parasite to evolve an egg colour and spotting pattern that accurately mimics eggs across the entire range of the host. David Lahti has tested this hypothesis using data from native and exotic populations of the village weaver *Ploceus cucullatus*.

Lahti compared the egg colour and spotting patterns of the village weaver in its native range in sub-Saharan Africa to two non-native populations, one on Mauritius and the other on Hispaniola. In its native range, the village weaver is a regular host to the diederik cuckoo *Chrysoccocyx caprius*, an obligate brood parasite. In its introduced range, the village weaver does not suffer interspecific brood parasitism. In response to diederik cuckoo brood parasitism in Africa, the village weaver lays eggs that are consistent in colour and spotting within clutches, but that are highly variable in these traits across individuals in the population (Lahti 2005; 2006). If the costs of maintaining those traits are relatively high, we would expect that non-native populations of village weavers would alter variation in egg appearance and lose the ability to recognize and reject parasite eggs in the absence of cuckoo parasitism. Lahti (2005) indeed found that village weaver eggs were more variable within clutches in the exotic relative to the native range, with a lower incidence of spotting. The increase in within-clutch variability suggests that village weavers in the non-native range experience reduced selection pressure to recognize foreign eggs. The reduction in spotting may also be a result of the relaxed evolutionary pressure to maintain across-population variability in egg appearance, as the incidence of spotting is thought to be a primary mechanism for creating this variability.

These changes in egg appearance may be accompanied by a loss of behavioural responses to eggs that look different. The ability of host females to discriminate between their eggs and that of a brood parasite is a product of how visually similar her eggs are to one another. The increase in within-clutch variability in village weaver egg appearance in the exotic populations should translate to a reduced ability to discriminate between these and a foreign egg. Lahti (2006) experimentally manipulated clutches of village weavers in Mauritius and Hispaniola to test this expectation. He found that in both exotic populations village weavers had a reduced capacity to identify and reject the eggs of the diederik cuckoo. However, if the changes in egg appearance Lahti documented in the non-native ranges were accounted for, village weavers rejected foreign eggs at a rate similar to that documented in their native range. In other words, weavers had not lost their ability to discriminate differences of the same magnitude as what would have been needed to reject cuckoo eggs in Africa. Instead, the baseline for comparison had changed between the native and non-native ranges, illustrating that the weaver's rejection behaviour was context specific.

Different responses of introduced species to an absence of brood parasites have been demonstrated by Hale and Briskie (2007) for species introduced to New Zealand from the United Kingdom. Hale and Briskie (2007) studied the responses of individual hosts in the exotic populations to the placement of mimetic or non-mimetic eggs in their nest, and to a taxidermic model of a European cuckoo. New Zealand blackbirds *Turdus merula* and song thrushes *T. philomelos* accepted mimetic eggs in their clutch more often than non-mimetic eggs, but rejected non-mimetic eggs more frequently in New Zealand than in similar experiments in the United Kingdom. Chaffinches also regularly rejected non-mimetic eggs, although there were not clear differences in rejection rates for mimetic and non-mimetic eggs in this species. These results suggest these exotic birds have not lost the ability to recognize and reject foreign eggs, despite roughly 130 years in the absence of brood parasites. In contrast, New Zealand individuals of these host species paid little attention to model adult cuckoos, even though these are vigorously attacked by individuals in the European source populations. Thus, it seems that different anti-parasite behaviours are lost at different rates in the same species (Hale and Briskie 2007), and, comparing these results with those for the village weaver (Lahti 2005; 2006), are differentially lost by different former host species.

There are several other examples of exotic birds that have probably escaped the effects of brood parasites within their non-native range. Since the hotspot of brood parasite radiation can be found in sub-Saharan Africa, the suite of well-established exotic passerines that have their source native populations in this area (>10 species) should be the primary focus for initial research. Nearly all continents have at least one species of brood parasite, however, indicating that there is a lot of fertile ground for the use of exotic bird populations to understand the evolutionary origins and mechanisms that drive host–parasite evolution.

The reverse situation, of the responses of species to a novel exotic brood parasite or a greater degree of parasitism, has yet to be explored, but this potential is provided by exotic bird introductions to Puerto Rico. The pin-tailed whydah *Vidua macroura*, an obligate brood parasite, was introduced successfully to Puerto Rico from Africa relatively recently and is now considered well-established there. In Puerto Rico, the whydah almost exclusively parasitizes orange-cheeked waxbill *Estrilda melpoda* nests. Orange-cheeked waxbills were introduced to Puerto Rico from Africa in the early 1900s and they can now be found across the island. These two species overlap in their native ranges in Africa, but pin-tailed whydahs have a larger native geographic range than orange-checked waxbills and are known to parasitize many other species there (Schuetz 2005). It is possible, although untested, that orange-cheeked waxbills in Puerto Rico suffer much higher parasitism rates from pin-tailed whydahs than they do in their native range simply because they are the only host of the whydah. A higher rate of parasitism of waxbill nests, and the reliance of the whydah exclusively on the waxbills as hosts, could set up an intensified co-evolutionary dynamic that will result in rapid changes to egg appearance in both species. These expectations could easily be tested via home-and-away comparisons of orange-cheeked waxbill and pin-tailed whydah populations in Puerto Rico and Africa.

9.3.4 Mutualism

Despite the historical preoccupation with the role of antagonistic interactions in non-native species establishment, there is an undeniable role for mutualistic or facultative interactions in determining the fate of exotic species (Simberloff and Von Holle 1999; Richardson, et al. 2000). Existing theory concerning mutualisms in general suggests that such interactions are established easily enough to be unlikely to provide a barrier to establishment success.

Here we explore the possibility that, once an exotic bird population has securely established a self-sustaining population, it will evolve in response to the novel selection forces generated by any mutualistic interaction it has established. We have good reason to believe that most exotic birds will establish some sort of mutualistic interaction with plants within their non-native range. In nearly all ecosystems, birds play a large role in the dispersal of plant seeds and the pollination of flowers. Indeed, the bird taxa most likely to participate in such mutualistic interactions are precisely the taxa that have been commonly introduced (e.g., passerines, columbiforms).

There are several examples of exotic birds providing pollination or seed dispersal functions for native and non-native plants. For example, Japanese white-eyes *Zosterops japonicus* on the island of Hawaii are the principal visitors to the invasive firetree *Myrica faya*, eating its berries and thus dispersing its seeds (Vitousek and Walker 1989). Japanese white-eyes on Maui visit the native shrub ohelo

Vaccinium calycinum almost to the exclusion of other nectar- and fruit-producing trees including the native and dominant ohia tree *Metrosideros polymorpha* and the invasive tree alfalfa *Cytisus proliferus* (Waring, et al. 1993). Given that most bird–plant mutualisms are highly generalized (Howe 1984; Jordano 1987), the fact that the Japanese white-eye, currently the most abundant bird on the Hawaiian Islands (Pratt, et al. 1987), interacts with a variety of native and non-native plants should not be surprising. Perhaps the more important observation relative to the probability of evolutionary response in Japanese white-eyes (or the plants it interacts with) is that none of the plants listed above can be found within the white-eye's native range, and thus these interactions have been generated in Hawaii *de novo*.

The degree to which the exotic birds that participate in mutualistic interactions with plants will evolve in response to that interaction is an open question. Most research to date has been on the effect of the exotic bird on the plants and not vice versa (e.g., Richardson, et al. 2000; Gosper, et al. 2005; Traveset and Richardson 2006). Still, this research suggests that birds in general have such broad foraging behaviours that interactions with native or exotic plants will be diffuse and relatively weak, and thus we should expect no evolutionary response in either participant (plant or bird; Traveset and Richardson 2006). This conclusion may be even more relevant to exotic birds specifically, since the possession of generalistic foraging behaviours is one of the life-history traits that have been shown consistently to increase the probability that an exotic bird population will establish (Chapter 4).

Nevertheless, there are instances where we might expect that the fitness of exotic birds may be tightly tied to their interaction with a native or exotic plant. The work of Peter and Rosemary Grant on Galapagos finches suggests that even a temporary change in seed size will precipitate a measurable change in bill morphology (Grant and Grant 2008). This observation suggests that, even in cases where an exotic bird does not establish a strong mutualistic interaction with its plant food source, it may evolve better to access or more efficiently to process that food.

Such interactions may be especially strong if the plant is pre-adapted to be ornithophilous. These plants rely exclusively on birds for pollination services, and seem to have evolved commonly among island endemics. They have also suffered from the mass extinction of native bird species that once served as their principal pollinators or seed-dispersers (Robertson, et al. 1999; Cox and Elmqvist 2000). In several cases, however, exotic birds have filled the role vacated by these extinct natives. Examples include the Hawaiian ie'ie vine *Freycinetia arborea*, which relies almost exclusively on Japanese white-eyes for pollination (Cox and Elmqvist 2000), and the rare Mauritius endemic plant *Nesocodon mauritianus*, whose flowers are more often visited by the exotic red-whiskered bulbul *Pycnonotus jocosus* than by all of the remaining native birds of Mauritius (Olesen, et al. 1988). In these instances, an exotic bird may be fulfilling the role of a lost native co-evolved mutualist, and we may thus expect that the plant will evolve traits to increase the reliance of these exotic birds on the resources it provides (nectar or fruit). Whether the exotic bird evolves morphological or behavioural traits to take further advantage of these increased resources, and thus increase its own fitness, is a question of theoretical and practical interest.

We may also expect some evolutionary responses of exotic birds to their mutualistic partners if they are constrained to include only a subset of all available plants. The assembly and co-evolution of such mutualistic complexes is a topic of interest that goes well beyond concern for the impacts of exotic species. A component of this research centres on the likelihood that exotic species preferentially interact with each other thus creating 'invader complexes' (D'Antonio and Dudley 1993). Since the participants in these invader complexes are typically species that are living in sympatry for the first time, their evolutionary response to one another should be of conservation interest. However, the evidence that invader complexes are regularly formed is decidedly mixed (Olesen, et al. 2002). Indeed, it appears equally as likely that exotic birds will form mutualistic networks with native plants as with other exotics (e.g., Foster and Robinson 2007).

9.4 Conclusions

There is a growing recognition that non-native species can and do evolve rather quickly within their new range (Vellend, et al. 2007), and we show in this chapter that exotic birds are no exception. The evidence that exotic birds evolve in response to abiotic conditions is well founded, although the particular mechanisms are certainly open to debate and further study. By dint of their status, exotic birds provide the opportunity to explore mechanistic evolutionary questions using methods that would be harder to justify on threatened species within their native range. This is not to advocate unethical practices, but rather to acknowledge that such considerations will tend to be less strict on species that in many cases have the official or unofficial status of pests.

The evidence for contemporary evolution within exotic bird populations in response to biotic factors is scant. This situation reflects a lack of research effort instead of a failure to observe substantial shifts in morphology or life history. Given the number of established exotic bird populations, there are multiple opportunities to answer basic questions in evolutionary ecology related to competition, predation, parasitism, and mutualism. By the happenchance of bird introductions, some species have been removed from biotic stressors in some situations, and they have experienced increased biotic stressors in others. In a relatively few but compelling cases the same species will have exotic populations spread across several such situations. These 'experiments in nature' have been drastically under-utilized by evolutionary biologists.

10
Lessons from Exotic Birds

> The more one learns of this whole subject the plainer it becomes that definitive rules for establishing a new bird or extending the range of a native one can not be laid down.
>
> J. C. Phillips (1928)

10.1 Introduction

Exotic birds have been considered fodder for testing ecological and evolutionary patterns for at least half a century. Although there is some indication that they are comparatively understudied within the broader invasion ecology literature (Pyšek, et al. 2008), exotic bird populations have been considered in hundreds of studies, as witnessed by the length of our (by no means comprehensive) bibliography. As should now be clear, the exotic bird literature is not evenly distributed with regard to various stages on the invasion pathway, or indeed to different topics relevant to any given stage. The emphasis we have given to issues often reflects the weight of previous work, which is a consequence of the interests of (and controversy between) earlier researchers, more than it reflects what we would consider to be the importance of different topics in the invasion process. Given this, in the final chapter it makes sense to us to address what lessons we believe have emerged from our synthesis of the major patterns and processes in the bird invasion literature.

Some lessons are recapitulations of already well-argued and accepted points. For example, we gain considerable insight into which factors influence exotic bird establishment success by explicitly considering invasions as a pathway instead of a single event. Williamson (1996) and Kolar and Lodge (2001) made this point, and since these publications the invasion pathway model has become a standard organizing principle within invasion ecology (Lockwood, et al. 2007). However, other lessons emerge that are worth closer examination because they are not well integrated across invasion ecology, and indeed highlight shortcomings to standard approaches in the discipline even after taking into account the influence of the structured (nested) nature of the invasion process. Here, we review these in the context of the wider study of biological invasions, highlighting particularly relevant sections from the previous chapters.

10.2 'Ockham's Razor': All Else being Equal, the Simplest Solution is the Best

Invasion ecology is replete with hypotheses about what determines invasion success and why there are more invasive species in one location than another. However, the simplest explanation for most of these patterns is that they are the direct product of the activities that brought the exotic species to the novel location in the first place, and thus require no (substantial) ecological knowledge to explain. To adopt this point of view is to acknowledge that the first two stages of the invasion process—transport and introduction—shape the subsequent stages, and that these early stages are distinctly non-ecological. Exotic birds illustrate this very well in that they reveal highly non-random patterns (1) in the taxonomic composition of the species that were transported and introduced; (2) from where in their native range those species were entrained for transport; (3) in the locations into which they were introduced; and (4) in how many individuals and species were introduced in one or more events. These non-random patterns can be attributed directly to fashionable and economic trends in the pet bird trade, economic and social incentives to introduce birds, and global patterns in commerce. This point is not to suggest that these patterns are unpredictable, but instead to suggest that they require little if any knowledge of the ecology or evolution of birds to understand.

Despite the acknowledgement of the important role that these initial stages play in the invasion process, they are badly under-represented in the invasion literature. In a large-scale review, Puth and Post (2005) showed that almost 90% of peer-reviewed publications on biological invasions have been on the establishment stage. These authors attribute some of this bias to the fact that, 'the vast number of potential [ecological] interactions involved with establishment, and the consequences of establishment, make considerable grist for the empirical research mill'. On the other hand, the study of patterns in transport and introduction is difficult to handle empirically, is nearly impossible to explore experimentally, and involves limited numbers of ecological interactions. In other words, the study of the early invasion stages is ecologically boring and nearly impossible to obtain funds to explore from basic science initiatives that value experimental scientific approaches.

Because we have comparatively good records of transport and introduction patterns in exotic birds, and the dynamics that drive them, we can see how the failure to account for these patterns confounds conclusions drawn. This issue shows up in every chapter of this book, except of course the first two, which describe how these patterns are created. We consistently find that a large portion of the variance in which species establish, and where they establish, can be explained by patterns in transport and introduction. Following Colautti, et al. (2006), we advocate the use of transport/introduction patterns as 'null hypotheses' in invasion ecology, and suggest that it is the deviation from these human-mediated patterns that require ecological explanation (see also Lonsdale 1999). This issue is so pervasive that we have divided up the discussion of it that follows into three sections, which are broadly congruent with event-level, species-level, and location-level effects.

10.2.1 Event-level Null

The evidence in favour of propagule pressure, defined as the number of individuals introduced or the number of independent introduction events, playing a defining role in which species establish is overwhelming. As we have reviewed in Chapter 3, a good portion of this evidence comes from exotic birds in large part because of the relatively good records kept by acclimatization societies and game clubs. Propagule pressure is by definition a product of trends that manifest at the transport and introduction stages, and for this reason can be used as a null hypothesis in invasion research. If so adopted, it is the patterns that persist once propagule pressure is accounted for that require ecological explanation (see also Williamson 1996).

Furthermore, it is becoming increasingly apparent that propagule pressure must be accounted for if robust conclusions are to be drawn about the effects of other variables. Yet, as Colautti et al. (2006) show, few studies control for the effects of propagule pressure (Figure 10.1). Propagule pressure is correlated with a broad range of location- and species-level factors, such that the results of comparative analyses of establishment success will be seriously biased without it (Cassey, et al. 2004c). In particular, we run the risk of associating a series of factors with establishment probability that in fact have no discriminatory power at all. If these factors are used in risk assessment, the rate of Type I error will be greatly inflated (i.e., identifying a species as 'risky', or a location as 'susceptible', when in fact it is not). In invasion ecology more broadly, we run the risk of forever being stuck with a muddied view of what determines establishment success in invasive species (e.g., Smallwood and Salmon 1992). This is especially worrisome given that data on propagule pressure are as rare as hen's teeth for most taxa other than birds.

Not all variance in establishment success can be explained by propagule pressure, however, and there appears to be an asymptote to the relationship beyond which increasing the number of individuals introduced does not have a strong effect on establishment probability. We suggest that the set of species- and location-level factors that deserve the closest attention are those that follow naturally from consideration of the mechanisms that drive the relationship between propagule

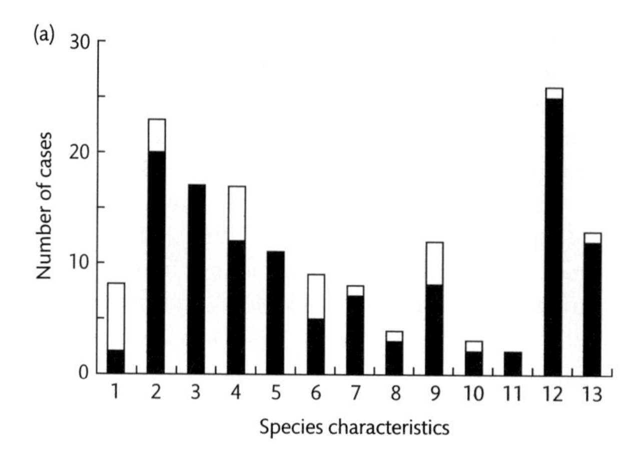

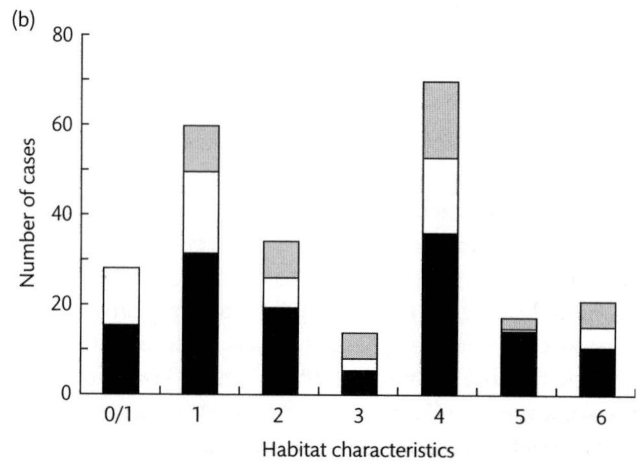

Fig. 10.1 The total number of comparisons identifying characteristics of successful invaders or habitats susceptible to invasion, that either measured (white), controlled for (grey), or failed to consider (black) the potentially confounding effects of propagule pressure. (a) Summary of studies of the characteristics of successful invaders (1. invasion history/widespread; 2. physiological tolerance; 3. consumption efficiency; 4. body size/biomass; 5. individual growth rate; 6. lifespan/generation time; 7. germination success/rate; 8. seed size; 9. reproductive output; 10. length of growing/breeding season; 11. hermaphroditic/asexual reproduction; 12. niche/habitat separation; 13. effects of herbivores/ predators); (b) Summary of studies of the characteristics of habitats susceptible to invasion (0/1. anthropogenic activity; 1. disturbance; 2. resource/food availability or quality; 3. light intensity; 4. species richness or diversity; 5. species abundance or density; 6. effects of herbivores/predators). From Colautti, et al. (2006).

pressure and establishment. These are the factors most likely to modify this relationship. For example, propagule pressure may be particularly important in helping an incipient exotic bird population avoid the trap of Allee effects, while we find evidence that species-level traits associated with Allee effects have stronger effects on establishment success than do traits associated with population growth rates (Chapter 4). Even if they have the ability to achieve positive population growth rates, small populations of exotic birds will still fail to establish if they become trapped in an extinction vortex.

Similarly, Duncan and Forsyth (2006) achieve some resolution concerning the role of interspecific competition in exotic bird establishment success by showing the interrelationship between competition and small population size. Competition evidently does play a role in establishment success. However, it does so only when the competitor has a relatively large effect and is abundant, and when the incipient exotic bird population is introduced in large enough numbers to avoid stochastic extinction.

10.2.2 Species-level Null

The search for species traits that predict establishment success or invasiveness in the non-native range goes back at least to Darwin (Proçhes, et al. 2008). There persists in the invasion literature a notion that evolution has 'built' some species to be good invaders (*sensu* Baker 1974) by virtue of traits that confer high competitive ability and the capacity to exist across a broad range of climates, habitats, and population sizes. Recently there has been a strong focus on understanding why some invasive species reach numerical dominance in their non-native range, or cause an ecological or economic impact. Most of these studies fail to consider the potential for previous invasion stages to colour their results (Colautti, et al. 2006; Hayes and Barry 2008). We highlighted above the importance of considering propagule pressure as a primary cause of establishment success, as it can often provide the more parsimonious explanation for observed patterns. However, the potential for the transport and introduction stages to confound efforts to identify species-level traits associated with establishment or invasion success go well beyond that observation.

The detailed record of which exotic birds were introduced, and from where they were taken, draws a stark image of just how non-random the transport process is relative to species' traits (Chapter 2). We show that five avian families contain more than half of all introduced bird species, despite comprising fewer than 15% of all extant bird species. In so much as species' traits are conserved within phylogenetic lineages (Owens and Bennett 1995), the variety of traits represented by these families is also very limited in comparison to all bird species. This variety is even further constrained by the non-random manner in which species are chosen for transport out of their native range. Species with large geographical range sizes and high native abundances are more often transported as exotics. These traits are relatively phylogenetically labile but are possessed by only a small subset of species in the avian families favoured for introduction. Commonness is itself associated with a restricted set of other traits and trait values (Kunin and Gaston 1997), so that common species are a non-random subset of regional species pools. In addition, species that are local to the home country of human colonists are more likely to be introduced into wherever these humans settle. Since regional avifaunas share an evolutionary history, the exotics transported from these regions will show the imprint of this history. In sum, the species that emerge from the transport stage of the invasion process are very far from a random subset of all species in terms of their traits. Colautti, et al. (2006) call this effect 'propagule bias'.

It seems that most invasion ecologists consider propagule bias interesting, but largely irrelevant to their experiments on what determines invasiveness. Yet, as Colautti, et al. (2006) demonstrate clearly, propagule bias can, if unaccounted for, influence the interpretation of a variety of invasion studies. At the heart of this issue is that the traits possessed by the limited set of species that were transported are the only ones that had the opportunity to be 'tested' in terms of conferring establishment or invasive spread advantages. Logically this means that, to take an extreme example, we can say nothing about the advantages that having hummingbird traits confers to establishment success since no hummingbirds have ever been introduced into a novel environment. It also means that we need to be cautious in inferring that exotic bird species that are abundant and widespread in their native range have a special propensity to have high geographical spread rates based solely on observations or experiments conducted on the set of birds that are now well established and spreading. Since few rare bird species have ever had the chance to spread, we have no idea whether or not being rare in its native environment disadvantages a species for rapid range expansion in non-native environments.

Propagule bias enters into the discussion of appropriate null expectations in two ways, one of which is relevant to experiments and the other to observational contrasts. Experimental approaches to invasion questions are now quite common, especially when considering exotic species that are (largely) sessile such as plants or marine invertebrates. These are also groups for which we have relatively little information on the set of species that were transported and introduced but failed to establish. Perhaps for this reason, most experimental research on these species deals with traits of invaders that cause ecological or economic impacts, or with traits that assist the attainment of widespread dispersal and numerical dominance. Many of these experiments have recently controlled for, or directly tested, the influence of propagule pressure in their experiments (e.g., Davis and Pelsor 2001; Von Holle and Simberloff 2005).

Propagule bias does not directly interfere with experimental outcomes as does propagule pressure, but it should constrain inferences made from experimental results. For example, many experiments have shown that exotic plants that efficiently fix nitrogen can have a strong influence on the native plant community in which they are embedded (e.g., Ehrenfeld 2003; Mack and D'Antonio 2003). However, this result does not imply that nitrogen-fixing plants are generally those that will cause a large ecological impact, even if this result is repeated across several independent experimental tests. For reasons related to agricultural markets, nitrogen-fixing plants may be more commonly transported as exotics (e.g., Daehler 1998) and we therefore may have disproportionately many chances to conduct experiments on such species. Non-nitrogen-fixing plants may cause just as many problems, albeit in a different way, if only they had more opportunity to do so. Moreover, the nitrogen-fixing plants important to agriculture may be so precisely because they have a large ecological impact. The conclusions of such experiments need to be tempered in the face of propagule bias.

Invasion studies that rely on observations of past events usually gain statistical power through comparisons of a set of exotic species to a species pool. Here, propagule bias needs to be considered in the context of this species pool. In terms of identifying traits that confer some advantage to establishment or geographic spread, the species pool should represent the set of traits that could influence these events, which logically is the set of species that successfully transited an invasion stage plus those that had the chance but failed. Species pools have been defined in at least two other ways, including using the set of all species from the region where the exotics were drawn (e.g., Jeschke and Strayer 2006; 2008 for birds) or using the set of native species at the location of introduction, within which the exotics are embedded (Williamson and Fitter 1996). Colautti, et al. (2006) illustrate how these approaches can be confounded by propagule bias (Figure 10.2).

Cassey, et al. (2004a) show that mis-specification of species pools for birds can lead to serious biases in conclusions about traits that influence establishment success. For example, if the set of parrot species that have established exotic populations is compared to the set that was transported (Stage III comparison to Stage 0 in Figure 10.2), a variety of traits show some ability to explain the observed variance in success. However, when the correct pool is specified (Stage III to Stage II in Figure 10.2), only two traits show any explanatory power: diet breadth and migratory tendency. Neither of these traits was recognized as important when the incorrect pool was specified. If this result generalizes (and so far we have read no other study that quantifies this effect in invasion ecology), then we think it is likely that there will be several traits that are purported to be important in establishment success that in fact show associations primarily due to propagule bias.

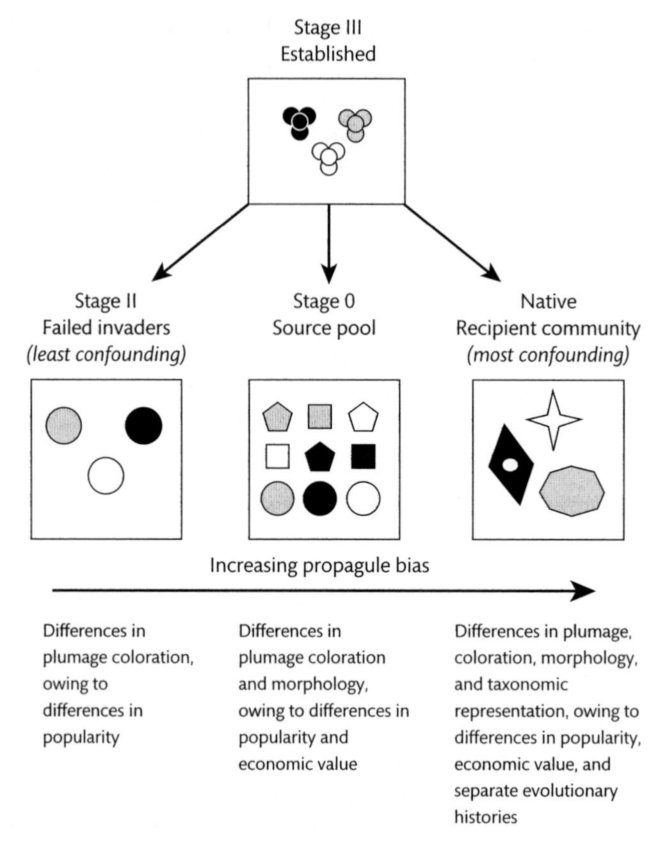

Fig. 10.2 Examples of comparison groups used to identify characteristics of invasiveness, and the likely effects of those comparison groups on the conclusions of studies of exotic species establishment. The characteristics of successfully established species may be contrasted with native species (native to the invaded region), with 'stage 0' species (species from the same source region - concatenating species that failed to be transported, failed to be released after transportation, or failed to establish following release; see Fig. 1. 2), or with 'stage II' species (species introduced but not established). Differences between contrast groups are expected to increase with the number of potential propagule biases, as exemplified by plumage coloration, morphology, and taxonomic representation. Modified from Colautti, et al. (2006).

10.2.3 Location-level Null

Another issue that has received a great deal of attention within invasion ecology through the years is what makes one site relatively more susceptible to invasion than another. Elton (1958) used differences in the number of invader species across geographical locations as evidence for several of his hypotheses concerning establishment success (e.g., that islands are more susceptible to invasion than continents). Differences in the number of exotic species across locations has been formalized in the concept of invasibility, which posits that some sites are prone to invasion by virtue of location-level factors such as human disturbance, availability of empty niches, or lack of competition from native species. In Chapter 7, we considered large-scale patterns in exotic bird species richness in relation to known biogeographical principles; these can also be considered questions of invasibility but where the mechanisms involved manifest themselves at larger spatial scales.

The influence of non-random patterns of transport and introduction on trends in invasibility was clearly acknowledged by Case (1996). He suggested that 'clearly there can only be a few introduced species in the avifauna of today if only a few species were introduced'. This concept is often referred to as 'propagule pressure' in the invasion ecology literature (e.g., Lonsdale 1999; Colautti, et al. 2006) and it is measured as the number of species that are introduced into one location over time. Note that this is a distinctly different use of this term than that adopted in this book and our published papers. Our definition (see above) concerns the population-level process of introducing more individuals *of a single species* either in one event, or across several independent events (Williamson 1996). To avoid confusion, we will define the number of species introduced into a single location as 'colonization pressure'.

Lonsdale (1999) illustrated in a simple equation how colonization pressure directly influences the number of exotic species that establish in a focal area. He posited that $E = IS$, where E is number of exotic species in a region, I is the number of species introduced (colonization pressure), and S is the establishment (or survival) rate of species once they have arrived in the new range. Lonsdale suggested that most ecological studies consider how differences in S across locations affect differences in E, but it is clear from his equation that we must also bear in mind I. Thus, any real differences in invasibility require looking at the residuals from the relationship between invasion success and colonization pressure. In this sense, colonization pressure becomes a kind of null hypothesis for differences in invasibility since no ecological explanation need be invoked to characterize I.

Exotic birds are again one of the few groups for which we can directly test the influence of I on E, and thus indicate the extent to which the proper null hypothesis can clarify patterns in invasibility. Thus, Blackburn, et al. (2008) showed that the number of exotic bird species established on islands worldwide can be explained by the number introduced on each island. Simply, islands with many exotic bird species (high E) are the islands that had many birds introduced there (high I). There is no reason to invoke any ecological or biogeographical characteristic of the islands themselves, such as land area, habitat complexity, or number of competitor species, to explain differences in invasibility ($S \approx$ constant; see also Case 1996; Chiron, et al. 2008).

When the invasibility of islands is compared with that of mainlands, very similar conclusions are reached. The susceptibility of islands to non-native species invasions is notorious and the proposed reasons for this pattern are varied and deeply engrained in the invasion ecology literature (e.g., Proçhes, et al. 2008). However, there are very few examples where this pattern persists once colonization pressure has been accounted for (cf. Lonsdale 1999). For birds there is no evidence that islands are easier to invade than mainlands and instead the preponderance of exotic bird species on islands seems driven almost exclusively by colonization pressure (Chapter 7).

It is very difficult to judge the extent to which colonization pressure influences our overall perception of invasibility patterns because for the vast majority of taxa we do not know how many species were introduced but failed. Nevertheless, we suggest that there are ways of assessing the influence of colonization pressure on invasibility using information gained from a close examination of transport patterns. One potential approach is to incorporate information on human population size. More people bring more exotic species, and higher numbers of individuals of some of these species, into a location. Thus, any location that harbours a high human population is likely also to show high propagule *and* colonization pressures. (People also tend to increase disturbance, suggesting that links between disturbance and human population size/density may be confounded with transport patterns.)

Any effect of human population size on propagule and colonization pressures is likely to derive from the relationship of all three to the volume of trade. A second promising approach is thus to characterize commercial traffic in order to locate networks of trade that are likely to exchange species, and from these patterns deduce a null expectation in terms of the number of exotics likely to be transported across network hubs. Tatem and Hay (2007) illustrate how traffic networks can be identified using commercial airline data. They showed strong traffic connections between particular locations that are climatically matched and thus are likely to result in the introduction of species that are well-suited to establish (see Chapter 5). One of their results is directly relevant to the influence of colonization pressure on island invasibility. Tatem and Hay (2007) showed that the Hawaiian Islands have high incoming traffic on flight routes originating from a diverse range of destinations, all of which are climatically matched to Hawaii during some part of the year. This result provides a compelling explanation for why Hawaii is overrun with exotic species. Tatem and Hay suggest that several network hubs like Hawaii exist. We suggest that these sites will appear particularly susceptible to invasions, when in reality international commerce has bestowed upon them a nearly limitless species pool of potential invaders. The methods employed by Tatem and Hay (2007) can easily be modified to other transportation networks such as shipping or trucking routes. Similar approaches have been used in ecology to account for the effects of propagule pressure (e.g., gravity models) and these methods should be amenable for use in characterizing colonization pressure as well.

10.3 False Dichotomies

Ecological and evolutionary questions are by their very nature complex. A strong linear relationship between two ecological variables is as uncommon as a pristine ecological environment. Given this complexity, large questions are often broken down into smaller, manageable sub-questions, which are then considered independently of each other (Gaston and Blackburn 1999). This situation typifies invasion ecology, and the result has often been that we become trapped into thinking those pieces are alternative possibilities instead of parts of a whole. Below we consider two of these false dichotomies and how exotic birds help to clarify the interrelationship between the two 'opposing' viewpoints.

10.3.1 Species or Locations?

In the Preface to the penultimate SCOPE book by Drake, et al. (1989), the editors set forth three key questions about biological invasions (reprinted in Williamson 1996) that have shaped the field ever since. They are:

1) What factors determine whether a species will become an invader or not?
2) What site properties determine whether an ecological system will be prone to, or resistant to, invasions?
3) How should management systems be developed to best advantage, given the knowledge gained from attempting to answer questions 1 and 2?

These questions were clearly an attempt to break down a complex issue into manageable pieces, but seem to have sown the seeds for a debate in invasion ecology as to whether the answer to question 1 (concerning species-level traits) or question 2 (concerning location-level effects) is more important in predicting invasion success (and hence structuring the response to question 3). We fall into this trap to some extent by considering species- and location-level effects in separate chapters (and separate sections in this chapter). However, our review is a good illustration of how these two effects are sub-components of a larger process.

Nowhere is this more obvious than when we consider the roles of ecological breadth of a species and climatic suitability of a location in determining exotic bird establishment success. In Chapter 4, we see that one of the few consistent species-level predictors of establishment success is the ecological flexibility or niche breadth of a bird. Species with more catholic tastes are more likely to find the recipient non-native locale suitable for their needs. In Chapter 5, we review evidence that a close climatic match between a species' native range and the location into which it is introduced will increase the likelihood of establishment (see also Hayes and Barry 2008). These two elements are clearly intertwined since an ecologically flexible species will by default have a higher likelihood of finding a close match (or at least some match) between conditions it can exploit in its native range and conditions it experiences in its introduced range. Which is the more important result? We suggest neither: rather, the two results should not be considered as separate, as they represent different ways of measuring a fundamental relationship between a species and its abiotic environment. Thus, we consider species-by-habitat interactions to be of prime importance in determining the establishment success of an exotic bird population.

10.3.2 Stochastic or Deterministic?

Somewhat related to the above is the question as to whether stochastic or deterministic forces are more important determinants of establishment success. Again, we fall into this trap to some extent by considering chance events, and mechanisms that reduce the effects of these events, in Chapter 3, and then in Chapters 4 and 5 considering the more deterministic factors of species traits and location-level factors separately. However, a running theme in these chapters is how little we gain from considering deterministic forces in isolation from the influence of propagule pressure, the principal variable related to stochastic causes of establishment failure.

Richard Duncan and colleagues provide the two best examples of how stochastic and deterministic forces combine to influence exotic bird establishment in their work on climatic suitability and competition relative to propagule pressure within the exotic birds of Australia and New Zealand respectively (Duncan, et al. 2001; Duncan and Forsyth 2006). Their research suggests that a species can be perfectly matched to the recipient location, be ecologically flexible, and have few natural enemies once introduced, but that the introduction will still fail if the initial number of individuals was too small to escape a stochastic extinction event. Furthermore, we should expect to see the interplay of stochastic and deterministic forces vary not only across locations but also across non-native species introduced to the same location. Thus, for example, some species introduced to New Zealand will fail due to stochastic extinction, some will fail due to competition, and yet others will fail due to the inability to survive in that climate. Finding a clear signal for any one deterministic variable will be rare, especially if we cannot partition out the effects of stochasticity, or indeed the effect of patterns of introduction, on establishment.

10.4 'Somewhere, Something Incredible is Waiting to be Known' (Carl Sagan)

Exotic species provide a unique perspective on a wide variety of ecological and evolutionary issues (Sax, et al. 2005b). We should thus expect that there will always be new ways to incorporate exotic species into basic research. Invasive species also provide a mechanism to explore the dynamics of small populations outside the context of the conservation of endangered species (Briskie 2006). Nevertheless, in writing this book we were surprised at the range of topics in which exotic birds have played a role. In some instances, the focus on exotic birds was held just long enough to provide a glimpse of the tantalizing possibilities they afford to our understanding of natural processes, but not nearly long enough to provide clear answers. We thus end by highlighting two areas in which the study of exotic birds would seem likely to pay high dividends.

10.4.1 Dispersal and Genetic Structure in Exotic Bird Populations

An abundance of avid birdwatchers around the globe has enabled some impressive datasets on the geographic spread of exotic birds to be compiled. Three species in particular seem to dominate the literature on this subject: house sparrow *Passer domesticus*, house finch *Carpodacus mexicanus*, and European starling *Sturnus vulgaris*. All three species were introduced into the eastern USA between 1840 and 1940, and all now occupy more or less the whole of North America (although the house finch was already native to the western half of the continent). Inventive use of data deriving from these introductions has greatly informed our understanding of invasive species spread. However, the use of exotic birds to understand spread dynamics, and the spatial structure of populations more generally, is seriously underdeveloped.

It is now well accepted that the pattern of long-range dispersal events can drive the rate of geographic range expansion in exotic (and native) species. However, such events tend to be rare, and hence rarely observed. Birds seem to offer one of our better hopes for getting good empirical data on long-distance dispersal, and thus injecting a much-needed dose of empiricism into the development of range expansion models. The techniques for marking and following birds are well worked out, and include the potential use of satellite or similar tags that can trace the movements of a single individual over ecologically relevant time scales. We have no shortage of exotic bird populations that are currently showing rapid range expansion. These would seem to present an open invitation to document directly the dispersal patterns of individuals in these populations while also collecting basic information on their rate of range expansion and demography.

Two much-neglected aspects of the dynamics of exotic populations are why some never expand their distributional extents, and why some undergo a 'boom and bust' cycle (Chapter 6). Any avid birdwatcher can tell you where to find the small but persistent population of an exotic bird in their neighbourhood (see e.g. Jiguet 2007; Crochet 2008; Monticelli 2008). These species may not regularly find their way into large catalogues of invasive species, but they are of considerable interest if we want to produce empirical information on the barriers to invasion success, or better understand the dynamics of very small populations. Do these species suffer from inbreeding depression? Do they suffer from Allee effects, or the ill-effects of enemies (parasites, competitors, predators)? Do individuals from these populations ever disperse over long distances, and, if so, what happens to these individuals?

Equally compelling is the potential to use standard analyses of genetic diversity to document spatial structure in populations. The rate of dispersal between patches is a key parameter for understanding the evolution and viability of metapopulations. Although we have some information on the genetic distance (F_{st}) between populations of exotic birds, the methodological approaches adopted by these studies are past their sell-by dates, and were never geared to answer basic questions in metapopulation theory. Exotic birds are thus ripe for exploration in terms of the effects of dispersal barriers to genetic differentiation, trait variance, and population persistence. They are likely to be especially useful in these regards because we often have basic information on their history of introduction and subsequent geographic spread. These are factors we seldom, if ever, know for native species.

10.4.2 Contemporary Evolution in Exotic Bird Populations

Exotic species provide some of the best evidence that evolution by natural selection can proceed at such a fast pace that we can easily document substantial phenotypic and genetic change within the span of our adult lives (Sax, et al. 2007). Because exotic birds have successfully established across all the ice-free land masses on Earth, and several dozen species have established multiple exotic populations, they provide us with an unprecedented opportunity to explore contemporary evolution (Hardy 1973).

Avian life-history evolution has been a centrepiece of ornithological science for decades. We thus have a variety of expectations in terms of what it would mean, for example, to take a bird that evolved in the temperate zone and introduce it into the tropics. Amongst other things, this species should tend towards a longer breeding season with multiple broods being produced each year. Since the climate is presumably more benign, individuals of this species should live longer on average than their native counterparts. These changes should in turn lead to fewer eggs being laid per clutch, and perhaps to the young fledging at larger weights. Do temperate birds that were introduced to the tropics show these predicted changes? We have some evidence for a few exotic bird populations (Chapter 9) that breeding seasons are extended and clutch sizes reduced in response to more benign climates (although not that climate itself is necessarily the driver of these changes), but in general we have no clue what has happened to the life histories of exotic bird species following establishment.

There is a similar story to be told in terms of the effects of community interactions on the direction and rate of evolution. Exotic birds may, or may not, escape from predators, competitors, and parasites in their non-native range. Some exotic birds are coming into contact with entirely novel assemblages of potential enemies. How do these species respond in the face of what almost certainly is a drastic change in selection regimes? As yet, no one knows.

Beyond the appeal of providing evidence for contemporary evolution stemming from these selection agents, the study of exotic birds provides us with some additional advantages over the study of native species. For example, species with multiple exotic populations allow us the rare chance to see if evolution repeats itself given largely the same set of conditions. The scant evidence on this topic presented in Chapter 9 suggests the answer is yes, but the particular mechanisms driving these changes may be more idiosyncratic than evolutionary biologists might suspect. Similarly, the often detailed knowledge of the specific point of origin for several exotic bird populations, and on the numbers in which these species were introduced, provide unprecedented background information upon which to base evolutionary expectations. We can empirically test the assumption that populations started at very small propagule sizes are constrained in their evolutionary response to novel selection regimes because they lost substantial genetic variation via the founding event. The possibilities we can present in this regard seem limited more by our understanding of evolutionary biology than by an inherent lack of opportunity afforded by exotic birds.

10.5 Conclusions

At this point we would refer the reader back to the quote that heads up this chapter, written eighty years ago by John C. Phillips. Eighty years on, it is fair to say that Phillips' observation still holds true, and indeed can be applied to the invasion process in general. However, lacking definitive rules does not serve to distinguish invasion biology from other fields of ecological research, where the existence or otherwise of hard and fast laws has been the subject of debate (e.g., Lawton 1999; Colyvan and Ginzburg 2003; O'Hara 2005). Moreover, it also does not mean that we have failed to make progress in understanding avian invasions.

We embarked on this book with the aim of constructing a coherent and informative picture of how a bird species goes from native to exotic, and beyond. Despite the limits set by the state of available knowledge, and by our own limitations in interpreting it, we think that a coherent and informative picture is starting to emerge. The guiding hand of humanity in the invasion process, the interplay between event, species, and location, and the importance of both the stochastic and the deterministic in establishment and spread, are all emerging as important and consistent themes. Nevertheless, we believe that we are only just beginning to scratch the surface in terms of what we can learn about bird invasions, and what bird invasions can tell us about how the natural world works. Whether invasions by birds are materially different from those by other taxa also remains to be seen. It would be helpful if they were not, but we suspect that the forces driving invasions by parrots, plankton, and plants are sufficiently diverse and different that they will be.

Bibliography

Able, K. P. and Belthoff, J. R. (1998). Rapid 'evolution' of migratory behaviour in the introduced house finch of eastern North America. *Proceedings of the Royal Society, London, B* **265**, 2063–2071.

Agoramoorthy, G. and Hsu, M. J. (2007). Ritual releasing of wild animals threatens island ecology. *Human Ecology* **35**, 251–254.

Alexander, D. J., Wilson, G. W. C., Russell, P. H., Lister, S. A. and Parsons, G. (1985). Newcastle disease outbreaks in fowl in Great Britain during 1984. *Veterinary Record* **117**, 429–434.

Allee, W. C. (1931). *Animal aggregations: A study in general sociology*. University of Chicago Press, Chicago.

Allee, W. C. (1938). *The social life of animals*. Norton, New York.

Allen, C. R. (2006). Predictors of introduction success in the South Florida avifauna. *Biological Invasions* **8**, 491–500.

Allen, J., Gosden, C. and White, J. P. (1989). Human Pleistocene adaptations in the tropical island Pacific: Recent evidence from New Ireland, a greater Australian outlier. *Antiquity* **63**, 548–561.

American Birding Association (2002). ABA checklist of North American birds. 7th edition. American Ornithologists Union, Washington, DC.

Anderson, A. (1991). The chronology of colonization in New Zealand. *Antiquity* **65**, 767–795.

Anderson, S. H., Kelly, D., Robertson, A. W., Ladley, J. J. and Innes, J. G. (2006). Birds as pollinators and dispersers: A case study from New Zealand. *Acta Zoologica Sinica* **52S**, 112–115.

Andow, D. A., Kareiva, P. M., Levin, S. A. and Okubo, A. (1990). Spread of invading organisms. *Landscape Ecology* **4**, 177–188.

Anon. (1805). *A description of the island of St Helena; containing observations on its singular structure and formation; and an account of its climate, natural history, and inhabitants.* R. Phillips, London.

Apanius, V., Yorinks, N., Bermingham, E. and Ricklefs, R. E. (2000). Island and taxon effects in parasitism and resistance of lesser Antillean birds. *Ecology* **81**, 1959–1969.

Araújo, M. B. (2003). The coincidence of people and biodiversity in Europe. *Global Ecology and Biogeography* **12**, 5–12.

Arrhenius, O. (1921). Species and area. *Journal of Ecology* **9**, 95–99.

Arrhenius, O. (1923). On the relation between species and area: A reply. *Ecology* **4**, 90–91.

Ashmole, N. P. (1963). The regulation of numbers of tropical oceanic birds. *Ibis* **103b** 458–473.

Ashton, K. G. (2002). Do amphibians follow Bergmann's rule? *Canadian Journal of Zoology* **80**, 708–716.

Atkinson, C. T., Lease, J. K., Dusek, R. J. and Samuel, M. D. (2005). Prevalence of pox-like lesions and malaria in forest bird communities on leeward Mauna Loa Volcano, Hawaii. *Condor* **107**, 537–546.

Avise, J. C. (2004). *Molecular markers, natural history, and evolution*, 2nd edition. Sinauer Associates Inc., Sunderland, MA.

Badyaev, A. V. (2005). Maternal inheritance and rapid evolution of sexual size dimorphism: Passive effects or active strategies? *American Naturalist* **166**, S17–S30.

Badyaev, A. V. (2009). Evolutionary significance of phenotypic accommodation in novel environments: An empirical test of the Baldwin effect. *Philosophical Transactions of the Royal Society, London, B.* **364**, 1125–1141.

Badyaev, A. V. and Hill, G. E. (2000). The evolution of sexual dimorphism in the house finch. I. Population divergence in morphological covariance structure. *Evolution* **54**, 1784–1794.

Badyaev, A. V. and Oh, K. P. (2008). Environmental induction and phenotypic retention of adaptive maternal effects. *BMC Evolutionary Biology* 8:3 doi: 10.1186/1471–2148-8–3.

Badyaev, A. V., Hill, G. E., Stoehr, A. M., Nolan. P. M. and McGraw, K. J. (2000). The evolution of sexual dimorphism in the house finch: II. Population divergence in relation to local selection. *Evolution* **54**, 2134–2144.

Badyaev, A. V., Oh, K. P. and Mui, R. (2006). Evolution of sex-biased maternal effects in birds: II. Contrasting sex-specific oocyte clustering in native and recently established populations. *Journal of Evolutionary Biology* **19**, 909–921.

Baker, A. J. (1980). Morphometric differentiation in New Zealand populations of the house sparrow (*Passer domesticus*). *Evolution* **34**, 638–653.

Baker, A. J. (1992). Genetic and morphometric divergence in ancestral European and descendent New Zealand populations of chaffinches (*Fringilla coelebs*). *Evolution* **46**, 1784–1800.

Baker, A. J. and Moeed, A. (1979). Evolution in the introduced New Zealand populations of the common myna, *Acridotheres tristis* (Aves: Sturnidae). *Canadian Journal of Zoology* **57**, 570–584.

Baker, A. J. and Moeed, A. (1987). Rapid genetic differentiation and founder effect in colonizing populations of common mynas (*Acridotheres tristis*). *Evolution* **41**, 525–538.

Baker, H. G. (1974). The evolution of weeds. *Annual Review of Ecology and Systematics*, **5**, 1–24.

Baker, I., Gale, N. H. and Simons, J. (1967). Geochronology of the St. Helena volcanoes. *Nature* **215**, 1451–1456.

Balmford, A., Moore, J. L., Brooks, T., Burgess, N., Hansen, L. A., Williams, P. and Rahbek, C. (2001). Conservation conflicts across Africa. *Science* **291**, 2616–2619.

Barrowclough, G. F. (1983). Biochemical studies of microevolutionary processes. In *Perspectives in ornithology: Essays presented for the centennial of the American Ornithologists Union*, eds A. H. Brush and G. A. C. Jr, pp. 223–261. Cambridge University Press, Cambridge.

Begon, M., Harper, J. L. and Townsend, C. R. (1996). *Ecology: Individuals, populations and communities*. Blackwell Scientific Publications, Oxford.

Bell, C. P. (1996). The relationship between geographic variation in clutch size and migration pattern in the yellow wagtail. *Bird Study* **43**, 333–341.

Belthoff, J. R. and Gauthreax Jr, S. A. (1991). Partial migration and differential winter distribution of house finches in the eastern United States. *Condor* **93**, 374–382.

Bennett, P. M. (1986). Comparative studies of morphology, life history and ecology among birds. PhD thesis, University of Sussex.

Bennett, P. M. and Harvey, P. H. (1987). Active and resting metabolism in birds: Allometry, phylogeny and ecology. *Journal of Zoology, London* **213**, 327–363.

Bennett, P. M. and Owens, I. P. F. (2002). *The evolutionary ecology of birds*. Oxford University Press, Oxford.

Bergmann, C. (1847). Ueber die Verhältnisse der Wärmeökonomie der Thiere zu ihrer Grösse. *Gottinger studien* **3**, 595–708.

Berthold, P. (2003). Genetic basis and evolutionary aspects of bird migration. *Advances in the Study of Behavior* **33**, 175–229.

Bielby, J., Mace, G. M., Bininda-Emonds, O. R. P., Cardillo, M., Gittleman, J. L., Jones, K. E., Orme, C. D. L. and Purvis, A. (2007). The fast–slow continuum in mammalian life history: An empirical reevalution. *American Naturalist* **169**, 748–757.

BirdLife International (2000). *Threatened birds of the world*. Lynx Edicions and BirdLife International, Barcelona and Cambridge.

Blackburn, T. M., and Cassey, P. (2007). Patterns of non-randomness in the exotic avifauna of Florida. *Diversity and Distributions* **13**, 519–526.

Blackburn, T. M., and Duncan, R. P. (2001a). Determinants of establishment success in introduced birds. *Nature* **414**, 195–197.

Blackburn, T. M. and Duncan, R. P. (2001b). Establishment patterns of exotic birds are constrained by non-random patterns in introduction. *Journal of Biogeography* **28**, 927–939.

Blackburn, T. M. and Gaston, K. J. (1994). The distribution of body sizes of the world's bird species. *Oikos* **70**, 127–130.

Blackburn, T. M. and Gaston, K. J. (1996a). Spatial patterns in the body sizes of bird species in the New World. *Oikos* 77, 436–446.

Blackburn, T. M. and Gaston, K. J. (1996b). Spatial patterns in the geographic range sizes of bird species in the New World. *Philosophical Transactions of the Royal Society, London, B* **351**, 897–912.

Blackburn, T. M. and Gaston, K. J. (2002). Extrinsic factors and the population sizes of threatened birds. *Ecology Letters* **5**, 568–576.

Blackburn, T. M. and Gaston, K. J. (2005). Biological invasions and the loss of birds on islands: Insights into the idiosyncrasies of extinction. In *Species invasions: Insights into ecology, evolution and biogeography*, eds D. Sax, S. D. Gaines and J. J. Stachowicz, pp. 85–110. Academic Press, Sunderland, MA.

Blackburn, T. M. and Jeschke, J. M. (2009). Invasion success and threat status: Two sides of a different coin? *Ecography* **32**, 83–88.

Blackburn, T. M., Cassey, P. and Gaston, K. J. (2006). Variations on a theme: Sources of heterogeneity in the form of the interspecific relationship between abundance and distribution. *Journal of Animal Ecology* **75**, 1426–1439.

Blackburn, T. M., Cassey, P. and Lockwood, J. L. (2008). The island biogeography of exotic bird species. *Global Ecology and Biogeography* **17**, 246–251.

Blackburn, T. M., Cassey, P. and Lockwood, J. L. (2009). The role of species traits in the establishment success of exotic birds. *Global Change Biology*, in press.

Blackburn, T. M., Cassey, P., Duncan, R. P., Evans, K. L. and Gaston, K. J. (2004). Avian extinction and mammalian introductions on oceanic islands. *Science* **305**, 1955–1958.

Blackburn, T. M., Cassey, P., Duncan, R. P., Evans, K. L. and Gaston, K. J. (2005a). Causes of avian extinction on islands. *Science* **307**, 1412b.

Blackburn, T. M., Gaston, K. J. and Duncan, R. P. (2001). Population density and geographic range size in the introduced and native passerine faunas of New Zealand. *Diversity and Distributions* 7, 209–221.

Blackburn, T. M., Gaston, K. J. and Lawton, J. H. (1998). Patterns in the geographic ranges of the world's woodpeckers. *Ibis* **140**, 626–638.

Blackburn, T. M., Gaston, K. J. and Loder, N. (1999). Geographic gradients in body size: A clarification of Bergmann's rule. *Diversity and Distributions* **5**, 165–174.

Blackburn, T. M., Lawton, J. H. and Gregory, R. D. (1996). Relationships between abundances and life histories of British birds. *Journal of Animal Ecology* **65**, 52–62.

Blackburn, T. M., Petchey, O. L., Cassey, P. and Gaston, K. J. (2005b). Functional diversity of mammalian predators and extinction in island birds. *Ecology* **86**, 2916–2923.

Blakers, N., Davies, S. J. J. F. and Reilly, P. N. (1984). *The atlas of Australian birds.* Melbourne University Press, Carlton.

Blanvillain, C., Salducci, J. M., Tutururai, G. and Maeura, M. (2003). Impact of introduced birds on the recovery of the Tahiti Flycatcher (*Pomarea nigra*), a critically endangered forest bird of Tahiti. *Biological Conservation* **109**, 197–205.

Böhning-Gaese, K. and Oberrath, R. (1999). Phylogenetic effects on morphological, life-history, behavioural and ecological traits of birds. *Evolutionary Ecology Research* **1**, 347–364.

Böhning-Gaese, K., Caprano, T., van Ewjik, K. and Veith, M. (2006). Range size: Disentangling current traits and phylogenetic and biogeographic factors. *American Naturalist* **167**, 555–567.

Bomford, M. and Hart, Q. (2002). Non-indigenous vertebrates in Australia. In *Biological Invasions: Environmental and economic costs of alien plant, animal and microbe species.* ed D. Pimentel, pp. 25–44. CRC Press, Boca Raton, FL.

Bonier, F., Martin, P. R. and Wingfield, J. C. (2007). Urban birds have broader environmental tolerance. *Biology Letters* **3**, 670–673.

Bortolini, M. C., Salzano, F. M., Thomas, M. G., Stuart, S., Nasanen, S. P. K., Bau, C. H. D., Hutz, M. H., Layrisse, Z., Petzl-Erler, M. L., Tsuneto, L. T., Hill, K., Hurtado, A. M., Castro-de-Guerra, D., Torres, M. M., Groot, H., Michalski, R., Nymadawa, P., Bedoya, G., Bradman, N., Labuda, D. and Ruiz-Linares, A. (2003). Y-chromosome evidence for differing ancient demographic histories in the Americas. *American Journal of Human Genetics* **73**, 524–539.

Bouzat, J. L. (1998). Genetic evaluation of a demographic bottleneck in the Greater Prairie Chicken. *Conservation Biology* **12**, 836–843.

Brisbin, I. L. Jr. (1995). Conservation of the wild ancestors of domestic animals. *Conservation Biology* **9**, 1327–1328.

Brisbin, I. L. Jr. and Peterson, A. T. (2007). Playing chicken with red junglefowl: Identifying phenotypic markers of genetic purity in *Gallus gallus*. *Animal Conservation* **10**, 429–435.

Briskie, J. V. (2006). Introduced birds as model systems for the conservation of endangered native birds. *Auk* **123**, 949–957.

Briskie, J. V. and Mackintosh, M. (2004). Hatching failure increases with severity of population bottlenecks in birds. *Proceedings of the National Academy of Sciences, USA* **101**, 558–561.

Brook, B. W. (2004). Australian bird invasions: Accidents of history? *Ornithological Science* **3**, 33–42.

Brook, B. W. (2008). Demographics versus genetics in conservation biology. In *Conservation Biology: Evolution in Action*, eds S.P. Carroll and C.W. Fox, pp. 35–49. Oxford University Press, Oxford.

Brook, B. W., Sodhi, N. S. and Bradshaw, C. J. A. (2008). Synergies among extinction drivers under global change. *Trends in Ecology & Evolution* **23**, 453–460.

Brook, B. W., Traill, L. W. and Bradshaw, C. J. A. (2006). Minimum viable population size and global extinction risk are unrelated. *Ecology Letters* **9**, 375–382.

Brooke, R. K., Lockwood, J. L. and Moulton, M. P. (1995). Patterns of success in passeriform bird introductions on Saint Helena. *Oecologia* **103**, 337–342.

Brooks, T. (2001). Are unsuccessful avian invaders rarer in their native range than successful invaders? In *Biotic homogenization*, eds J. L. Lockwood and M. L. McKinney, pp. 125–155. Kluwer Academic/Plenum Publishers, New York.

Brown, J. H. (1981). Two decades of homage to Santa Rosalia: Toward a general theory of diversity. *American Zoologist* **21**, 877–888.

Brown, J. H. (1984). On the relationship between abundance and distribution of species. *American Naturalist* **124**, 255–279.

Brown, J. H. (1989). Patterns, modes and extents of invasions by vertebrates. In *Biological invasions, a global perspective*, eds J. A. Drake, F. di Castri, R. H. Groves, F. J. Kruger, M. Rejmánek and M. Williamson, pp. 85–109. John Wiley & Sons, Chichester.

Brown, J. H., and Sax, D. (2004). An essay on some topics concerning invasive species. *Austral Ecology* **29**, 530–536.

Brown, J. H., Gillooly, J. F., West, G. B. and Savage, V. M. (2003). The next step in macroecology: From general empirical patterns to universal ecological laws. In *Macroecology: Concepts and consequences*, eds T. M. Blackburn and K. J. Gaston, pp. 408–423. Blackwell Science, Oxford.

Bull, P. C., Gaze, P. D. and Robertson, C. J. R. (1985). *The atlas of bird distribution in New Zealand.* The Ornithological Society of New Zealand, Wellington.

Bumpus, H. (1899). The elimination of the unfit as illustrated by the introduced sparrow, *Passer domesticus*. *Biological Lectures, Marine Biology Lab, Woods Hole*, 209–226.

Butler, C. J. (2005). Feral parrots in the continental United States and United Kingdom: Past, present and future. *Journal of Avian Medicine and Surgery* **19**, 142–149.

Cabe, P. R. (1998). The effects of founding bottlenecks on genetic variation in the European starling (*Sturnus vulgaris*). in North America. *Heredity* **80**, 519–525.

Cade, T. J. (1983). Hybridization and gene exchange among birds in relation to conservation. In *Genetics and conservation*, eds C. M. Schonewald-Cox, S. M. Chambers, B. MacBryde and W. L. Thomas, pp. 288–309. Benjamin/Cummings, Menlo Park, CA.

Cadotte, M. W., McMahon, S. M. and Fukami, T. eds. (2006). *Conceptual ecology and invasion biology: Reciprocal approaches to nature.* Springer, Dordrecht.

Calhoun, J. B. (1947). The role of temperature and natural selection in the variations of size of the English sparrow in the United States. *American Naturalist* **81**, 203–228.

Cardillo, M. (2002). The life-history basis of latitudinal diversity gradients: How do species traits vary from the poles to the equator? *Journal of Animal Ecology* **71**, 79–87.

Carlton, J. T. (1996). Pattern, process, and prediction in marine invasion ecology. *Biological Conservation* **78**, 97–106.

Carrete, M. and Tella, J. L. (2008). Wild-bird trade and exotic invasions: A new link of conservation concern? *Frontiers in Ecology and the Environment*, **6**.

Carroll, S. P., Hendry, A. P., Reznick, D. N. and Fox, C. W. (2007). Evolution on ecological time-scales. *Functional Ecology* **21**, 387–393.

Case, T. J. (1996). Global patterns in the establishment and distribution of exotic birds. *Biological Conservation* **78**, 69–96.

Cassey, P. (2001a). Are there body size implications for the success of globally introduced land birds? *Ecography* **24**, 413–420.

Cassey, P. (2001b). Determining variation in the success of New Zealand land birds. *Global Ecology and Biogeography* **10**, 161–172.

Cassey, P. (2002a). Comparative analysis of successful establishment among introduced landbirds. PhD thesis, Griffith University, Australia.

Cassey, P. (2002b). Life history and ecology influences establishment success of introduced land birds. *Biological Journal of the Linnean Society* **76**, 465–480.

Cassey, P. (2003). A comparative analysis of the relative success of introduced landbirds on islands. *Evolutionary Ecology Research* **5**, 1011–1021.

Cassey, P., Blackburn, T. M. and Evans, K. L. (2005d). Correlated changes in reproductive effort of exotic passerines. *Notornis* **52**, 243–246.

Cassey, P., Blackburn, T. M., Duncan, R. P. and Chown, S. L. (2005a). Concerning invasive species: A reply to Brown and Sax. *Austral Ecology* **30**, 475–480.

Cassey, P., Blackburn, T. M. Duncan, R. P. and Gaston, K. J. (2005b). Causes of exotic bird establishment across oceanic islands. *Proceedings of the Royal Society, London B*, **272**, 2059–2063.

Cassey, P., Blackburn, T. M. Duncan, R. P. and Lockwood, J. L. (2005c). Lessons from the establishment of exotic species: A meta-analytical case study using birds. *Journal of Animal Ecology* **74**, 250–258.

Cassey, P., Blackburn, T. M., Duncan, R. P. and Lockwood, J. L. (2008). Lessons from introductions of exotic species as a possible information source for managing translocations of birds. *Wildlife Research* **35**, 193–201.

Cassey, P., Blackburn, T. M., Jones, K. E. and Lockwood, J. L. (2004a). Mistakes in the analysis of exotic species establishment: Source pool designation and correlates of introduction success among parrots (Psittaciformes) of the world. *Journal of Biogeography* **31**, 277–284.

Cassey, P., Blackburn, T. M., Lockwood, J. L. and Sax, D. F. (2006). The shape of biotic homogenisation: A stochastic model for integrating changes in species richness and identity. *Oikos* **115**, 207–218.

Cassey, P., Blackburn, T. M., Russell, G. J., Jones, K. E. and Lockwood, J. L. (2004b). Influences on the transport and establishment of exotic bird species: An analysis of the parrots (Psittaciformes) of the world. *Global Change Biology* **10**, 417–426.

Cassey, P., Blackburn, T. M., Sol, D. Duncan, R. P. and Lockwood, J. L. (2004c). Introduction effort and establishment success in birds. *Proceedings of the Royal Society, London, B (supplement)* **271**, S405–S408.

Cassey, P., Lockwood, J. L., Blackburn, T. M. and Olden, J. D. (2007). Spatial scale and evolutionary history determine the degree of taxonomic homogenization across island bird assemblages. *Diversity and Distributions* **13**, 458–466.

Caswell, H., Lensink, R., and Neubert, M. G. (2003). Demography and dispersal: Life table response experiments for invasion speed. *Ecology* **84**, 1968–1978.

Caughley, G. (1994). Directions in conservation biology. *Journal of Animal Ecology* **63**, 215–244.

Caughley, G. and Gunn, A. (1996). *Conservation biology in theory and practice.* Blackwell Science, Oxford.

Cawthorne, R. A. and Marchant, J. H. (1980). The effects of the 1978/79 winter on British bird populations. *Bird Study* **27**, 163–172.

Chamberlain, D. E., Fuller, R. J., Bunce, R. G. H. Duckworth, J. C. and Shrubb, M. (2000). Changes in the abundance of farmland birds in relation to the timing of agricultural intensification in England and Wales. *Journal of Applied Ecology* **37**, 771–778.

Chase, J. M. and Leibold, M. A. (2003). *Ecological niches: Linking classical and contemporary approaches* University of Chicago Press, Chicago.

Chave, J. (2004). Neutral theory and community ecology. *Ecology Letters* 7, 241–253.

Cheke, A. and Hume, J. (2008). *Lost land of the dodo: An ecological history of Mauritius, Réunion & Rodrigues*. T & A. D. Poyser, London.

Chiron, F., Shirley, S. and Kark, S. (2008). Human-related processes drive the richness of exotic birds in Europe. *Proceedings of the Royal Society, London, B* **276**, 47–53.

Chown, S. L., Gremmen, N. J. M. and Gaston, K. J. (1998). Ecological biogeography of southern ocean islands: Species-area relationships, human impacts, and conservation. *American Naturalist* **152**, 562–575.

Clark, J. S. (1998). Why trees migrate so fast: Confronting theory with dispersal biology and the paleorecord. *American Naturalist* **152**, 204–224.

Clark, J. S., Lewis, M., McLachlan, J. S. and HilleRisLambers, J. (2003). Estimating population spread: What can we forecast and how well? *Ecology* **84**, 1979–1988.

Clavero, M. and García-Berthou, E. (2005). Invasive species are a leading cause of animal extinctions. *Trends in Ecology & Evolution* **20**, 110.

Clegg, S. M., Degnan, S. M., Kikkawa, J., Moritz, C., Estoup, A. and Owens, I. P. F. (2002). Genetic consequences of sequential founder events by an island-colonizing bird. *Proceedings of the National Academy of Sciences, USA* **99**, 8127–8132.

Clements, F. E. (1916). Plant succession: An analysis of the development of vegetation. *Carnegie Institution of Washington Publication 242*.

Clergeau, P., and Mandon-Dalger, I. (2001). Fast colonization of an introduced bird: The case of *Pycnonotus jocosus* on the Mascarene Islands. *Biotropica* **33**, 542–546.

Clergeau, P., Jokimäki, J. and Savard, J.-P. L. (2001). Are urban bird communities influenced by the bird diversity of adjacent landscapes? *Journal of Applied Ecology* **38**, 1122–1134.

Colautti, R. I., Grigorovich, I. A. and MacIsaac, H. J. (2006). Propagule pressure: A null model for biological invasions. *Biological Invasions* **8**, 1023–1037.

Colautti, R. I., Muirhead, J. R., Biswas, R. N. and MacIsaac, H. J. (2005). Realized vs apparent reduction in enemies of the European starling. *Biological Invasions* 7, 723–732.

Colautti, R. I., Ricciardi, A., Grigorovich, I. A. and MacIsaac, H. J. (2004). Is invasion success explained by the enemy release hypothesis? *Ecology Letters* 7, 721–733.

Cole, F. R., Loope, L. L., Medeiros, A. C., Raikes, J. A. and Wood, C. S. (1995). Conservation implications of introduced game birds in high-elevation Hawaiian shrubland. *Conservation Biology* **9**, 306–313.

Colyvan, M. and Ginzburg, L. R. (2003). Laws of nature and laws of ecology. *Oikos* **101**, 649–653.

Conner, J. K. and Hartl, D. L. (2004). *A primer of ecological genetics*. Sinauer Associates Inc., Sunderland, MA.

Connor, E. F. and McCoy, E. D. (1979). The statistics and biology of the species-area relationship. *American Naturalist* **113**, 791–833.

Courchamp, F, Berec, L. and Gascoigne, J. (2008). Allee effects in ecology and conservation. Oxford University Press, Oxford.

Courchamp, F., Clutton-Brock, T. and Grenfell, B. (1999). Inverse density dependence and the Allee effect. *Trends in Ecology & Evolution* **14**, 405–410.

Cousins, S. H. (1989). Species richness and the energy theory. *Nature* **340**, 350–351.

Cox, G. W. (2004). *Alien species and evolution*. Island Press, Washington, DC.

Cox, P. A. and Elmqvist, T. (2000). Pollinator extinction in the Pacific Islands. *Conservation Biology* **14**, 1237–1239.

Cramp, S. and Perrins, C. M. eds. (1994). *Birds of the Western Palaearctic*. Volume IX. *Buntings and New World warblers*. Oxford University Press, Oxford.

Cramp, S. ed. (1988). *Birds of the Western Palaearctic*. Volume V. *Tyrant flycatchers to thrushes*. Oxford University Press, Oxford.

Crochet, P.-A. (2008). Birding Algeria for Algerian Nuthatch and other specialities. *Birding World* **21**, 19–25.

Crooks, J. A. (2005). Lag times and exotic species: The ecology and management of biological invasions in slow-motion. *Ecoscience* **12**, 316–329.

Crosby, A. W. (1993). *Ecological imperialism: The biological expansion of Europe, 900–1900*. Cambridge University Press, Cambridge.

Cruz, F. B., Fitzgerald, L. A., Espinoza, R. E. and Schulte, J. A. (2005). The importance of phylogenetic scale in tests of Bergmann's and Rapoport's rules: Lessons from a clade of South American lizards. *Journal of Evolutionary Biology* **18**, 1559–1574.

Currie, D. J. and Fritz, J. T. (1993). Global patterns of animal abundance and species energy use. *Oikos* **67**, 56–68.

D'Antonio, C. M. and Dudley, T. L. (1993). Alien species: How inadvertent immigrants affect habitats. *Pacific Discovery* **46**, 8–11.

Daehler, C. C. (1998). The taxonomic distribution of invasive angiosperm plants: Ecological insights and comparison to agricultural weeds. *Biological Conservation* **84**, 167–180.

Daehler, C. C. (2001). Darwin's naturalization hypothesis revisited. *American Naturalist* **158**, 324–330.

Daehler, C. C. and Strong, D. R. (1993). Prediction and biological invasions. *Trends in Ecology & Evolution* **8**, 380.

Darwin, C. (1859). *On the origin of species*. Murray, London.

Davies, N. B. and Brooke, M. de L. (1989). An experimental study of co-evolution between the cuckoo *Cuculus canorus*, and its hosts. I. Host egg discrimination. *Journal of Animal Ecology* **58**, 207–224.

Davies, R. G., Orme, C. D. L., Storch, D., Olson, V., Thomas, G. H., Bennett, P. M., Blackburn, T. M., Owens, I. P. F. and Gaston, K. J. (2007). Topography, temperature and the global distribution of bird species richness. *Proceedings of the Royal Society, London, B* **274**, 1189–1197.

Davis, M. A. (2003). Biotic globalization: Does competition from introduced species threaten biodiversity? *Bioscience* **53**, 481–489.

Davis, M. A. and Pelsor, M. (2001). Experimental support for a resource-based mechanistic model of invasibility. *Ecology Letters* **4**, 421–428.

Dawson, J. C. (1984). A statistical analysis of species characteristics affecting the success of bird introductions. BSc thesis, University of York.

Day, T. M. (1995). Bird species composition and abundance in relation to native plants in urban gardens, Hamilton, New Zealand. *Notornis* **42**, 175–186.

Dayan, T. and Simberloff, D. (1994). Character displacement, sexual dimorphism, and morphological variation among British and Irish mustelids. *Ecology* **75**, 1063–1073.

Dean, W. R. J. (2000). Alien birds in southern Africa: What factors determine success? *South African Journal of Science* **96**, 9–14.

Dennis, B. (1989). Allee effects: Population growth, critical density, and the chance of extinction. *Natural Resource Modelling* **3**, 481–538.

Dennis, B. (2002). Allee effects in stochastic populations. *Oikos* **96**, 389–401.

DeSalle, R. (2005). Genetics at the brink of extinction. *Heredity* **94**, 386–387.

Devictor, V., Julliard, R., Clavel, J., Jiguet, F., Lee, A. and Couvet, D. (2008). Functional biotic homogenization of bird communities in disturbed landscapes. *Global Ecology and Biogeography* **17**, 252–261.

Devictor, V., Julliard, R., Couvet, D., Lee A. and Jiguet, F. (2007). Functional homogenization effect of urbanization on bird communities. *Conservation Biology* **21**, 741–751.

Dhondt, A. A. (2006). Dynamics of mycoplasmal conjunctivitis in the native and introduced range of the host. *Ecohealth* **3**, 95–102.

di Castri, F. (1989). History of biological invasions with special emphasis on the Old World. In *Biological invasions, a global perspective*, eds J. A. Drake, F. di Castri, R. H. Groves, F. J. Kruger, M. Rejmánek and M. Williamson, pp. 1–30. John Wiley & Sons, Chichester.

Diamond, A. ed. (1987). *Studies of Mascarene Island birds*. British Ornithologists' Union and Cambridge University Press, Cambridge.

Diamond, J. (1986). Overview: Laboratory experiments, field experiments, and natural experiments. In *Community ecology*, eds J. Diamond and T. J. Case, pp. 3–22. Harper & Row, New York.

Diamond, J. (1998). *Guns, germs and steel: A short history of everybody for the last 13,000 years*. Vintage, London.

Diamond, J. M. (1984). Historic extinctions: A Rosetta Stone for understanding prehistoric extinctions. In *Quaternary Extinctions: A Prehistoric Revolution*, eds P. S. Martin and R. Klein, pp. 824–862. University of Arizona Press, Tuscon.

Diamond, J. M. (1989). The present, past and future of human-caused extinctions. *Philosophical Transactions of the Royal Society, London, B* **325**, 469–476.

Diamond, J. M. and Veitch, C. R. (1981). Extinctions and introductions in the New Zealand avifauna: Cause and effect? *Science* **211**, 499–501.

Didham, R. K., Tylianakis, J. M., Gemmell, N. J., Rand, T. A. and Ewers, R. M. (2007). Interactive effects of habitat modification and species invasion on native species decline. *Trends in Ecology & Evolution* **22**, 489–496.

Diez, J. M., Sullivan, J. J., Hulme, P. E., Edwards, G. and Duncan, R. P. (2008). Darwin's naturalization conundrum: Dissecting taxonomic patterns of species invasions. *Ecology Letters* **11**, 674–681.

Diniz-Filho, J. A. and Bini, L. M. (2008). Macroecology, global change and the shadow of forgotten ancestors. *Global Ecology and Biogeography* **17**, 11–17.

Domènech, J., Carrillo, J. and Senar, J. C. (2003). Population size of the Monk Parakeet *Myiopsitta monachus* in Catalonia. *Revista Catalana d'Ornitologia*, **20** 1–9.

Donze, J., Moulton, M. P., Labisky, R. F. and Jetz, W. (2004). Sexual plumage differences and the outcome of game bird (Aves: Galliformes). introductions on oceanic islands. *Evolutionary Ecology Research* **6**, 595–606.

Dormann, C. F., McPherson, J. M., Araujo, M. B., Bivand, R., Bolliger, J., Carl, G., Davies, R. G., Hirzel, A., Jetz, W., Kissling, W. D., Kuhn, I., Ohlemuller, R., Peres-Neto, P. R., Reineking, B., Schroder, B., Schurr, F. M. and Wilson, R. (2007). Methods to account for spatial autocorrelation in the analysis of species distributional data: A review. *Ecography* **30**, 609–628.

Drake, J. A., di Castri, F., Groves, R. H., Kruger, F. J., Rejmánek, M. and Williamson, M. eds. (1989). *Biological invasions, a global perspective*. John Wiley & Sons, Chichester.

Drake, J. M. (2003). The paradox of the parasites: Implications for biological invasion. *Proceedings of the Royal Society, London, B (supplement)* **270**, 133–135.

Drake, J. M. (2006). Heterosis, the catapult effect, and establishment success of a colonizing bird. *Biology Letters* **2**, 304–307.

Drake, J. M. (2007). Parental investment and fecundity, but not brain size, are associated with establishment success in introduced fishes. *Functional Ecology* **21**, 963–968.

Drake, J. M. and Lodge, D. M. (2006). Allee effects, propagule pressure and the probability of establishment: Risk analysis for biological invasions. *Biological Invasions* **8**, 365–375.

Drake, J. M., Baggenstos, P. and Lodge, D. M. (2005). Propagule pressure and persistence in experimental populations. *Biology Letters* **1**, 480–483.

Draycott, R. A., Hoodless, A. N. and Sage, R. B. (2008). Effects of pheasant management on vegetation and birds in lowland woodlands. *Journal of Applied Ecology* **45**.

Drummond, J. (1907). On introduced birds. *Transactions & Proceedings of the New Zealand Institute* **39**, 227–252.

Duckworth, R. A. (2008). Adaptive dispersal strategies and the dynamics of a range expansion. *American Naturalist* **172**, S4–S17.

Duncan, R. P. (1997). The role of competition and introduction effort in the success of passeriform birds introduced to New Zealand. *American Naturalist* **149**, 903–915.

Duncan, R. P. and Blackburn, T. M. (2002). Morphological over-dispersion in game birds (Aves: Galliformes). successfully introduced to New Zealand was not caused by interspecific competition. *Evolutionary Ecology Research* **4**, 551–561.

Duncan, R. P. and Forsyth, D. M. (2006). Competition and the assembly of introduced bird communities. In *Conceptual ecology and invasion biology: Reciprocal approaches to nature,* eds M. W. Cadotte, S. M. McMahon and T. Fukami, 415–431. Springer, Dordrecht.

Duncan, R. P. and Young, J. R. (1999). The fate of passeriform introductions on oceanic islands. *Conservation Biology* **13**, 934–936.

Duncan, R. P., Blackburn, T. M. and Cassey, P. (2006). Factors affecting the release, establishment and spread of introduced birds in New Zealand. In *Biological invasions in New Zealand*, eds R. B. Allen and W. G. Lee, pp. 137–154. Springer, Dordrecht.

Duncan, R. P., Blackburn, T. M. and Sol, D. (2003). The ecology of bird introductions. *Annual Review of Ecology, Evolution and Systematics* **34**, 71–98.

Duncan, R. P., Blackburn, T. M. and Veltman, C. J. (1999). Determinants of geographical range sizes: A test using introduced New Zealand birds. *Journal of Animal Ecology* **68**, 963–975.

Duncan, R. P., Bomford, M., Forsyth, D. M. and Conibear, L. (2001). High predictability in introduction outcomes and the geographical range size of introduced Australian birds: A role for climate. *Journal of Animal Ecology* **70**, 621–632.

Dunn, P. O., Thusius, K. J., Kimber, K. and Winkler, D. W. (2000). Geographic and ecological variation in clutch size of tree swallows. *Auk* **117**, 211–215.

Dunning, J. B. (1992). *CRC handbook of avian body masses*. CRC Press, Boca Raton.

Eguchi, K. and Amano, H. E. (2004). Spread of exotic birds in Japan. *Ornithological Science* **3**, 3–11.

Ehrenfeld, J. G. (2003). Effects of exotic plant invasions on soil nutrient cycling processes. *Ecosystems* **6**, 503–523.

Ehrlich, P. R. (1986). Which animals will invade? In *Ecology of biological invasions of North America and Hawaii*, eds H. A. Mooney and J. A. Drake, 79–95. Springer-Verlag, New York.

Ehrlich, P. R. (1989). Attributes of invaders and the invading processes: Vertebrates. In *Biological invasions, a global perspective*, eds J. A. Drake, F. di Castri, R. H. Groves, F. J. Kruger, M. Rejmánek and M. Williamson, pp. 315–328. John Wiley & Sons, Chichester.

Elliott, J. J. and Arbib, R. S. (1953). Origin and status of the house finch in the eastern United States. *Auk* **70**, 31–37.

Ellis, M. M. and Elphick, C. S. (2007). Using a stochastic model to examine the ecological, economic and ethical consequences of population control in a charismatic invasive species: Mute swans in North America. *Journal of Applied Ecology* **44**, 312–322.

Ellstrand, N. C. (2000). Hybridization as a stimulus for the evolution of invasiveness in plants? *Proceedings of the National Academy of Sciences, USA* **97**, 7043–7050.

Elton, C. S. (1958). *The ecology of invasions by animals and plants.* Methuen, London.

Eraud, C., Boutin, J. M., Roux, D. and Faivre, B. (2007). Spatial dynamics of an invasive bird species assessed using robust design occupancy analysis: The case of the Eurasian collared dove (*Streptopelia decaocto*). in France. *Journal of Biogeography* **34**, 1077–1086.

Evans, K. L. and Gaston, K. J. (2005). People, energy and avian species richness. *Global Ecology and Biogeography* **14**, 187–196.

Evans, K. L., Duncan, R. P., Blackburn, T. M. and Crick, H. Q. P. (2005a). Investigating geographic variation in clutch size using a natural experiment. *Functional Ecology* **19**, 616–624.

Evans, K. L., Warren, P. H. and Gaston, K. J. (2005b). Does energy availability influence classical patterns of spatial variation in exotic species richness? *Global Ecology and Biogeography* **14**, 57–65.

Fagan, W. F. and Holmes, E. E. (2006). Quantifying the extinction vortex. *Ecology Letters* **9**, 51–60.

Falk-Petersen, J., Bøhn, T. and Sandlund, O. T. (2006). On the numerous concepts in invasion biology. *Biological Invasions* **8**, 1409–1424.

Feare, C. J. (1984). *The starling.* Oxford University Press, Oxford.

Felsenstein, J. (1985). Phylogenies and the comparative method. *American Naturalist* **125**, 1–15.

Filin, I., Holt, R. D. and Barfield, M. (2008). The relation of density regulation to habitat specialization, evolution of a species' range, and the dynamics of biological invasions. *American Naturalist* **172**, 233–247.

Fischer, J. R., Stallknecht, D. E., Luttrell, M. P., Dhondt, A. A. and Converse, K. A. (1997). Mycoplasma conjunctivitis in wild songbirds: The spread of a new contagious disease in a mobile host population. *Emerging Infectious Diseases* **3**, 69–72.

Fisher, R. (1937). The advance of advantageous genes. *Annals of Eugenics* (*London*) **7**, 355–369.

Fisher, R. J. and Wiebe, K. L. (2006). Nest site attributes and temporal patterns of northern flicker nest loss: Effects of predation and competition. *Oecologia* **147**, 744–753.

Flux, J. E. C. (2006). No evidence for a reduction in egg size in introduced populations of European starlings (*Sturnus vulgaris*) and song thrushes (*Turdus philomelos*) in New Zealand. *Notornis* **53**, 383–384.

Forshaw, J. M. (1973). *Parrots of the world.* Lansdowne Press, Melbourne.

Forstmeier, W. (2007). Genetic variation and differentiation in captive and wild zebra finches (*Taeniopygia guttata*). *Molecular Ecology* **16**, 4039–4050.

Forys, E. A. and Allen, C. R. (1999). Biological invasions and deletions: Community change in south Florida. *Biological Conservation* **87**, 341–347.

Foster, J. T. and Robinson, S. K. (2007). Introduced birds and the fate of Hawaiiian rainforest. *Conservation Biology* **21**, 1248–1257.

Frankham, R. (1996). Relationship of genetic variation to population size in wildlife. *Conservation Biology* **10**, 1500–1508.

Frankham, R., Ballou, J. D. and Briscoe, D. A. (2002). *Introduction to conservation genetics*. Cambridge University Press, Cambridge.

Frankham, R., Ballou, J. D., Briscoe, D. A. and McInnes, K. H. (2004). *A primer of conservation genetics*. Cambridge University Press, Cambridge.

Frantzen, J. and van den Bosch, F. (2000). Spread of organisms: Can travelling and dispersive waves be distinguished? *Basic and Applied Ecology* **1**, 83–91.

Freckleton, R. P., Dowling, P. M. and Dulvy, N. K. (2006a). Stochasticity, nonlinearity and instability in biological invasions. In *Conceptual ecology and invasion biology: Reciprocal approaches to nature*, eds M. W. Cadotte, S. M. McMahon and T. Fukami, pp. 125–146. Springer, Dordrecht.

Freckleton, R. P., Gill, J. A., Noble, D. and Watkinson, A. R. (2005). Large-scale population dynamics, abundance-occupancy relationships and the scaling from local to regional population size. *Journal of Animal Ecology* **74**, 353–364.

Freckleton, R. P., Harvey, P. H. and Pagel, M. (2002). Phylogenetic analysis and comparative data: A test and review of the evidence. *American Naturalist* **160**, 712–726.

Freckleton, R. P., Noble, D. and Webb, T. J. (2006b). Distributions of habitat suitability and the abundance-occupancy relationship. *American Naturalist* **167**, 260–275.

Fridley, J. D., Stachowicz, J. J., Naeem, S., Sax, D. F., Seabloom, E. W., Smith, M. D., Stohlgren, T. J., Tilman, D. and Von Holle, B. (2007). The invasion paradox: Reconciling pattern and process in species invasions. *Ecology* **88**, 3–17.

Fritts, T. H. and Rodda, G. H. (1998). The role of introduced species in the degradation of island ecosystems: A case history of Guam. *Annual Review of Ecology, Evolution and Systematics* **29**, 113–140.

Gaillard, J.-M., Pontier, D., Allainé, D., Lebreton, J. D., Trouvilliez, J. and Clobert, J. (1989). An analysis of demographic tactics in birds and mammals. *Oikos* **56**, 59–76.

Garamszegi, L. Z., Møller, A. P. and Erritzoe, J. (2003). The evolution of immune defense and song complexity in birds. *Evolution* **57**, 905–912.

Garden, J., McAlpine, C., Peterson, A., Jones, D. and Possingham, H. (2006). Review of the ecology of Australian urban fauna: A focus on spatially explicit processes. *Austral Ecology* **31**, 126–148.

Garrett, K. L. (1997). Population status and distribution of naturalized parrots in southern California. *Western Birds* **28**, 181–195.

Garrick, A. (1981). Diet of Pipits and Skylarks at Huiarua Station, Tokomaru Bay, North Island, New Zealand. *New Zealand Journal of Ecology* **4**, 106–114.

Gaston, K. J. (1994). *Rarity*. Chapman & Hall, London.

Gaston, K. J. (1996). The multiple forms of the interspecific abundance-distribution relationship. *Oikos* **76**, 211–220.

Gaston, K. J. (2003). *The structure and dynamics of geographic ranges*. Oxford University Press, Oxford.

Gaston, K. J. (2006). Biodiversity and extinction: Macroecological patterns and people. *Progress in Physical Geography* **30**, 258–269.

Gaston, K. J. and Blackburn, T. M. (1995). The frequency distribution of bird body weights: Aquatic and terrestrial species. *Ibis*, **137**, 237–240.

Gaston, K. J. and Blackburn, T. M. (1996). Global scale macroecology: Interactions between population size, geographic range size and body size in the Anseriformes. *Journal of Animal Ecology* **65**, 701–714.

Gaston, K. J. and Blackburn, T. M. (1997). Age, area and avian diversification. *Biological Journal of the Linnean Society* **62**, 239–253.

Gaston, K. J. and Blackburn, T. M. (1999). A critique for macroecology. *Oikos* **84**, 353–368.

Gaston, K. J. and Blackburn, T. M. (2000). *Pattern and process in macroecology*. Blackwell Science, Oxford.

Gaston, K. J. and Blackburn, T. M. (2003). Macroecology and conservation biology. In *Macroecology: Concepts and consequences*, eds T. M. Blackburn and K. J. Gaston, pp. 345–367. Blackwell Science, Oxford.

Gaston, K. J. and Chown, S. L. (2005). Neutrality and the niche. *Functional Ecology* **19**, 1–6.

Gaston, K. J. and Evans, K. L. (2004). Birds and people in Europe. *Proceedings of the Royal Society, London, B* **271**, 1649–1655.

Gaston, K. J. and Fuller, R. A. (2008). Commonness, population depletion and conservation biology. *Trends in Ecology & Evolution* **23**, 14–19.

Gaston, K. J., Blackburn, T. M. and Lawton, J. H. (1997). Interspecific abundance-range size relationships: An appraisal of mechanisms. *Journal of Animal Ecology* **66**, 579–601.

Gaston, K. J., Blackburn, T. M., Greenwood, J. J. D., Gregory, R. D., Quinn, R. M. and Lawton, J. H. (2000). Abundance-occupancy relationships. *Journal of Applied Ecology* **37**, 39–59.

Gaston, K. J., Davies, R. G., Orme, C. D. L., Olson, V., Thomas, G. H. Ding, T.-S., Rasmussen, P. C., Lennon, J. J., Bennett, P. M., Owens, I. P. F. and Blackburn, T. M. (2007). Spatial turnover in the global avifauna. *Proceedings of the Royal Society, London, B* **274**, 1567–1574.

Gaston, K. J., Jones, A. G., Hanel, C. and Chown, S. L. (2003). Rates of species introduction to a remote oceanic island. *Proceedings of the Royal Society, London, B* **270**, 1091–1098.

Gause, G. F. (1932). Experimental studies on the struggle for existence. I. Mixed population of two yeast species. *Journal of experimental biology* **9**, 389–402.

Gibbons, D. W., Reid, J. B. and Chapman, R. A. (1993). *The new atlas of breeding birds in Britain and Ireland:1988–1991*. T. & A. D. Poyser, London.

Gilardi, J. D. (2006). Captured for conservation: Will cages save wild birds? A response to Cooney & Jepson. *Oryx* **40**, 24–26.

Gilchrist, G. W. and Lee, C. E. (2006). All stressed out and nowhere to go: Does evolvability limit adaptation in invasive species? *Genetica* **129**, 127–132.

Gill, B. J. (1999). A myna increase—notes on introduced mynas (*Acridotheres*) and bulbuls (*Pycnonotus*) in Western Samoa. *Notornis* **46**, 268–269.

Gill, B. J., Lovegrove, T. G. and Hay, J. R. (1993). More myna matters—notes on introduced passerines in Western Samoa. *Notornis* **40**, 300–302.

Gilpin, M. (1990). Ecological prediction. *Science* **248**, 88–89.

Gilpin, M. E. and Soulé, M. E. (1986). Minimum viable populations: Processes of extinction. In *Conservation biology: The science of scarcity and diversity*, ed M. E. Soulé, pp. 19–34. Sinauer Associates Inc., Sunderland, MA.

Gittleman, J. L. and Kot, M. (1990). Adaptation: Statistics and a null model for estimating phylogenetic effects. *Systematic Zoology* **39**, 227–241.

Gleason, H. A. (1922). On the relation between species and area. *Ecology* **3**, 158–162.

Gleason, H. A. (1925). Species and area. *Ecology* **6**, 66–74.

Gleason, H. A. (1926). The individualistic concept of the plant association. *Bulletin of the Torrey Botanical Club* **53**, 7–26.

Gobster, P. H. (1994). The urban savanna: Reuniting ecological preference and function. *Restoration and management notes* **12**, 64–71.

Goldstein, H. (1995). *Multilevel statistical models*. Edward Arnold, London.

Gosper, C. R., Stansbury, C. D. and Vivian-Smith, G. (2005). Seed dispersal of fleshy-fruited invasive plants by birds: Contributing factors and management options. *Diversity and Distributions* **11**, 549–558.

Gotelli, N. J. and Graves, G. R. (1996). *Null models in ecology*. Smithsonian Institution Press, Washington, DC.

Grant, B. R. (2002). Lack of premating isolation at the base of a phylogenetic tree. *American Naturalist* **160**, 1–19.

Grant, P. R. (1972). Centripetal selection and the house sparrow. *Systematic Zoology* **21**, 23–30.

Grant, P. R. (1986). *Ecology and evolution of Darwin's finches*. Princeton University Press, Princeton, NJ.

Grant, P. R. and Grant, B. R. (2008). Pedigrees, assortative mating and speciation in Darwin's finches. *Proceedings of the Royal Society, London*, B **275**, 661–668.

Graur, D. and Li, W. H. (2000). *Fundamentals of molecular evolution*. Sinauer Associates Inc., Sunderland, MA.

Graves, G. R. and Gotelli, N. J. (1983). Neotropical land-bridge avifaunas: New approaches to null hypotheses in biogeography. *Oikos* **41**, 322–333.

Grayson, D. K. (2001). The archaeological record of human impacts on animal populations. *Journal of World Prehistory* **15**, 1–68.

Green, R. E. (1997). The influence of numbers released on the outcome of attempts to introduce exotic bird species to New Zealand. *Journal of Animal Ecology* **66**, 25–35.

Green, R. J. (1984). Native and exotic birds in a suburban habitat. *Australian Wildlife Research* **11**, 181–190.

Gregory, R. D. and Blackburn, T. M. (1998). Macroecological patterns in British breeding birds: Covariation of species' geographical range sizes at differing spatial scales. *Ecography* **21**, 527–534.

Gregory, R. D. and Gaston, K. J. (2000). Explanations of commonness and rarity in British breeding birds: Separating resource use and resource availability. *Oikos* **88**, 515–526.

Greig-Smith, P. W., Feare, C. J., Freeman, E. M. and Spencer, P. L. (1988). Causes and consequences of egg size variation in the European starling *Sturnus vulgaris*. *Ibis* **130**, 1–10.

Grevstad, F. (1999). Factors influencing the chance of population establishment: Implications for release strategies in biocontrol. *Ecological Applications* **9**, 1439–1447.

Grinnell, J. (1906). Foolish introductions of foreign birds. *Condor* **8**, 58.

Grosholz, E. D. (1996). Contrasting rates of spread for introduced species in terrestrial and marine systems. *Ecology* **77**, 1680–1686.

Guild, E. (1938). Tahitian aviculture: Acclimation of foreign birds. *Avicultural Magazine* **3**, 8–11.

Guild, E. (1940). Western bluebirds in Tahiti. *Avicultural Magazine* **5**, 284–285.

Haccou, P. and Iwasa, Y. (1996). Establishment probability in fluctuating environments: A branching process model. *Theoretical Population Biology* **50**, 254–280.

Haccou, P. and Vatunin, V. (2003). Establishment success and extinction risk in autocorrelated environments. *Theoretical Population Biology* **64**, 303–314.

Hadfield, M. G., Miller, S. E. and Carwisle, A. H. (1993). The decimation of endemic Hawaiian tree snails by alien predators. *American Zoologist* **33**, 610–622.

Haemig, P. D. (1978). Aztec Emperor Auitzotl and the great-tailed grackle. *Biotropica* **10**, 11–17.

Haemig, P. D. (1979). Secret of the painted jay. *Biotropica* **11**, 81–87.

Hagemeijer, W. J. M. and Blair, M. J. (1997). *The EBCC atlas of European breeding birds.* T. & A. D. Poyser, London.

Hale, K. A. and Briskie, J. V. (2007). Decreased immunocompetence in a severely bottlenecked population of an endemic New Zealand bird. *Animal Conservation* **10**, 2–10.

Hamilton, W. D. and Zuk, M. (1982). Heritable true fitness and bright birds: A role for parasites? *Science* **218**, 384–387.

Hanski, I. (1989). Metapopulation dynamics: Does it help to have more of the same? *Trends in Ecology & Evolution* **4**, 113–114.

Hanski, I. (1999). *Metapopulation ecology.* Oxford University Press, Oxford.

Hanski, I. and Gaggiotti, O. E. (2004). Metapopulation biology: Past, present, and future. In *Ecology, genetics and evolution of metapopulations*, eds I. Hanski and O. E. Gaggiotti, pp. 3–22. Elsevier Academic Press, Amsterdam.

Hardy, J. W. (1973). Feral exotic birds in southern California. *Wilson Bulletin* **85**, 506–512.

Harper, M. J., McCarthy, M. A. and van der Ree, R. (2005). The use of nest boxes in urban natural vegetation remnants by vertebrate fauna. *Wildlife Research* **32**, 509–516.

Harris, J. A. (1911). A neglected paper on natural selection in the English sparrow. *American Naturalist* **45**, 314–318.

Harrison, T. (1971). Easter Island, a last outpost. *Oryx* **11**, 2–3.

Harvey, P. H. and Pagel, M. D. (1991). *The comparative method in evolutionary biology.* Oxford University Press, Oxford.

Hastings, A. (1996). Models of spatial spread: Is the theory complete? *Ecology* **77**, 1675–1679.

Hastings, A., Cuddington, K., Davies, K. F., Dugaw, C. J., Elmendorf, S., Freestone, A., Harrison, S., Holland, M., Lambrinos, J., Malvadkar, U., Melbourne, B. A., Moore, K., Taylor, C. and Thomson, D. (2005). The spatial spread of invasions: New developments in theory and evidence. *Ecology Letters* **8**, 91–101.

Haugen, T. O. and Vollestad, L. A. (2001). A century of life-history evolution in grayling. *Genetica* **112**, 475–491.

Hawkins, B. A. (2001). Ecology's oldest pattern? *Trends in Ecology & Evolution* **16**, 470.

Hawkins, B. A. and Diniz-Filho, J. A. F. (2006). Beyond Rapoport's rule: Evaluating range size patterns of New World birds in a two-dimensional framework. *Global Ecology and Biogeography* **15**, 461–469.

Hawkins, B. A., Field, R., Cornell, H. V., Currie, D. J., Guégan, J.-F., Kaufman, D. M., Kerr, J. T., Mittelbach, G. G., Oberdorff, T., O'Brien, E. M., Porter, E. E. and Turner, J. R. G. (2003). Energy, water, and broad-scale geographic patterns of species richness. *Ecology* **84**, 3105–3117.

Hawley, D. M., Hanley, D., Dhondt, A. A. and Lovette, I. J. (2006). Molecular evidence for a founder effect in invasive house finch (*Carpodacus mexicanus*) populations experiencing an emergent disease epidemic. *Molecular Ecology* **15**, 263–275.

Hawley, D. M., Sydenstricker, K. V., Kollias, G. V. and Dhondt, A. A. (2005). Genetic diversity predicts pathogen resistance and cell-mediated immunocompetence in house finches. *Biology Letters* **1**, 326–329.

Hayes, K. R. and Barry, S. C. (2008). Are there consistent predictors of invasion success? *Biological Invasions* **10**, 483–506.

Heath, L., Martin, D. P., Warburton, L., Perrin, M., Horsfield, W., Kingsley, C., Rybicki, E. P. and Williamson, A. L. (2004). Evidence of unique genotypes of beak and feather disease virus in southern Africa. *Journal of Virology* **78**, 9277–9284.

Heather, B. and Robertson, H. (1997). *The field guide to the birds of New Zealand*. Oxford University Press, Oxford.

Hengeveld, R. (1989). *Dynamics of biological invasions*. Chapman & Hall, London.

Hengeveld, R. (1994). Small-step invasion research. *Trends in Ecology & Evolution* **9**, 339–342.

Hengeveld, R. and Van den Bosch, F. (1991). The expansion velocity of the collared dove *Streptopelia decaocto* population in Europe. *Ardea* **79**, 67–72.

Hesse, R., Allee, W. C. and Schmidt, K. P. (1937). *Ecological animal geography*. Wiley, New York.

Higgins, S. I., Nathan, R. and Cain, M. L. (2003). Are long-distance dispersal events in plants usually caused by nonstandard means of dispersal? *Ecology* **84**, 1945–1956.

Higham, T., Anderson, A. and Jacomb, C. (1999). Dating the first New Zealanders: The chronology of Wairau Bar. *Antiquity* **73**, 420–427.

Hill, G. E. (1993). Geographic variation in the carotenoid plumage pigmentation of male house finches (*Carpodacus mexicanus*). *Biological Journal of the Linnean Society* **49**, 63–86.

Hill, G. E. (2002). *A red bird in a brown bag*. Oxford University Press, Oxford.

Hochachka, W. M. and Dhondt, A. A. (2000). Density-dependent decline of host abundance resulting from a new infectious disease. *Proceedings of the National Academy of Sciences, USA* **97**, 5303–5306.

Hoffmann, A. A. and Parsons, P. A. (1991). *Evolutionary genetics and environmental stress*. Oxford University Press, Oxford.

Holmes, E. E. (1993). Are diffusion models too simple: A comparison with telegraph models of invasion. *American Naturalist* **142**, 779–795.

Holt, A. R. and Gaston, K. J. (2003). Interspecific abundance-occupancy relationships of British mammals and birds: Is it possible to explain the residual variation? *Global Ecology and Biogeography* **12**, 37–46.

Holt, R. D. (1977). Predation, apparent competition, and the structure of prey communities. *Theoretical Population Biology* **12**, 197–229.

Holt, R. D. (1984). Spatial heterogeneity, indirect interactions, and the coexistence of prey species. *American Naturalist* **124**, 377–406.

Holt, R. D. and Lawton, J. H. (1994). The ecological consequences of shared natural enemies. *Annual Review of Ecology Evolution and Systematics* **25**, 495–520.

Holt, R. D., Barfield, M. and Gomulkiewicz, R. (2005). Theories of niche conservatism and evolution: Could exotic species be potential tests? In *Species invasions: Insights into ecology, evolution and biogeography*, eds D. F. Sax, J. J. Stachowicz and S. D. Gaines, pp. 259–290. Sinauer Associates Inc., Sunderland, MA.

Holyoak, D. T. and Thibault, J.-C. (1984). Contribution à l'étude des oiseaux de Polynésie orientale. *Memoirs du Museum National d'Histoire Naturelle, Nouvelle Serie A. Zoologie* **127**, 1–209.

Hooten, M. B. and Wikle, C. K. (2008). A hierarchical Bayesian non-linear spatio-temporal model for the spread of invasive species with application to the Eurasian Collared-Dove. *Environmental and Ecological Statistics* **15**, 59–70.

Hooten, M. B., Wikle, C. K., Dorazio, R. M. and Royle, J. A. (2007). Hierarchical spatio-temporal matrix models for characterizing invasions. *Biometrics* **63**, 558–567.

Hopper, K. R. and Roush, R. T. (1993). Mate finding, dispersal, number released, and the success of biological control introductions. *Ecological Entomology* **18**, 321–331.

Howe, H. F. (1984). Constraints on the evolution of mutualisms. *American Naturalist* **123**, 764–777.

Hubbell, S. P. (2001). *The unified neutral theory of biodiversity and biogeography.* Princeton University Press, Princeton, NJ.

Hughes, B. (1996). The ruddy duck *Oxyura jamaicensis* in the Western Palaearctic and the threat to the white-headed duck *Oxyura leucocephala*. In *The introduction and naturalisation of birds*, eds J. S. Holmes and J. R. Simons, pp. 79–86. HMSO, London.

Hughes, B. J., Martin, G. R. and Reynolds, S. J. (2008). Has eradication of feral cats *Felis silvestris* halted the decline in the Sooty Tern *Onychoprion fuscata* population on Ascension Island, South Atlantic? *Ibis* **150** (suppl. 1), 122–131.

Hughes, J. B. (2000). The scale of reseource specialization and the distribution and abundance of lycaenid butterflies. *Oecologia* **123**, 375–383.

Hurlbert, S. H. (1984). Pseudoreplication and the design of ecological field experiments. *Ecological Monographs* **54**, 187–211.

Hurles, M. E., Matisso-Smith, E., Gray, R. D. and Penny, D. (2003). Untangling oceanic settlement: The edge of the knowable. *Trends in Ecology & Evolution* **18**, 531–540.

Hursthouse, C. (1857). *New Zealand or Zealandia: The Britain of the south.* Stanford, London.

Hutson, A. M. (1975). Observations on the birds of Diego Garcia, Chagos Archipelago, with notes on other vertebrates. *Atoll Research Bulletin* **175**, 1–25.

Ingold, D. J. (1998). The influence of starlings on flicker reproduction when both naturally excavated cavities and artificial nest boxes are available. *Wilson Bulletin* **110**, 218–225.

Ishtiaq, F., Beadell, J. S., Baker, A. J., Rahmani, A. R., Jhala, Y. V. and Fleischer, R. C. (2006). Prevalence and evolutionary relationships of haematozoan parasites in native versus introduced populations of common myna *Acridotheres tristis*. *Proceedings of the Royal Society, London, B* **273**, 587–594.

James, F. C. (1970). Geographic size variation in birds and its relationship to climate. *Ecology* **51**, 365–390.

Jarvi, S. I., Atkinson, C. T. and Fleischer, R. C. (2001). Immunogenetics and resistance to avian malaria in Hawaiian Honeycreepers (Drepanidinae). In *Evolution Ecology Conservation, and Management of Hawaiian Birds: A Vanishing Avifauna*, eds J. M. Scott, S. Conant and C. van Riper III, pp. 254–263. Studies in Avian Biology, No. 22.

Jarvi, S. I., Farias, M. E. M., Baker, H., Freifeld, H. B., Baker, P. E., Van Gelder, E., Massey, J. G. and Atkinson, C. T. (2003). Detection of avian malaria (*Plasmodium* spp.) in native land birds of American Samoa. *Conservation Genetics* **4**, 629–637.

Jeschke, J. M. and Strayer, D. L. (2005). Invasion success of vertebrates in Europe and North America. *Proceedings of the National Academy of Sciences, USA* **102**, 7198–7202.

Jeschke, J. M. and Strayer, D. L. (2006). Determinants of vertebrate invasion success in Europe and North America. *Global Change Biology* **12**, 1608–1619.

Jeschke, J. M. and Strayer, D. L. (2008). Are threat status and invasion success two sides of the same coin? *Ecography* **31**, 124–130.

Jiguet, F. (2007). Finding naturalised birds in France. *Birding World* **20**, 216–219.

Johnston, R. F. (1973). Evolution in the house sparrow. IV. Replicate studies in phenetic covariation. *Systematic Zoology* **22**, 219–226.

Johnston, R. F. (1990). Variation in size and shape in pigeons, *Columba livia. Wilson Bulletin* **102**, 213–225.

Johnston, R. F. (1992). Evolution in the rock dove: Skeletal morphology. *Auk* **109**, 530–542.

Johnston, R. F. (1994). Geographic variation of size in feral pigeons. *Auk* **111**, 398–404.

Johnston, R. F. (2001). The synanthropic birds of North America. In *Avian ecology and conservation in an urbanizing world*, eds J. M. Marzluff, R. Bowman and R. Donnelly, pp. 49–67. Kluwer Academic, Norwell, MA.

Johnston, R. F. and Fleisher, R. C. (1981). Overwinter mortality and sexual size dimoprhism in the house sparrow. *Auk* **98**, 503–511.

Johnston, R. F. and Garrett, K. L. (1994). Population trends of introduced birds in western North America. *Studies in Avian Biology* **15**, 221–231.

Johnston, R. F. and Selander R. K. (1964). House sparrows: Rapid evolution of races in North America. *Science* **144**, 548–550.

Johnston, R. F. and Selander, R. K. (1971). Evolution in the house sparrow. II. Adaptive differentiation in North American populations. *Evolution* **25**, 1–28.

Johnston, R. F. and Selander, R. K. (1973). Evolution of the house sparrow. III. Variation in size and sexual dimorphism in Europe and North and South America. *American Naturalist* **107**, 373–390.

Johnston, R. F., Niles, D. M. and Rohwer, S. A. (1972). Hermon Bumpus and natural selection in the house sparrow *Passer domesticus. Evolution* **26**, 20–31.

Jordano, P. (1987). Patterns of mutualistic interactions in pollination and seed dispersal: Connectance, dependence asymmetries and coevolution. *American Naturalist* **119**, 657–677.

Jozan, M., Evans, R., McLean, R., Hall, R., Tangredi, B., Reed, L. and Scott, J. (2003). Detection of West Nile Virus infection in birds in the United States by blocking ELISA and immunohistochemistry. *Vector Borne Zoonotic Diseases* **3**, 99–110.

Juniper, A. and Parr, M. (1998). *Parrots: A guide to the parrots of the world*. Christopher Helm, London.

Kalmar, A. and Currie, D. J. (2006). A global model of island biogeography. *Global Ecology and Biogeography* **15**, 72–81.

Karels, T. J., Dobson, F. S., Trevino, H. S. and Skibiel, A. L. (2008). The biogeography of avian extinction on oceanic islands. *Journal of Biogeography* **35**, 1106–1111.

Kark, S. and Sol, D. (2005). Establishment success across convergent Mediterranean ecosystems: An analysis of bird introductions. *Conservation Biology* **19**, 1519–1527.

Kawakami, K. and Higuchi, H. (2003). Interspecific interactions between the native and introduced White-eyes in the Bonin Islands. *Ibis* **145**, 583–592.

Kawakami, K. and Yamaguchi, Y. (2004). The spread of the introduced Melodious Laughing Thrush *Garrulax canorus* in Japan. *Ornithological Science* **3**, 13–21.

Keane, R. M. and Crawley, M. J. (2002). Exotic plant invasions and the enemy release hypothesis. *Trends in Ecology & Evolution* **17**, 164–170.

Keddy, P. A. (2001). *Competition*. 2nd edition. Kluwer Academic Publishers, Dordrecht, The Netherlands.

Keitt, T. H. and Marquet, P. A. (1996). The introduced Hawaiian avifauna reconsidered: Evidence for self-organised criticality? *Journal of theoretical Biology* **182**, 161–167.

Keitt, T. H., Lewis, M. A. and Holt, R. D. (2001). Allee effects, invasion pinning, and species' borders. *American Naturalist* **157**, 203–216.

Keller, L. F. and Waller, D. M. (2002). Inbreeding effects in wild populations. *Trends in Ecology & Evolution* **17**, 230–241.

Kennedy, C. R. and Bush, A. O. (1994). The relationship between pattern and scale in parasite communities: A stranger in a strange land. *Parasitology* **109**, 187–196.

Kessel, B. (1979). Starlings become established at Fairbanks, Alaska. *Condor* **81**, 437–438.

Kilpatrick, A. M., Kramer, L. D., Jones, M. J., Marra, P. P. and Daszak, P. (2006). West Nile Virus epidemics in North America are driven by shifts in mosquito feeding behavior. *PLoS Biology* **4**, e82.

King, M. (2003). *The Penguin History of New Zealand*. Penguin Books, Rosedale, Auckland.

Kirkpatrick, M., Price, T. and Arnold, S. J. (1990). The Darwin–Fisher theory of sexual selection in monogamous birds. *Evolution* **44**, 180–193.

Klein, J. (1990). *Immunology*. Oxford University Press, Oxford.

Koenig, W. D. (2003). European Starlings and their effect on native cavity-nesting birds. *Conservation Biology* **17**, 1134–1140.

Kolar, C. S. and Lodge, D. M. (2001). Progress in invasion biology: Predicting invaders. *Trends in Ecology & Evolution* **16**, 199–204.

Kolbe, J. J., Glor, R. E., Schettino, L. R., Lara, A. C., Larson, A. and Losos, J. B. (2004). Genetic variation increases during biological invasion by a Cuban lizard. *Nature* **431**, 177–181.

Koleff, P., Gaston, K. J. and Lennon, J. J. (2003). Measuring beta diversity for presence–absence data. *Journal of Animal Ecology* **72**, 367–382.

Komdeur, J. (1996). Breeding of the Seychelles Magpie Robin *Copsychus sechellarum* and implications for its conservation *Ibis* **138**, 485–498.

Kot, M., Lewis, M. A. and van den Driessche, P. (1996). Dispersal data and the spread of invading organisms. *Ecology* **77**, 2027–2042.

Kowarik, I. (1990). Some responses of flora and vegetation to urbanization in Central Europe. In *Urban ecology*, eds H. Sukopp, S. Hejny and I. Kowarik, pp. 45–74. SPB Academic Publishers, The Hague.

Kulesza, G. (1990). An analysis of clutch-size in New World passerine birds. *Ibis* **132**, 407–422.

Kunin, W. E. (1997). Introduction: On the causes and consequences of rare–common differences. In *The biology of rarity: Causes and consequences of rare–common differences*, eds W. E. Kunin and K. J. Gaston, pp. 3–11. Chapman & Hall, London.

Kunin, W. E. and Gaston, K. J. eds. (1997). *The biology of rarity: Causes and consequences of rare–common differences*. Chapman & Hall, London.

Labra, F. A., Abades, S. and Marquet, P. A. (2005). Distribution and abundance: Scaling patterns in exotic and native bird species. In *Species invasions: Insights into ecology, evolution and biogeography*, eds D. F. Sax, J. J. Stachowicz and S. D. Gaines, pp. 421–446. Sinauer Associates Inc., Sunderland, MA.

Lack, D. (1947). The significance of clutch-size. Part I. Intraspecific variations. *Ibis* **89**, 302–352.

Lack, D. (1948). The significance of clutch-size. Part III. Some interspecific comparisons. *Ibis* **90**, 25–45.

Lack, D. (1968). *Ecological adaptations for breeding in birds*. Methuen, London.

Lahti, D. C. (2005). Evolution of bird eggs in the absence of cuckoo parasitism. *Proceedings of the National Academy of Sciences, USA* **102**, 18057–18062.

Lahti, D. C. (2006). Persistence of egg recognition in the absence of cuckoo brood parasitism: Pattern and mechanism. *Evolution* **60**, 157–168.

Lambrinos, J. G. (2004). How interactions between ecology and evolution influence contemporary invasion dynamics. *Ecology* **85**, 2061–2070.

Lanciotti, R. S., Roehrig, J., Deubel, V., Smith, J., Parker, M., Steele, K., Crise, B., Volpe, K. E., Crabtree, M. B., Scherret, J. H., Hall, R. A., MacKenzie, J. S., Cropp, C. B., Panigrahy, B., Ostlund, E., Schmitt, B., Malkinson, M., Banet, C., Weissman, J., Komar, N., Savage, H. M., Stone, W., McNamara, T. and Gubler, D. J. (1999). Origin of the West Nile virus responsible for an outbreak of encephalitis in the northeastern United States. *Science* **286**, 2333–2337.

Lande, R. (1993). Risks of population extinction from demographic and environmental stochasticity and random catastrophes. *American Naturalist* **142**, 911–927.

Lande, R., Engen, S. and Sæther, B.-E. (2003). *Stochastic population dynamics in ecology and conservation*. Oxford University Press, Oxford.

Landsborough Thomson, A. (1964). *A new dictionary of birds*. Thomas Nelson, London.

Lang, A. L. and Barlow, J. C. (1997). Cultural evolution in the Eurasian Tree Sparrow: Divergence between introduced and ancestral populations. *Condor* **99**, 413–423.

Lawton, J. H. (1999). Are there general laws in ecology? *Oikos* **84**, 177–192.

Leavesley, M. G. (2005). Prehistoric hunting strategies in New Ireland, Papua New Guinea: The evidence of the cuscus (*Phalanger orientalis*) remains from Buang Merabak cave. *Asian Perspectives* **44**, 207–218.

Lee, C. E. (2002). Evolutionary genetics of invasive species. *Trends in Ecology & Evolution* **17**, 386–391.

Lee, K. A., Martin, L. B. and Wikelski, M. C. (2006). Responding to inflammatory challenges is less costly for a successful avian invader, the house sparrow (*Passer domesticus*) than its less-invasive congener. *Oecologia* **145**, 244–251.

Lefebvre, L., Gaxiola, A., Dawson, S., Rozsa, L. and Kabai, P. (1998). Feeding innovations and forebrain size in Australasian birds. *Behaviour* **135**, 1077–1097.

Lefebvre, L., Whittle, P., Lascaris, E. and Finkelstein, A. (1997). Feeding innovations and forebrain size in birds. *Animal Behaviour* **53**, 549–560.

Legendre, S., Clobert, J., Møller, A. P. and Sorci, G. (1999). Demographic stochasticity and social mating system in the process of extinction of small populations: The case of passerines introduced to New Zealand. *American Naturalist* **153**, 449–463.

Lensink, R. (1997). Range expansion of raptors in Britain and the Netherlands since the 1960s: Testing an individual-based diffusion model. *Journal of Animal Ecology* **66**, 811–826.

Lensink, R. (1998). Temporal and spatial expansion of the Egyptian goose *Alopochen aegyptiacus* in The Netherlands, 1967–94. *Journal of Biogeography* **25**, 251–263.

Leprieur, F., Beauchard, O., Blanchet, S., Oberdorff, T. and Brosse, S. (2008). Fish invasions in the world's river systems: When natural processes are blurred by human activities. *PLoS Biology* **6**, e28.

Leung, B., Drake, J. M. and Lodge, D. M. (2004). Predicting invasions: Propagule pressure and the gravity of Allee effects. *Ecology* **85**, 1651–1660.

Leven, M. R. and Corlett, R. T. (2004). Invasive birds in Hong Kong, China. *Ornithological Science* **3**, 43–55.

Lever, C. (1987). *Naturalized birds of the world*. Longman Scientific and Technical, New York.

Lever, C. (1992). *They dined on eland: The story of the acclimatisation societies*. Quiller Press, London.

Lever, C. (2005). *Naturalised birds of the world*. T. & A. D. Poyser, London.

Levin, D. A. (2002). Hybridization and extinction. *American Scientist* **90**, 254–261.

Lewin, V. and Lewin, G. (1984). The Kalij pheasant, a newly established game bird on the island of Hawaii. *Wilson Bulletin* **96**, 634–646.

Lewis, M. A. (1997). Variability, patchiness, and jump dispersal in the spread of an invading population. In *Spatial ecology: The role of space in population dynamics and interspecific interactions*, eds D. Tilman and P. M. Kareiva, pp. 46–69. Princeton University Press, Princeton, NJ.

Lewis, M. A. and Kareiva, P. (1993). Allee dynamics and the spread of invading organisms. *Theoretical Population Biology* **43**, 141–158.

Lewis, M. A. and van den Driessche, P. (1993). Waves of extinction from sterile insect release. *Mathematical Biosciences* **116**, 221–247.

Lim, H. C., Sodhi, N. S., Brook, B. W. and Soh, M. C. K. (2003). Undesirable aliens: Factors determining the distribution of three invasive bird species in Singapore. *Journal of Tropical Ecology* **19**, 685–695.

Lindholm, A. K. (2005). Invasion success and genetic diversity of introduced populations of guppies *Poecilia reticulata* in Australia. *Molecular Ecology* **14**, 3671–3682.

Lockwood, J. L. (1999). Using taxonomy to predict success among introduced avifauna: Relative importance of transport and establishment. *Conservation Biology* **13**, 560–567.

Lockwood, J. L. (2005). Insights into biogeography. In *Species invasions: Insights into ecology, evolution and biogeography*, eds D. F. Sax, J. J. Stachowicz and S. D. Gaines, pp. 309–313. Sinauer Associates Inc., Sunderland, MA.

Lockwood, J. L. (2006). Life in a double-hotspot: The transformation of Hawaiian passerine bird diversity following invasion and extinction. *Biological Invasions* **8**, 49–457.

Lockwood, J. L. and McKinney, M. L. eds. (2001). *Biotic homogenization*. Kluwer, New York.

Lockwood, J. L. and Moulton, M. P. (1994). Ecomorphological pattern in Bermuda birds: The influence of competition and implications for nature preserves. *Evolutionary Ecology* **8**, 53–60.

Lockwood, J. L., Brooks, T. M. and McKinney, M. L. (2000). Taxonomic homogenization of the global avifauna. *Animal Conservation* **3**, 27–35.

Lockwood, J. L., Cassey, P. and Blackburn, T. (2005). The role of propagule pressure in explaining species invasions. *Trends in Ecology & Evolution* **20**, 223–228.

Lockwood, J. L., Hoopes, M. F. and Marchetti, M. P. (2007). *Invasion ecology*. Blackwell Publishing, Oxford.

Lockwood, J. L., Moulton, M. P. and Anderson, S. K. (1993). Morphological assortment and the assembly of communities of introduced Passeriformes on oceanic islands: Tahiti versus Oahu. *American Naturalist* **141**, 398–408.

Lockwood, J. L., Moulton, M. P. and Brooke, R. K. (1996). Morphological dispersion of the introduced landbirds of Saint Helena. *Ostrich* **67**, 111–117.

Lodge, D. M. (1993). Biological invasions: Lessons for ecology. *Trends in Ecology & Evolution* **8**, 133–137.

Long, J. L. (1981). *Introduced birds of the world: The worldwide history, distribution and influence of birds introduced to new environments*. David & Charles, London.

Lonsdale, W. M. (1999). Global patterns of plant invasions and the concept of invasibility. *Ecology* **80**, 1522–1536.

Luniak, M. (2004). Synurbization: Adaptation of animal wildlife to urban development. In *Proceedings of the 4th International Urban Wildlife Symposium*, eds W. W. Shaw, L. K. Harris and L. VanDruff, pp. 50–55. University of Arizona Press, Tucson.

Lutz, F. E. (1921). Geographic average, a suggested method for the study of distribution. *American Museum Novitates* **5**, 1–7.

McAllan, I. A. W. and Hobroft, D. (2005). The further spread of introduced birds in Samoa. *Notornis* **52**, 16–20.

McArdle, B. H. (1996). Levels of evidence in studies of competition, predation, and disease. *New Zealand Journal of Ecology* **20**, 7–15.

MacArthur, R. H. (1972). *Geographical ecology: Patterns in the distribution of species.* Harper & Row, New York.

MacArthur, R. H. and Wilson, E. O. (1963). An equilibrium theory of insular zoogeography. *Evolution* **17**, 373–387.

MacArthur, R. H. and Wilson, E. O. (1967). *The theory of island biogeography.* Princeton University Press, Princeton, NJ.

McCarthy, E. M. (2006). *Handbook of avian hybrids of the world.* Oxford University Press, Oxford.

McClure, H. E. (1989). What characterizes an urban bird? *Journal of the Yamashina Institute of Ornithology* **21**, 178–192.

Macdonald, I. A., Thebaud, C., Strahm, W. A. and Strasberg, D. (1991). Effects of alien plant invasions on native vegetation remnants on La Réunion (Mascarene Islands, Indian Ocean). *Environmental Conservation* **18**, 51–61.

McDowall, R. M. (1994). *Gamekeepers for the nation: The story of New Zealand's acclimatisation societies 1861–1990.* Canterbury University Press, Christchurch, New Zealand.

Mace, G. M. and Lande, R. (1991). Assessing extinction threats: Toward a reevaluation of IUCN threatened species categories. *Conservation Biology* **5**, 148–157.

Mack, M. C. and D'Antonio, C. M. (2003). Exotic grasses alter controls over soil nitrogen dynamics in a Hawaiian woodland. *Ecological Applications* **13**, 154–166.

Mack, R. N., Simberloff, D., Lonsdale, W. M., Evans, H., Clout, M. and Bazzaz, F. A. (2000). Biotic invasions: Causes, epidemiology, global consequences, and control. *Ecological Applications* **10**, 689–710.

McKinney, M. L. (2004). Do exotics homogenize or differentiate? Roles of sampling and exotic species richness. *Biological Invasions* **6**, 495–504.

McKinney, M. L. (2005). Species introduced from nearby sources have a more homogenizing effect than species from distant sources: Evidence from plants and fishes in the USA. *Diversity and Distributions* **11**, 367–374.

McKinney, M. L. (2006a). Urbanization as a major cause of biotic homogenization. *Biological Conservation* **127**, 247–260.

McKinney, M. L. (2006b). Correlated non-native species richness of birds, mammals, herptiles and plants: Scale effects of area, human population and native plants. *Biological Invasions* **8**, 415–425.

McKinney, M. L. and Lockwood, J. L. (1999). Biotic homogenization: A few winners replacing many losers in the next mass extinction. *Trends in Ecology & Evolution* **14**, 450–453.

McKinney, M. L. and Lockwood, J. L. (2005). Community composition and homogenization: Evenness and abundance of native and exotic plant species. In *Species invasions: Insights into ecology, evolution and biogeography*, eds D. F. Sax, J. J. Stachowicz and S. D. Gaines, pp. 365–380. Sinauer Associates Inc., Sunderland, MA.

McLain, D. K., Moulton, M. P. and Redfearn, T. P. (1995). Sexual selection and the risk of extinction of introduced birds on oceanic islands. *Oikos* **74**, 27–34.

McLain, D. K., Moulton, M. P. and Sanderson, J. G. (1999). Sexual selection and extinction: The fate of plumage dimorphic and plumage-monomorphic birds introduced onto islands. *Evolutionary Ecology Research* **1**, 549–565.

McLean, R. G. (2003). The emergence of major avian diseases in North America: West Nile Virus and more. In *Proceedings of the 10th Wildlife Damage Management Conference*, eds K. A. Fagerstone and G. W. Witmer, pp. 300–305.

MacLeod, C. J., Duncan, R. P., Parish, D. M. B., Wratten, S. D. and Hubbard, S. F. (2005a). Can increased niche opportunities and release from enemies explain the success of introduced Yellowhammer populations in New Zealand? *Ibis* **147**, 598–607.

MacLeod, C. J., Parish, D. M. B., Duncan, R. P., Moreby, S. and Hubbard, S. F. (2005b). Importance of niche quality for Yellowhammer *Emberiza citrinella* nestling survival, development and body condition in its native and exotic ranges: The role of diet. *Ibis* **147**, 270–282.

McMahon, S. M., Cadotte, M. W. and Fukami, T. (2006). Tracking the tractable: Using invasions to guide the explortion of conceptual ecology. In *Conceptual ecology and invasion biology: Reciprocal approaches to nature*, eds M. W. Cadotte, S. M. McMahon and T. Fukami, pp. 3–14. Springer, Dordrecht.

McNutt, K. L. (1998). *Impacts of reduced bird densities on pollination and dispersal mutualisms in New Zealand forests.* Massey University, Palmerston North, New Zealand.

Madge, S. and Burn, H. (1989). *Wildfowl: An identification guide to ducks, geese and swans of the world.* Christopher Helm, London.

Mandon-Dalger, I., Clergeau, P., Tassin, J., Riviere, J. N. and Gatti, S. (2004). Relationships between alien plants and an alien bird species on Reunion Island. *Journal of Tropical Ecology* **20**, 635–642.

Marchetti, M. P., Light, T., Feliciano, J., Armstrong, T., Hogan, Z., Viers, J. and Moyle, P. B. (2001). Homogenization of California's fish fauna through abiotic change. In *Biotic homogenization*, eds J. L. Lockwood and M. L. McKinney, pp. 259–278. Kluwer Academic/Plenum Publishers, New York.

Martin, L. B. and Fitzgerald, L. (2005). A taste for novelty in invading house sparrows, *Passer domesticus. Behavioural Ecology* **16**, 702–707.

Martin, T. E. (1995). Avian life history evolution in relation to nest sites, nest predation, and food. *Ecological Monographs* **65**, 101–127.

Martin, T. E. (1996). Life history evolution in tropical and south temperate birds: What do we really know? *Journal of Avian Biology* **27**, 263–272.

Martuscelli, P. (1994). A parrot with a tiny distribution and a big problem. *PsittaScene* **6**, 3–7.

Marzluff, J. M. (2001). Worldwide increase in urbanization and its effects on birds. In *Avian ecology and conservation in an urbanizing world*, eds J. M. Marzluff, R. Bowman and R. Donnelly, pp. 19–47. Kluwer Academic, Norwell, MA.

Massaro, M., Starling-Windhof, A., Briskie, J. V. and Martin, T. E. (2008). Introduced mammalian predators induce behavioural changes in parental care in an endemic New Zealand bird. *PLoS One* **3**: e2331. doi:10. 1371/journal. pone. 0002331.

Matter, S. F., Hanski, I. and Gyllenberg, M. (2002). A test of the metapopulation model of the species-area relationship. *Journal of Biogeography* **29**, 977–983.

Mayr, E. (1963). *Animal species and evolution*. Belknap Press, Cambridge, MA.

Meiri, S. and Dayan, T. (2003). On the validity of Bergmann's rule. *Journal of Biogeography* **30**, 331–351.

Meiri, S. and Thomas, G. H. (2007). The geography of body size: Challenges of the interspecific approach. *Global Ecology and Biogeography* **16**, 689–693.

Meiri, S., Yom-Tov, Y. and Geffen, E. (2007). What determines conformity to Bergmann's rule? *Global Ecology and Biogeography* **16**, 788–794.

Melbourne, B. A. and Hastings, A. (2008). Extinction risk depends strongly on factors contributing to stochasticity. *Nature* **454**, 100–103.

Melbourne, B. A., Cornell, H. V., Davies, K. F., Dugaw, C. J., Elmendorf, S., Freestone, A. L., Hall, R. J., Harrison, S., Hastings, A., Holland, M., Holyoak, M., Lambrinos, J., Moore, K. and Yokomizo, H. (2007). Invasion in a heterogeneous world: Resistance, coexistence, or hostile takeover? *Ecology Letters* **10**, 77–94.

Méndez, V. and Camacho, J. (1997). Dynamics and thermodynamics of delayed population growth. *Physical Review E* **55**, 6476–6482.

Méndez, V., Fort, J. and Farjas, J. (1999). Speed of wave-front solutions to hyperbolic reaction-diffusion equations. *Physical Review E* **60**, 5231–5243.

Merilä, J. and Sheldon, B. C. (2001). Avian quantitative genetics. In *Current ornithology*, eds V. Nolan, V. Nolan Jr and C. F. Thompson, Vol. 16, pp. 179–255. Springer-Verlag, New York.

Merilä, J., Bjorklund, M. and Baker, A. J. (1996). The successful founder: Genetics of introduced *Carduelis chloris* (greenfinch) populations in New Zealand. *Heredity* 77, 410–422.

Milberg, P. and Tyrberg, T. (1993). Naïve birds and noble savages—A review of man-caused prehistoric extinctions of island birds. *Ecography* **16**, 229–250.

Minnis, P. E., Whalen, M. E., Kelley, J. H. and Stewart, J. D. (1993). Prehistoric macaw breeding in the North American southwest. *American Antiquity* **58**, 270–276.

Mitchell, C. E. and Power, A. G. (2003). Release of invasive plants from fungal and viral pathogens. *Nature* **421**, 625–627.

Moeed, A. (1975). Food of skylarks and pipits, finches and feral pigeons near Christchurch. *Notornis* **22**, 135–142.

Møller, A. P. and Cassey, P. (2004). On the relationship between T-cell mediated immunity in bird species and the establishment success of introduced populations. *Journal of Animal Ecology* **73**, 1035–1042.

Møller, A. P., Christie, P. and Lux, E. (1999). Parasitism, host immune function, and sexual selection. *Quarterly Review of Biology*, **74**, 3–20.

Møller, A. P., Martín-Vivaldi, M. and Soler, J. J. (2003). Parasitism, host immune defence and dispersal. *Journal of Evolutionary Biology* **17**, 603–612.

Monticelli, D. (2008). Finding introduced birds near Lisbon, Portugal. *Birding World* **21**, 203–206.

Moreau, R. E. (1944). Clutch-size: A comparative study, with special reference to African birds. *Ibis* **86**, 286–347.

Moulton, M. P. (1985). Morphological similarity and coexistence of congeners: An experimental test with introduced Hawaiian birds. *Oikos* **44**, 301–305.

Moulton, M. P. (1993). The all-or-none pattern in introduced Hawaiian passeriforms: The role of competition sustained. *American Naturalist* **141**, 105–119.

Moulton, M. P. and Pimm, S. L. (1983). The introduced Hawaiian avifauna: Biogeographic evidence for competition. *American Naturalist* **121**, 669–690.

Moulton, M. P. and Pimm, S. L. (1986a). Species introductions to Hawaii. In *The ecology of biological invasions of North America and Hawaii*, eds H. A. Mooney and J. A. Drake, pp. 231–249. Springer-Verlag, New York.

Moulton, M. P. and Pimm, S. L. (1986b). The extent of competition in shaping an introduced avifauna. In *Community ecology*, eds J. Diamond and T. J. Case, pp. 80–97. Harper & Row, New York.

Moulton, M. P. and Pimm, S. L. (1987). Morphological assortment in introduced Hawaiian passerines. *Evolutionary Ecology* **1**, 113–124.

Moulton, M. P. and Sanderson, J. G. (1997). Predicting the fates of passeriform introductions on oceanic islands. *Conservation Biology* **11**, 552–558.

Moulton, M. P. and Sanderson, J. G. (1999). The fate of passeriform introductions: Reply to Duncan and Young. *Conservation Biology* **13**, 937–938.

Moulton, M. P. and Scioli, M. E. T. (1986). Range sizes and abundances of passerines introduced to Oahu, Hawaii. *Journal of Biogeography* **13**, 339–344.

Moulton, M. P., Miller, K. and Tillman, E. A. (2001a). Patterns of success among introduced birds in the Hawaiian Islands. In *Evolution, ecology, conservation, and management of Hawaiian birds: A vanishing avifauna*, eds J. M. Scott, S. Conant and C. van Riper III, pp. 31–46. *Studies in Avian Biology* **22**.

Moulton, M. P., Sanderson, J. and Simberloff, D. (1996). Passeriform introductions to the Mascarenes (Indian Ocean): An assessment of the role of competition. *Ecologie* **27**, 143–152.

Moulton, M. P., Sanderson, J. G. and Labisky, R. F. (2001b). Patterns of success in game bird (Aves: Galliformes) introductions to the Hawaiian islands and New Zealand. *Evolutionary Ecology Research* **3**, 507–519.

Moulton, M.P., Pimm, S.L., and Krissinger, M.W. (1990) Nutmeg Mannikin (*Lonchura punctulata*): a comparison of abundances in Oahu vs Maui sugarcane: evidence for competitive exclusion? *Elepaio* **50**, 83–85

Muller-Landau, H. C., Levin, S. A. and Keymer, J. E. (2003). Theoretical perspectives on evolution of long-distance dispersal and the example of specialized pests. *Ecology* **84**, 1957–1967.

Narasue, M. and Obara, H. (1982). Feralization of cage birds, in and around Tokyo. In *Chiba Bay-Coast Project, IV, Tokyo*, ed M. Numata, pp. 82–86.

Nash, S. V. (1993). *Sold for a song: The trade in southeast Asian non-CITES birds.* Traffic International, Cambridge.

Nee, S., Read, A. F., Greenwood, J. J. D. and Harvey, P. H. (1991). The relationship between abundance and body size in British birds. *Nature* **351**, 312–313.

Nekola, J. C. and White, P. S. (1999). The distance decay of similarity in biogeography and ecology. *Journal of Biogeography* **26**, 867–878.

Neubert, M. G. and Caswell, H. (2000). Demography and dispersal: Calculation and sensitivity analysis of invasion speed for structured populations. *Ecology* **81**, 1613–1628.

Newsome, A. E. and Noble, I. R. (1986). Ecological and physiological characters of invading species. In *Ecology of biological invasions*, eds R. H. Groves and J. J. Burdon, pp. 1–20. Cambridge University Press, Cambridge.

Newton, I. (1998). *Population limitation in birds.* Academic Press, San Diego, CA.

Nicolakakis, N. and Lefebvre, L. (2000). Forebrain size and innovation rate in European birds: Feeding, nesting and confounding variables. *Behaviour* **137**, 1415–1429.

Niethammer, G. (1970). Clutch sizes of introduced European passeriformes in New Zealand. *Notornis* **17**, 214–222.

Nolan, P. M., Hill, G. E. and Stoehr, M. (1998). Sex, size and plumage redness predict House Finch survival in an epidemic. *Proceedings of the Royal Society, London, B* **265**, 961–965.

Novak, S. J. and Mack, R. N. (2005). Genetic bottlenecks in alien plant species: Influence of mating systems and introduction dynamics. In *Species invasions: Insights into ecology, evolution and biogeography*, eds D. F. Sax, J. J. Stachowicz and S. D. Gaines, pp. 201–228. Sinauer Associates Inc., Sunderland, MA.

Nyári, A., Ryall, C. and Peterson, A. T. (2006). Global invasive potential of the house crow *Corvus splendens* based on ecological niche modelling. *Journal of Avian Biology* **37**, 306–311.

O'Brien, S. J., Roelke, M. E., Marker, L., Newman, A., Winkler, C. A., Meltzer, K. D., Colly, L., Evermann, J. F., Bush, M. B. and Wildt, D. E. (1985). Genetic basis for species vulnerability in the cheetah. *Science* **227**, 1428–1434.

O'Connor, R. J. (1986). Biological characteristics of invaders among bird species in Britain. *Philosophical Transactions of the Royal Society, London, B* **314**, 583–598.

O'Hara, R. B. (2005). The anarchist's guide to ecological theory. Or, we don't need no stinkin' laws. *Oikos* **110**, 390–393.

Okubo, A. (1980). *Diffusion and ecological problems: Mathematical models.* Springer-Verlag, Berlin.

Okubo, A. (1988). Diffusion-type models for avian range expansion. In *Acta XIX Congress Internationalis Ornithologici. I,* ed H. Quellet, pp. 1038–1049. National Museum of Natural Sciences, University of Ottawa Press, Ottawa.

Olden, J. D. and Poff, N. L. (2003). Toward a mechanistic understanding and prediction of biotic homogenization. *American Naturalist* **162**, 442–460.

Olden, J. D. and Poff, N. L. (2004). Ecological processes driving biotic homogenization: Testing a mechanistic model using fish faunas. *Ecology* **85**, 1867–1875.

Olden, J. D. and Rooney, T. P. (2006). On defining and quantifying biotic homogenization. *Global Ecology and Biogeography* **15**, 113–120.

Olesen, J. M., Eskildsen, L. I. and Venkatasamy, S. (2002). Invasion of pollination networks on oceanic islands: Importance of invader complexes and endemic super generalists. *Diversity and Distributions* **8**, 181–192.

Olesen, J.M., Rønsted, N., Tolderlund, U., Cornett, C., Mølgaard, P., Madsen, J., Jones, C.G., and Olsen, C.E. (1998) Mauritian red nectar remains a mystery. *Nature* **393**, 529.

Olson, V., Davies, R. G., Orme, C. D. L., Thomas, G. H., Meiri, S., Blackburn, T. M., Gaston, K. J., Owens, I. P. F. and Bennett, P. M. (2009) Global biogeography and ecology of body size in birds. *Ecology Letters* **12**, 249–259.

Orme, C. D. L., Davies, R. G., Burgess, M., Eigenbrod, F., Pickup, N., Olson, V., Webster, A. J., Ding, T.-S., Rasmussen, P. C., Ridgely, R. S., Stattersfield, A. J., Bennett, P. M., Blackburn, T. M., Gaston, K. J. and Owens, I. P. F. (2005). Global biodiversity hotspots of species richness, threat and endemism are not congruent. *Nature* **436**, 1016–1019.

Orme, C. D. L., Davies, R. G., Olson, V., Thomas, G. H., Stattersfield, A. J., Bennett, P. M., Owens, I. P. F., Blackburn, T. M. and Gaston, K. J. (2006). Global patterns of geographic range size in birds. *PLoS Biology* **4**, e208. DOI: 10. 1371/journal. pbio. 0040208.

Ortega-Cejas, V., Fort, J. and Méndez, V. (2004). The role of the delay time in the modeling of biological range expansions. *Ecology* **85**, 258–264.

Ottens, G. and Ryall, C. (2003). House crows in the Netherlands and Europe. *Dutch Birding* **25**, 312–319.

Owens, I. P. F. and Bennett, P. M. (1995). Ancient ecological diversification explains life-history variation among living birds. *Proceedings of the Royal Society, London, B* **261**, 227–232.

Owens, I. P. F., Bennett, P. M. and Harvey, P. H. (1999). Species richness among birds: Body size, life history, sexual selection or ecology? *Proceedings of the Royal Society, London, B* **266**, 933–939.

Pain, D. J., Martins, T. L. F., Boussekey, M., Diaz, S. H., Downs, C. T., Ekstrom, J. M. M., Garnett, S., Gilardi, J. D., McNiven, D., Primot, P., Rouys, S., Saoumoé, M., Symes, C. T., Tamungang, S. A., Theuerkauf, J., Villafuerte, D., Verfailles, L., Widmann, P. and Widmann, I. D. (2006). Impact of protection on nest take and nesting success of parrots in Africa, Asia and Australasia. *Animal Conservation* **9**, 322–330.

Paradis, E., Baillie, S. R., Sutherland, W. J. and Gregory, R. D. (1998). Patterns of natal and breeding dispersal in birds. *Journal of Animal Ecology* **67**, 518–536.

Parker, I. M., Simberloff, D., Lonsdale, W. M., Goodell, K., Wonham, M., Kareiva, P. M., Williamson, M. H., Von Holle, B., Moyle, P. B., Byers, J. E. and Goldwasser, L. (1999). Impact: Towards a framework for understanding the ecological effects of invaders. *Biological Invasions* **1**, 3–19.

Parkin, D. T. and Cole, S. R. (1985). Genetic differentiation and rates of evolution in some introduced populations of the House Sparrow, *Passer domesticus* in Australia and New Zealand. *Heredity* **54**, 15–23.

Pass, D. A. and Perry, R. A. (1984). The pathology of psittacine beak and feather disease. *Australian Veterinary Journal* **61**, 69–74.

Paterson, A. M., Palma, R. L. and Gray, R. D. (1999). How frequently do avian lice miss the boat? Implications for coevolutionary studies. *Systematic Biology* **48**, 214–223.

Payne, R. B. (1977). The ecology of brood parasitism in birds. *Annual Review of Ecology and Systematics* **8**, 1–28.

Pease, C. M. and Grzybowski, J. A. (1995). Assessing the consequences of brood parastism and nest predation on seasonal fecundity in passerine birds. *Auk* **112**, 343–363.

Pederson, K., Clark, L., Andelt, W. F. and Salman, M. D. (2006). Prevalence of SHIGA toxin-producing *Escherichia coli* and *Salmonella enterica* in rock pigeons captured in Fort Collins, Colorado. *Journal of Wildlife Diseases* **42**, 46–55.

Petrovskii, S. V. and Li, B. L. (2005). *Exactly solvable models of biological invasions.* Chapman & Hall, London.

Phillips, B. L., Brown, G. P., Travis, J. M. J. and Shine, R. (2008). Reid's paradox revisited: The evolution of dispersal kernels during range expansion. *American Naturalist* **172**, S34–S48.

Phillips, J. C. (1928). Wild birds introduced or transplanted in North America. *United States Department of Agriculture Technical Bulletin* **61**, 1–63.

Pimentel, D., Lach, L., Zuniga, R. and Morrison, D. (2000). Environmental and economic costs of nonindigenous species in the United States. *BioScience* **50**, 53–65.

Pimentel, D., McNair, S., Janecka, J., Wightman, J., Simmonds, C., O'Connell, C., Wong, E., Russel, L., Zern, J., Aquino, T. and Tsomondo, T. (2001). Economic and environmental threats of alien plant, animal, and microbe invasions. *Agriculture Ecosystems & Environment* **84**, 1–20.

Pimm, S. L. (1989). Theories of predicting success and impact of introduced species. In *Biological Invasions a global perspective*, eds J. A. Drake, F. di Castri, R. H. Groves, F. J. Kruger, M. Rejmánek and M. Williamson, pp. 351–367. John Wiley & Sons, Chichester.

Pimm, S. L. (1991). *The balance of nature? Ecological issues in the conservation of species and communities.* University of Chicago Press, Chicago.

Pimm, S. L., Jones, H. L. and Diamond, J. (1988). On the risk of extinction. *American Naturalist* **132**, 757–785.

Pithon, J. A. and Dytham, C. (2002). Distribution and population development of introduced Ring-necked Parakeets *Psittacula krameri* in Britain between 1983 and 1998. *Bird Study* **49**, 110–117.

Pranty, B. (2001). The budgerigar in Florida: Rise and fall of an exotic psittacid. *North American Birds* **55**, 389–397.

Pranty, B. (2004). Florida's exotic avifauna: A preliminary checklist. *Birding* **36**, 362–372.

Pratt, D. H., Bruner, P. L. and Berrett, D. G. (1987). *A field guide to the birds of Hawaii and the tropical Pacific.* Princeton University Press, Princeton, NJ.

Prinzing, A., Durka, W., Klotz, S. and Brandl, R. (2002). Which species become aliens? *Evolutionary Ecology Research* **4**, 385–404.

Proçhes, S., Wilson, J. R. U., Richardson, D. M. and Rejmánek, M. (2008). Searching for phylogenetic pattern in biological invasions. *Global Ecology and Biogeography* **17**, 5–10.

Pruett-Jones, S., Newman, J. R., Newman, C. M., Avery, M. L. and Lindsay, J. R. (2007). Population viability analysis of monk parakeets in the United States and examination of alternative management strategies. *Human–Wildlife Conflicts* **1**, 35–44.

Pugesek, B. H. and Tomer, A. (1996). The Bumpus house sparrow data: A reanalysis using structural equation models. *Evolutionary Ecology* **10**, 387–404.

Purvis, A. and Rambaut, A. (1995). Comparative analysis by independent contrasts (CAIC): An Apple Macintosh application for analysing comparative data. *Computer Applications in the Biosciences* **11**, 247–251.

Puth, L. M. and Post, D. M. (2005). Studying invasion: Have we missed the boat? *Ecology Letters* **8**, 715–721.

Pyšek, P., Prach, K. and Mandák, B. (1998). Invasions of alien plants into habitats of Central European landscape: An historical pattern. In *Plant invasions: Ecological mechanisms and human responses*, eds U. Starfinger, K. Edwards, I. Kowarik and M. Williamson, pp. 23–32. Backhuys Publishers, Leiden.

Pyšek, P., Richardson, D. M., Pergl, J., Jarosik, V., Sixtova, Z. and Weber, E. (2008). Geographical and taxonomic biases in invasion ecology. *Trends in Ecology & Evolution* **23**, 237–244.

Quintana-Murci, L., Semino, O., Bandelt, H.-J., Passarion, G. McElreavey and Santachiara-Benerecetti, A. S. (1999). Genetic evidence of an early exit of *Homo sapiens sapiens* from Africa through eastern Africa. *Nature Genetics* **23**, 437–441.

Raffaele, H. (1989). *A guide to the birds of Puerto Rico and the Virgin Islands.* Princeton University Press, Princeton, NJ.

Raidal, S. R., McElnea, C. L. and Cross, G. M. (1993). Seroprevalence of psittacine beak and feather disease virus in wild psittacine birds in New South Wales. *Australian Veterinary Journal* **70**, 137–139.

Rangel, T., Diniz-Filho, J. A. F. and Bini, L. M. (2006). Towards an integrated computational tool for spatial analysis in macroecology and biogeography. *Global Ecology and Biogeography* **15**, 321–327.

Rapoport, E. H. (1982). *Areography: Geographical strategies of species.* Pergamon, Oxford.

Rayner, M. J., Hauber, M. E., Imber, M. J., Stamp, R. K. and Clout, M. N. (2008). Spatial heterogeneity of mesopredator release within an oceanic island. *Proceedings of the National Academy of Sciences, USA* **104**, 20862–20865.

Read, A. F. and Harvey, P. H. (1989). Life history differences among the eutherian radiations. *Journal of Zoology, London* **219**, 329–353.

Restall, R. (1997). *Munias and mannikins.* Yale University Press, New Haven, CT.

Rhymer, J. M. and D.S. Simberloff. 1996. Genetic extinction through hybridization and introgression. *Annual Review of Ecology and Systematics* **27**, 83–109.

Rhymer, J. M., Williams, M. J. and Braun, M. J. (1994). Mitochondrial analysis of gene flow between New Zealand mallards (*Anas platyrhynchos*) and grey ducks (*A. superciliosa*). *Auk* **111**, 970–978.

Richardson, D. M., Allsopp, N., D'Antonio, C. M., Milton, S. J. and Rejmanek, M. (2000). Plant invasions: The role of mutualisms. *Biological Reviews* 75, 65–93.

Richardson, D. M., Cambray, J. A., Chapman, R. A., Dean, W. R. J., Griffiths, C. L., Le Maitre, D. C., Newton, D. J. and Winstanley, T. J. (2003). Vectors and pathways of biological invasions in South Africa: Past, present and future. In *Invasive species: Vectors and management strategies*, eds G. Ruiz and J. Carlton, pp. 292–349. Island Press, Washington, DC.

Richardson, D. M., Rouget, M. and Rejmánek, M. (2004). Using natural experiments in the study of alien tree invasions. In *Experimental approaches to conservation biology*, eds M. S. Gordon and S. M. Bartol, pp. 180–200. University of California Press.

Ricklefs, R. E. and Travis, J. (1980). A morphological approach to the study of avian community organization. *Auk* 97, 321–333.

Robert, M. and Sorci, G. (1999). Rapid increase of host defence against brood parasites in a recently parasitized area: The case of village weavers in Hispaniola. *Proceedings of the Royal Society, London, B* **266**, 941–946.

Roberts, R. G., Flannery, T. F., Ayliffe, L. K., Yoshida, H., Olley, J. M., Prideaux, G. J., Laslett, G. M., Baynes, A., Smith, M. A., Jones, R. and Smith, B. L. (2001). New ages for the last Australian megafauna: Continent-wide extinction about 46,000 years ago. *Science* **292**, 1888–1892.

Robertson, A. W., Kelly, D., Ladley, J. J. and Sparrow, A. D. (1999). Effects of pollinator loss on endemic New Zealand mistletoes (Loranthaceae). *Conservation Biology* **13**, 499–508.

Robinson, J. M. (2001). The dynamics of avicultural markets. *Environmental Conservation* **28**, 76–85.

Robinson, R. A., Siriwardena, G. M. and Crick, H. Q. P. (2005). Size and trends of the House Sparrow *Passer domesticus* population in Great Britain. *Ibis* **147**, 552–562.

Robinson-Wolrath, S. I. and Owens, I. P. F. (2003). Large size in an island-dwelling bird: Intraspecific competition and the Dominance Hypothesis. *Journal of Evolutionary Biology* **16**, 1106–1114.

Roff, D. A. and Roff, R. J. (2003). Of rats and Maoris: A novel method for the analysis of patterns of extinction in the New Zealand avifauna before human contact. *Evolutionary Ecology Research* **5**, 759–779.

Romagosa, C. M. and Labisky, R. F. (2000). Establishment and dispersal of the Eurasian collared-dove in Florida. *Journal of Field Ornithology* **71**, 159–166.

Roman, J. and Darling, J. A. (2007). Paradox lost: Genetic diversity and the success of aquatic invasions. *Trends in Ecology & Evolution* **22**, 454–464.

Rosenzweig, M. L. (1995). *Species diversity in space and time*. Cambridge University Press, Cambridge.

Ross, H. A. (1983). Genetic differentiation of starling (*Sturnus vulgaris*: Aves). Populations in New Zealand and Great Britain. *Journal of Zoology, London* **201**, 351–362.

Ryall, C. (2003). Notes on ecology and behaviour of house crows at Hoek van Holland. *Dutch Birding* **25**, 167–172.

Sæther, B.-E. (1987). The influence of body weight on the covariation between reproductive traits in European birds. *Oikos* **48**, 79–88.

Sæther, B.-E. and Engen, S. (2002). Pattern of variation in avian population growth rates. *Philosophical Transactions of the Royal Society, London, B* **357**, 1185–1195.

Sæther, B.-E., Engen, S., Møller, A. P., Visser, M. E., Matthysen, E., Fiedler, W., Lambrechts, M. M., Becker, P. H., Brommer, J. E., Dickinson, J., Du Feu, C., Gehlbach, F. R., Merila, J., Rendell, W., Robertson, R. J., Thomson, D. and Torok, J. (2005). Time to extinction of bird populations. *Ecology* **86**, 693–700.

Sæther, B.-E., Engen, S., Møller, A. P., Weimerskirch, H., Visser, M. E., Fiedler, W., Matthysen, E., Lambrechts, M. M., Badyaev, A. V., Becker, P. H., Brommer, J. E., Bukacinski, D., Bukacinski, M., Christensen, H., Dickinson, J., du Feu, C., Gehlbach, F. R., Heg, D., Hötker, H., Merilä, J., Nielsen, J. T., Rendell, W., Robertson, R. J., Thomson, D. L., Török, J. and Van Hecke, P. (2004). Life-history variation predicts the effects of demographic stochasticity on avian population dynamics. *American Naturalist* **164**, 793–802.

Sahagun, B. de. ([1577] 1963). *Florentine codex, general history of things of New Spain, Book 11: Earthly things (Translated from Aztec by C. E. Dibble and A. J. O. Anderson).* University of Utah and School of American Research, Santa Fé.

Sakai, A. K., Allendorf, F. W. Holt, J. S., Lodge, D. M., Molofsky, J., With, K. A., Baughman, S., Cabin, R. J., Cohen, J. E., Ellstrand, N. E., McCauley, D. E., O'Neil, P., Parker, I. M., Thompson, J. N. and G, W. S. (2001). The population biology of invasive species. *Annual Review of Ecology, Evolution and Systematics* **32**, 305–332.

Salo, P., Korpimäki, E., Banks, P. B., Nordström, M. and Dickman, C. (2007). Alien predators are more dangerous than native predators to prey populations. *Proceedings of the Royal Society, London, B* **274**, 1237–1243.

Sauer, J. R., Hines, J. E. and Fallon, J. (2008). The North American Breeding Bird Survey, results and analysis 1966- 2007. USGS Patuxent Wildlife Research Center, Laurel, MD.

Sax, D. F. (2001). Latitudinal gradients and geographic ranges of exotic species: Implications for biogeography. *Journal of Biogeography* **28**, 139–150.

Sax, D. F. and Gaines, S. D. (2006). The biogeography of naturalised species and the species-area relationship: Reciprocal insights to biogeography and invasion biology. In *Conceptual ecology and invasion biology: Reciprocal approaches to nature,* eds M. W. Cadotte, S. M. McMahon and T. Fukami, pp. 449–480. Springer, Dordrecht.

Sax, D. F., Brown, J. H., White, E. P. and Gaines, S. D. (2005a). The dynamics of species invasions: Insights into the mechanisms that limit species diversity. In *Species invasions: Insights into ecology, evolution and biogeography,* eds D. Sax, S. D. Gaines and J. J. Stachowicz, pp. 447–465. Academic Press, Sunderland, Massachusetts.

Sax, D. F., Gaines, S. D. and Brown, J. H. (2002). Species invasions exceed extinctions on islands worldwide: A comparative study of plants and birds. *American Naturalist* **160**, 766–783.

Sax, D. F., Gaines, S. D. and Stachowicz, J. J. (2005b). *Species invasions. Insights into ecology, evolution and biogeography.* Academic Press, Sunderland, Massachussets.

Sax, D. F., Gaines, S. D. and Stachowicz, J. J. (2005c). Introduction. In *Species invasions: Insights into ecology, evolution and biogeography,* eds D. Sax, S. D. Gaines and J. J. Stachowicz, pp. 1–7. Academic Press, Sunderland, Massachusetts.

Sax, D. F., Stachowicz, J. J., Brown, J. H., Bruno, J. F., Dawson, M. N., Gaines, S. D., Grosberg, R. K., Hastings, A., Holt, R. D., Mayfield, M. M., O'Connor, M. I. and Rice, W. R. (2007). Ecological and evolutionary insights from species invasions. *Trends in Ecology & Evolution* **22**, 465–471.

Schoener, T. W. (1988). Testing for non-randomness in sizes and habitats of West Indian lizards: Choice of species pool affects conclusions from null models. *Evolutionary Ecology* **2**, 1–26.

Schuetz, J. G. (2005). Reduced growth but not survival of chicks with altered gape patterns: Implications for the evolution of nestling similarity in a parasitic finch. *Animal Behaviour* **70**, 839–848.

Scott, S. N., Clegg, S. M., Blomberg, S. P., Kikkawa, J. and Owens, I. P. F. (2003). Morphological shifts in island-dwelling birds: The roles of generalist foraging and niche expansion. *Evolution* 57, 2147–2156.

Seehausen, O. (2002). Patterns in fish radiation are compatible with Pleistocene desiccation of Lake Victoria and 14,600 year history for its cichlid species flock. *Proceedings of the Royal Society, London, B* **269**, 491–497.

Seehausen, O. (2004). Hybridization and adaptive radiation. *Trends in Ecology & Evolution* **19**, 198–207.

Selander, R. K. and Johnston, R. F. (1967). Evolution in the house sparrow. Intrapopulation variation in North America. *Condor* **69**, 217–258.

Severinghaus, L. L. and Chi, L. (1999). Prayer animal release in Taiwan. *Biological Conservation* **89**, 301–304.

Shea, K. and Chesson, P. (2002). Community ecology theory as a framework for biological invasions. *Trends in Ecology & Evolution* **17**, 170–176.

Shigesada, N. and Kawasaki, K. (1997). *Biological invasions: Theory and practice* Oxford University Press, Oxford.

Sibley, C. G. and Monroe, B. L. (1990). *Distribution and taxonomy of birds of the world.* Yale University Press, New Haven, CT.

Sibley, C. G. and Monroe, B. L. (1993). *Supplement to the distribution and taxonomy of birds of the world.* Yale University Press, New Haven, CT.

Siefferman, L. and Hill, G. E. (2003). Structural and melanin coloration indicate parental effort and reproductive success in male eastern bluebirds. *Behavioral Ecology* **14**, 855–861.

Silva, T., Reino, L. M. and Borralho, R. (2002). A model for range expansion of an introduced species: The common waxbill *Estrilda astrild* in Portugal. *Diversity and Distributions* **8**, 319–326.

Simberloff, D. (1992). Extinction, survival, and effects of birds introduced to the Mascarenes. *Acta Oecologica* **13**, 663–678.

Simberloff, D. (1995). Why do introduced species appear to devastate islands more than mainland areas? *Pacific Science* **49**, 87–97.

Simberloff, D. (2003). Confronting introduced species: A form of xenophobia? *Biological Invasions* **5**, 179–192.

Simberloff, D. (2004). Community ecology: Is it time to move on? *American Naturalist* **163**, 787–799.

Simberloff, D. and Boecklen, W. J. (1991). Patterns of extinction in the introduced Hawaiian avifauna: A reexamination of the role of competition. *American Naturalist* **138**, 300–327.

Simberloff, D. and Gibbons, L. (2004). Now you see them, now you don't!—Population crashes of established introduced species. *Biological Invasions* **6**, 161–172.

Simberloff, D. and Von Holle, B. (1999). Positive interactions of nonindigenous species: Invasional meltdown? *Biological Invasions* **1**, 21–32.

Simberloff, D., Dayan, T., Jones, C. and Ogura, G. (2000). Character displacement and release in the small Indian mongoose, *Herpestes javanicus*. *Ecology* **81**, 2086–2099.

Sinclair, A. R. E., Pech, R. P., Dickman, C. R., Hik, D., Mahon, P. and Newsome, A. E. (1998). Predicting effects of predation on conservation of endangered prey. *Conservation Biology* **12**, 564–575.

Skellam, J. G. (1951). Random dispersal in theoretical populations. *Biometrika* **38**, 196–218.

Skutch, A. F. (1949). Do tropical birds rear as many young as they can nourish? *Ibis* **91**, 430–455.

Slatkin, M. (1977). Gene flow and genetic drift in a species subject to frequent local extinctions. *Theoretical Population Biology* **12**, 253–262.

Smallwood, K. S. (1994). Site invasibility by exotic birds and mammals. *Biological Conservation* **69**, 251–259.

Smallwood, K. S. and Salmon, T. P. (1992). A rating system for potential exotic bird and mammal pests. *Biological Conservation* **62**, 149–159.

Smart, S. M., Thompson, K., Marrs, R. H., Le Duc, M. G., Maskell, L. C. and Firbank, L. G. (2006). Biotic homogenization and changes in species diversity across human-modified ecosystems. *Proceedings of the Royal Society, London, B* **273**, 2659–2665.

Smith, G. C., Henderson, I. S. and Robertson, P. A. (2005). A model of ruddy duck *Oxyura jamaicensis* eradication for the UK. *Journal of Applied Ecology* **42**, 546–555.

Soberón, J. (2007). Grinnellian and Eltonian niches and geographic distributions of species. *Ecology Letters* **10**, 1115–1123.

Sokal, R. R. and Rohlf, F. J. (1995). *Biometry*, 3rd edition. W. H. Freeman & Co, New York.

Sol, D. (2000a). Are islands more susceptible to be invaded than continents? Birds say no. *Ecography* **23**, 687–692.

Sol, D. (2000b). Introduced species: A significant component of global environmental change. Ph D thesis, Universitat de Barcelona, Barcelona.

Sol, D. (2007). Do successful invaders exist? Pre-adaptations to novel environments in terrestrial vertebrates. In *Biological Invasions*, ed W. Nentwig, pp. 127–144. Springer-Verlag, Berlin.

Sol, D., and Lefebvre, L. (2000). Behavioural flexibility predicts invasion success in birds introduced to New Zealand. *Oikos* **90**, 599–605.

Sol, D., Bacher, S., Reader, S. M. and Lefebvre, L. (2008). Brain size predicts the success of mammal species introduced into novel environments. *American Naturalist* **172**, S63–S71.

Sol, D., Blackburn, T. M., Cassey, P., Duncan, R. P. and Clavell, J. (2005a). The ecology and impact of non-indigenous birds. In *Handbook of the birds of the world*. X. *Cuckoo-shrikes to thrushes*, eds J. del Hoyo, A. Elliott and J. Sargatal, pp. 13–35. Lynx Ediçions and BirdLife International, Cambridge.

Sol, D., Duncan, R. P., Blackburn, T. M., Cassey, P. and Lefebvre, L. (2005b). Big brains, enhanced cognition, and response of birds to novel environments. *Proceedings of the National Academy of Sciences, USA* **102**, 5460–5465.

Sol, D., Lefebvre, L. and Rodríguez-Teijeiro, J. D. (2005c). Brain size, innovative propensity and migratory behaviour in temperate Palaearctic birds. *Proceedings of the Royal Society, London, B* **274**, 763–769.

Sol, D., Santos, D. M., Feria, E. and Clavell, J. (1997). Habitat selection by the Monk Parakeet during colonization of a new area in Spain. *Condor* **99**, 39–46.

Sol, D., Timmermans, S. and Lefebvre, L. (2002). Behavioural flexibility and invasion success in birds. *Animal Behaviour* **63**, 495–502.

Sol, D., Vilá, M. and Kühn, I. (2008). The comparative analysis of historical alien introductions. *Biological Invasions* **10**, 1119–1129.

Sorci, G., Møller, A. P. and Clobert, J. (1998). Plumage dichromatism of birds predicts introduction success in New Zealand. *Journal of Animal Ecology* **67**, 263–269.

Sossinka, R. (1982). Domestication in birds. *Studies in Avian Biology* **6**, 373–403.

South, J. M. and Pruett-Jones, S. (2000). Patterns of flock size, diet, and vigilance of naturalized monk parakeets in Hyde Park, Chicago. *Condor* **102**, 848–854.

Spear, D. and Chown, S. L. (2008). Taxonomic homogenization in ungulates: Patterns and mechanisms at local and global scales. *Journal of Biogeography* **35**, 1962–1975.

Spielman, D., Brook, B.W., & Frankham, R. (2004). Most species are not driven to extinction before genetic factors impact them. *Proceedings of the National Academy of Sciences*, U.S.A, **101**, 15261–15264.

St Louis, V. L. and Barlow, J. C. (1988). Genetic differentiation among ancestral and introduced populations of the Eurasian tree sparrow (*Passer montanus*). *Evolution* **42**, 266–276.

Starrfelt, J. and Kokko, H. (2008). Are the speeds of species invasions regulated? The importance of null models. *Oikos* **117**, 370–375.

Stattersfield, A. J., Crosby, M. J., Long, A. J. and Wege, D. C. (1998). *Endemic bird areas of the world: Priorities for bird conservation*. BirdLife International, Cambridge.

Steadman, D. W. (2006). *Extinction and biogeography of tropical Pacific birds*. Chicago University Press, Chicago.

Steadman, D. W., Greiner, E. C. and Wood, C. S. (1990). Absence of blood parasites in indigenous and introduced birds from the Cook Islands, South Pacific. *Conservation Biology* **4**, 398–404.

Stearns, S. C. (1983). The impact of size and phylogeny on patterns of covariation in the life-history traits of mammals. *Oikos* **41**, 173–187.

Steinitz, O., Heller, J., Tsoar, A., Rotem, D. and Kadmon, R. (2006). Environment, dispersal and patterns of species similarity. *Journal of Biogeography* **33**, 1044–1054.

Stephens, P. A. and Sutherland, W. J. (1999). Consequences of the Allee effect for behaviour, ecology and conservation. *Trends in Ecology & Evolution* **14**, 401–405.

Stephens, P. A., Sutherland, W. J. and Freckleton, R. P. (1999). What is the Allee effect? *Oikos* **87**, 185–190.

Stevens, G. C. (1989). The latitudinal gradient in geographical range: How so many species co-exist in the tropics. *American Naturalist* **133**, 240–256.

Stockwell, C. A., Hendry, A. P. and Kinnison, M. T. (2003). Contemporary evolution meets conservation biology. *Trends in Ecology & Evolution* **18**, 94–101.

Stohlgren, T. J., Barnett, D., Flather, C., Fuller, P., Peterjohn, B., Kartesz, J. and Master, L. L. (2006). Species richness and patterns of invasion in plants, birds, and fishes in the United States. *Biological Invasions* **8**, 443–463.

Suarez, A. V., Holway, D. A. and Case, T. J. (2001). Patterns of spread in biological invasions dominated by long-distance jump dispersal: Insights from Argentine ants. *Proceedings of the National Academy of Sciences, USA* **98**, 1095–1100.

Swinnerton, K. J., Greenwood, A. G., Chapman, R. E. and Jones, C. G. (2005). The incidence of the parasitic disease trichomoniasis and its treatment in reintroduced and wild pink pigeons *Columba mayeri*. *Ibis* **147**, 772–782.

Tatem, A. J. and Hay, S. I. (2007). Climatic similarity and biological exchange in the worldwide airline transportation network. *Proceedings of the Royal Society, London, B* **274**, 1489–1496.

Taylor, B. W. and Irwin, R. E. (2004). Linking economic activities to the distribution of exotic plants. *Proceedings of the National Academy of Sciences, USA* **101**, 17725–17730.

Taylor, C. M. and Hastings, A. (2005). Allee effects in biological invasions. *Ecology Letters* **8**, 895–908.

Terborgh, J. and Winter, B. (1980). Some causes of extinction. In *Conservation Biology: An evolutionary-ecological perspective*, eds M. E. Soule and B. A. Wilcox, pp. 119–133. Sinauer Associates Inc., Sunderland, MA.

Thibault, J.-C. (1988). Menaces et conservation des oiseaux de Polynésie Française. In *Livre rouge des oiseaux menacés des régions françaises d'outre-mer*, eds J.-C. Thibault and I. Guyot, pp. 87–124. Conseil International pour la Protection des Oiseaux, Saint-Cloud.

Thibault, J.-C. and Cibois, A. (2006). Une situation favorable pour le Rupe de Makatea. *Te Manu*, **54**, 2–3.

Thomas, G. H., Orme, C. D. L., Davies, R. G., Olson, V., Stattersfield, A. J., Bennett, P. M., Gaston, K. J., Owens, I. P. F. and Blackburn, T. M. (2008). Regional variation in the historical components of global avian species richness. *Global Ecology and Biogeography* **17**, 340–351.

Thomsen, J. B. and Brautigam, A. (1991). Sustainable use of Neotropical parrots. In *Neotropical wildlife: Use and conservation*, eds J. G. Robinson and K. H. Redford, pp. 359–379. University of Chicago Press, Chicago.

Thomsen, J. B. and Mulliken, T. A. (1992). Trade in Neotropical psittacines and its conservation implications. In *Neotropical parrots in crises*, eds S. R. Beissinger and N. F. R. Snyder, pp. 221–241. Smithsonian Institution Press, Washington, DC.

Thomsen, J. B., Edwards, S. R. and Mulliken, T. A. (1995). *Perceptions, conservation and management of wild birds in trade*. Traffic International, Cambridge.

Thomson, G. M. (1922). *The naturalisation of plants and animals in New Zealand*. Cambridge University Press, Cambridge.

Tilman, D. (2004). Niche tradeoffs, neutrality, and community structure: A stochastic theory of resource competition, invasion, and community assembly. *Proceedings of the National Academy of Sciences, USA* **101**, 10854–10861.

Tobin, P. C., Liebhold, A. M. and Roberts, E. A. (2007). Comparison of methods for estimating the spread of a non-indigenous species. *Journal of Biogeography* **34**, 305–312.

Torchin, M. E., Lafferty, K. D. and Kuris, A. M. (2001). Release from parasites as natural enemies: Increased performance of a globally introduced marine crab. *Biological Invasions* **3**, 333–345.

Torchin, M. E., Lafferty, K. D., Dobson, A. P., McKenzie, V. J. and Kuris, A. M. (2003). Introduced species and their missing parasites. *Nature* **421**, 628–630.

Traill, L. W., Bradshaw, C. J. A. and Brook, B. W. (2007). Minimum viable population size: A meta-analysis of 30 years of published estimates. *Biological Conservation* **139**, 159–166.

Traveset, A. and Richardson, D. M. (2006). Biological invasions as disruptors of plant reproductive mutualisms. *Trends in Ecology & Evolution* **21**, 208–216.

Trevelyan, R., Harvey, P. H. and Pagel, M. D. (1990). Metabolic rates and life histories in birds. *Functional Ecology* **4**, 135–141.

Tsutsui, N. D., Suarez, A. V., Holway, D. A. and Case, T. J. (2000). Reduced genetic variation and the success of an invasive species. *Proceedings of the National Academy of Sciences, USA* **97**, 5948–5953.

van Bael, S. and Pruett-Jones, S. (1996). Exponential population growth of Monk Parakeets in the United States. *Wilson Bulletin* **108**, 584–588.

van den Bosch, F., Hengeveld, R. and Metz, J. A. J. (1992). Analyzing the velocity of animal range expansion. *Journal of Biogeography* **19**, 135–150.

van den Bosch, F., Metz, J. A. J. and Diekmann, O. (1990). The velocity of spatial population expansion. *Journal of Mathematical Biology* **28**, 529–565.

van den Bosch, F., Zadoks, J. C. and Metz, J. A. J. (1994). Continental expansion of plant disease: A survey of some recent results. In *Predictability and non-linear modelling in natural science and economics*, ed J. Grasman and G. van Straten, pp. 274–281. Kluwer Academic Publishers, Dordrecht, The Netherlands.

Van Kirk, R. W. and Lewis, M. A. (1997). Integrodifference models for persistence in fragmented habitats. *Bulletin of Mathematical Biology* **59**, 107–137.

van Riper III, C., van Riper, S. G. and Hansen, W. R. (2002). Epizootiology and effect of avian pox on Hawaiian forest birds. *Auk* **119**, 929–942.

van Riper III, C., van Riper, S. G., Goff, M. L. and Laird, M. (1986). The epizootiology and ecological significance of malaria in Hawaiian land birds. *Ecological Monographs* **56**, 327–344.

van Zyl, A. V. (1999). Breeding biology of the Common Kestrel in southern Africa (32 degrees S) compared to studies in Europe (53 degrees N). *Ostrich* **70**, 127–132.

Vázquez, D. P. (2006). Exploring the relationship between niche breadth and invasion success. In *Conceptual ecology and invasion biology: Reciprocal approaches to nature*, eds M. W. Cadotte, S. M. McMahon, and T. Fukami, pp. 307–322. Springer, Dordrecht.

Veit, R. R. and Lewis, M. A. (1996). Dispersal, population growth, and the Allee effect: Dynamics of the house finch invasion of eastern North America. *American Naturalist* **148**, 255–274.

Vellend, M., Harmon, L. J., Lockwood, J. L., Mayfield, M. M., Hughes, A. R., Wares, J. P. and Sax, D. F. (2007). Effects of exotic species on evolutionary diversification. *Trends in Ecology & Evolution* **22**, 481–488.

Veltman, C. J., Nee, S. and Crawley, M. J. (1996). Correlates of introduction success in exotic New Zealand birds. *American Naturalist* **147**, 542–557.

Vitousek, P. M. and Walker, L. R. (1989). Biological invasion by *Myrica faya* in Hawaii- Plant demography, nitrogen-fixation, ecosystem effects. *Ecological Monographs* **59**, 247–265.

von Haast, H. F. (1948). *The life and times of Sir Julius von Haast*. Avery, New Plymouth, New Zealand.

Von Holle, B. and Simberloff, D. (2005). Ecological resistance to biological invasion overwhelmed by propagule pressure. *Ecology* **86**, 3212–3218.

Waclawiw, M. A. and Liang, K.-Y. (1993). Prediction of random effects in the generalized linear model. *Journal of the American Statistical Association* **88**, 171–178.

Wakelin, D. (1996). *Immunology to parasites*. Cambridge University Press, Cambridge.

Walsh, M. R. and Reznick, D. N. (2008). Interactions between the direct and indirect effects of predators determine life history evolution in a killifish. *Proceedings of the National Academy of Sciences, USA* **105**, 594–599.

Wang, M. H. and Kot, M. (2001). Speeds of invasion in a model with strong or weak Allee effects. *Mathematical Biosciences* **171**, 83–97.

Wang, M. H., Kot, M. and Neubert, M. G. (2002). Integrodifference equations, Allee effects, and invasions. *Journal of Mathematical Biology* **44**, 150–168.

Wang, Z. S., Baker, A. J. Hill, G. E. and Edwards, S. V. (2003). Reconciling actual and inferred population histories in the house finch (*Carpodacus mexicanus*) by AFLP analysis. *Evolution* **57**, 2852–2864.

Wares, J. P., Hughes, A. R. and Grosberg, R. K. (2005). Mechanisms that drive evolutionary change: Insights from species introductions and invasions. In *Species invasions: Insights into ecology, evolution and biogeography*, eds D. F. Sax, J. J. Stachowicz and S. D. Gaines, pp. 229–258. Sinauer Associates Inc., Sunderland, MA.

Waring, G. H., Loope, L. L. and Medeiros, A. C. (1993). Study on use of alien versus native plants by nectarivorous forest birds on Maui, Hawaii. *Auk* **110**, 917–920.

Warner, R. E. (1968). The role of introduced diseases in the extinction of the endemic Hawaiian avifauna. *Condor* **70**, 101–120.

Webb, T. J. and Gaston, K. J. (2003). On the heritability of geographic range sizes. *American Naturalist* **161**, 553–566.

Webb, T. J. and Gaston, K. J. (2005). Heritability of range sizes revisited: A reply to Hunt, et al. *American Naturalist* **166**, 136–143.

Webb, T. J., Kershaw, M. and Gaston, K. J. (2001). Rarity and phylogeny in birds. In *Biotic homogenization*, eds J. L. Lockwood and M. L. McKinney, pp. 57–80. Kluwer Academic/ Plenum Publishers, New York.

Wellwood, J. M. (1968). *Hawkes Bay Acclimatisation Society centenary 1868–1968*. Hawkes Bay Acclimatisation Society, Hastings, New Zealand.

West, B. and Zhou, B.-X. (1989). Did chickens go north? New evidence for domestication. *World's Poultry Science Journal* **45**, 205–218.

West, G. B. and Brown, J. H. (2005). The origin of allometric scaling laws in biology from genomes to ecosystems: Towards a quantitative unifying theory of biological structure and organization. *Journal of experimental biology* **208**, 1575–1592.

Westphal, M. I., Browne, M. MacKinnon, K. and Noble, I. (2008). The link between international trade and the global distribution of invasive alien species. *Biological Invasions* **10**, 391–398.

Wetmore, A. (1918). Bones of birds collected by Theodor de Booy from kitchen midden deposits in the islands of St. Thomas and St. Croix. *Proceedings of the U S National Museum* **54**, 513–523.

White, C. R., Blackburn, T. M. Butler, P. J. and Martin, G. R. (2007). The basal metabolic rate of birds is associated with habitat temperature, not productivity. *Proceedings of the Royal Society, London, B* **274**, 287–293.

Whittaker, R. J. and Fernandez-Palacios, J. M. (2007). *Island biogeography: Ecology evolution and conservation*. Oxford University Press, Oxford.

Whittaker, R. J., Willis, K. J. and Field, R. (2003). Climate-energetic explanations of diversity: A macroscopic perspective. In *Macroecology: Concepts and consequences*, eds T. M. Blackburn and K. J. Gaston, pp. 107–129. Blackwell Science, Oxford.

Wiebe, K. L. (2003). Delayed timing as a strategy to avoid nest-site competition: Testing a model using data from starlings and flickers. *Oikos* **100**, 291–298.

Wikle, C. K. (2003). Hierarchical Bayesian models for predicting the spread of ecological processes. *Ecology* **84**, 1382–1394.

Williams, P. A. and Karl, B. J. (1996). Fleshy fruits of indigenous and adventive plants in the diet of birds in forest remnants, Nelson, New Zealand. *New Zealand Journal of Ecology* **20**, 127–145.

Williams, R. N. and Giddings, L. V. (1984). Differential range expansion and population-growth of bulbuls in Hawaii. *Wilson Bulletin* **96**, 647–655.

Williamson, M. (1988). Relationship of species number to area, distance and other variables. In *Analytical biogeography: An integrated approach to the study of animal and plant distributions*, eds A. A. Myers and P. S. Giller, pp. 91–115. Chapman & Hall, London.

Williamson, M. (1996). *Biological invasions*. Chapman & Hall, London.

Williamson, M. (1999). Invasions. *Ecography* **22**, 5–12.

Williamson, M. (2001). Can the impacts of invasive species be predicted? In *Weed risk assessment*, eds R. H. Groves, F. D. Panetta and J. G. Virtue, pp. 20–33. CSIRO Publishing, Collingwood, Australia.

Williamson, M. (2006). Explaining and predicting the success of invading species at different stages of invasion. *Biological Invasions* **8**, 1561–1568.

Williamson, M. and Brown, K. C. (1986). The analysis and modelling of British invasions. *Philosophical Transactions of the Royal Society, London, B* **314**, 505–522.

Williamson, M. and Fitter, A. (1996). The varying success of invaders. *Ecology* 77, 1661–1666.

Wilson, J. R. U., Richardson, D. M., Rouget, M., Proches, S., Amis, M. A., Henderson, L. S. and Thuiller, W. (2007). Residence time and potential range: Crucial considerations in modelling plant invasions. *Diversity and Distributions* **13**, 11–22.

Wing, L. (1943). Spread of the starling and English sparrow. *Auk* **60**, 74–87.

With, K. A. (2002). The landscape ecology of invasive spread. *Conservation Biology* **16**, 1192–1203.

With, K. A. (2004). Assessing the risk of invasive spread in fragmented landscapes. *Risk Analysis*, **24**, 803–815.

Wood, C. A. (1924). The starling family at home and abroad. *Condor* **26**, 123–136.

Woolnough, A. P., Lowe, T. J. and Rose, K. (2006). Can the Judas technique be applied to pest birds? *Wildlife Research* **33**, 449–455.

Yap, C. A.-M., Sodhi, N. S. and Brook, B. W. (2002). Roost characteristics of invasive mynas in Singapore. *Journal of Wildlife Management* **66**, 1118–1127.

Yom-Tov, Y., Christie, M. I. and Iglesias, G. J. (1994). Clutch size in passerines of southern South America. *Condor* **96**, 170–177.

Index

Note: page numbers in *italics* refer to Figures and Tables.